24 0309718 9

AF616334

ENVIRONMENTAL MODELING
Vol. II

Computer Methods and Software for Simulating Environmental Pollution and its Adverse Effects

Acknowledgement is made to C. Seigneur for the use of Figure 1 on page 229, Volume 1, which has been used as a basis for the front cover of this book.

ENVIRONMENTAL MODELING
Vol. II

Computer Methods and Software for Simulating Environmental Pollution and its Adverse Effects

Editor:

P. Zannetti

Senior Managing Scientist
Failure Analysis Associates, Inc., California, USA
Visiting Professor
Wessex Institute of Technology, Southampton, UK

Computational Mechanics Publications
Southampton Boston

P. Zannetti
Failure Analysis Associates., Inc.
149 Commonwealth Drive
P.O. Box 3015
Menlo Park, California 94025
USA

Published by

Computational Mechanics Publications
Ashurst Lodge, Ashurst, Southampton, SO40 7AA, UK
Tel: 44 (0) 703 293223; Fax: 44 (0) 703 292853
E-mail: cmi@ib.rl.ac.uk

For USA, Canada and Mexico:

Computational Mechanics Inc
25 Bridge Street, Billerica, MA 01821, USA
Tel: (508) 667-5841; Fax: (508) 667-7582

British Library Cataloguing-in-Publication Data

A Catalogue record for this book is available
from the British Library

ISBN 1-85312-281-5 Computational Mechanics Publications, Southampton
ISBN 1-56252-205-1 Computational Mechanics Publications, Boston

Library of Congress Catalog Card Number 94-70418

Printed and bound in Great Britain by Bell & Bain Ltd., Glasgow

CONTENTS

TABLE OF CONTENTS OF PREVIOUS VOLUMES

VOLUME I

PREFACE

Environmental Modeling, volume 2 is the second volume of an edited series of publications on computer methods for simulating environmental pollution and its adverse effects. It presents a careful selection of invited review papers covering environmental modeling topics. Each chapter was authored by a leading scientist in the field and was written to provide the reader with an organized and consistent discussion on the field of mathematical and numerical simulation of environmental phenomena. In addition to the discussion of the mathematical, numerical, physical, chemical, biological, and ecological aspects of environmental phenomena, *Environmental Modeling*, volume 2 provides a critical review of software available for environmental simulations.

This *Environmental Modeling* series represents an effort that is unique in its contents and especially in its long-term goals. Volume 2, which contains nine chapters on different environmental subjects plus an introductory chapter, deals both with major environmental topics such as groundwater pollution and with some special issues such as the modeling of the aerial spray drift associated with the application of agricultural pesticides. Volumes 1 and 2 of *Environmental Modeling* present a total of 17 technical chapters that can assist the reader in achieving a comprehensive knowledge of environmental modeling issues and computer simulation of anthropogenic pollution. An introductory chapter in volume 2 provides the reader with a guide to both volumes.

Subsequent volumes will expand the coverage of environmental topics by including new material and rearranging or rewriting previously published chapters. Therefore, each new volume will change the structure of the entire publication series and provide the reader with an updated, expanded, and reorganized review. An introductory chapter in each new volume will assist and guide the reader through the different volumes.

I would like to conclude by encouraging the environmental scientific community to provide me with input, suggestions, contributions, and constructive criticism. I will do my best to incorporate new ideas into future volumes.

Paolo Zannetti
December 1993

About the Authors – Volumes I and II

John W. Barry graduated from the University of Cincinnati in 1956, and then served five years in the U.S. Army. His service with the U.S. Army Chemical Corps included evaluating ecological and epidemiological impact of human and crop diseases, and conducting field tests of biological application and detection systems. In 1961 Mr. Barry served as a test officer at U.S. Army Dugway Proving Ground, Utah, conducting engineering systems and atmospheric dispersion tests. Since joining the USDA Forest Service, Washington Office Staff in 1975, Mr Barry has been Program Manager of the dispersion and deposition model program and sits as chairperson for four national steering committees. He was Program Manager of Program WIND (Winds In Non-uniform Domains) a multi agency four year field study in northern California during the mid-1980s.

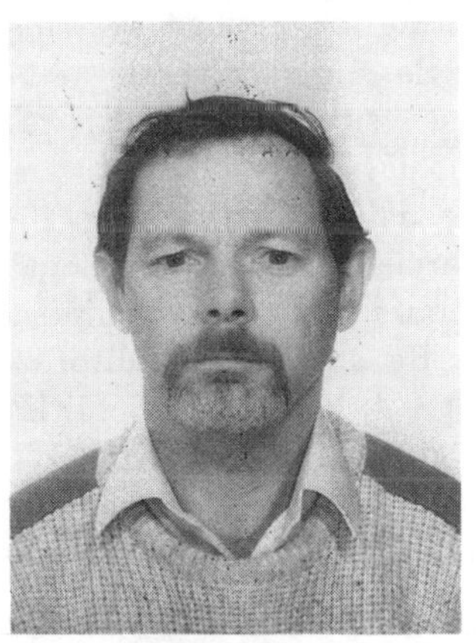

James R. Davis initially trained as a physicist in New Zealand, but moved to the use of computers for environmental planning and management in the early 1980s at CSIRO, Australia. He has integrated such technologies as GIS, expert systems, databases and models into decision support systems for a variety of problems. As part of his work, he and colleagues have developed a spatial expert system, ARX. He is particularly concerned with the developing of the support needed to ensure that the software is used in practice. He has visited the major research institutions in Europe and the USA working in this area and has been visiting Professor of Expert Systems for Environmental Management at the Technical University of Milan, Italy.

Elizabeth M. Donley received her B.S. in Physics from Virginia Polytechnic and State University and her M.S. in Mathematical Sciences from East Tennessee State University. In her professional career she has worked for Cambridge Research Associates, Inc. as a senior analyst and manager, for Comarco, Inc. as a technical analyst, systems engineer and administrator. She is currently executive director of Donley Technology. She is publisher/editor of the Environmental Software Report (official publication of the Environment & Safety Data Exchange – ESDX) and the Environmental Software Directory. She is also a member of the editorial board of the *Environmental Software Journal* and of *Environmental Manager*.

Giuseppe Gambolati is Professor of Numerical Methods at the School of Engineering, University of Padova, Italy. He received his Doctor degree with honors in Mechanical Engineering from the Technical University of Turin in 1968. He joined the IBM Scientific Center of Venice in 1969 where he became chairman of the hydrology modeling program in 1976. He has been a Professor at the University of Padova since 1980. Dr. Gambolati has long been a consultant to national and international environmental programs, including those developed by the Italian Regional Governments, the National Electricity Board (ENEL), and the Commission of the European Communities (CEC). He is the author of more than 100 papers on groundwater flow, contaminant transport, land subsidence prediction and related modeling topics, and serves as a member of the editorial board for several international journals.

Giorgio Guariso obtained a Doctor Degree in Systems Engineering in 1976, and has been a researcher with the Italian National Research Council in Milan, Italy, the Egyptian Academy of Scientific Research and Technology in Cairo, Egypt, and the International Institute for Applied Systems Analysis (IIASA), Laxenburg, Austria. He is presently associate Professor at the Technical University of Milan. His research activities have focused on the development of software systems for environmental modeling and decision making. He has published papers in this field through many international journals and conferences, and two books particularly centered on environmental decision support. He is Associate Editor of the *Environmental Software* Journal and chairman of the IFIP Working Group 5.11 'Computers and Environment'.

Giovanni Graziani received his Doctor Degree in Physics at Naples University, Italy. In 1964 he joined the OECD Dragon Project (Winfrith, UK) in charge of the nuclear fuel cycle studies. In 1967 he seconded to the division of Reactor Theory and Calculus at JRC Ispra (Establishment of Commission of European Communities, Italy). He joined the unit of System Analysis in 1974, and in 1980 moved to the division of Systems Engineering and Reliability. In 1986 Dr. Graziani contributed to the conception of the JRC specific research activity REM (Radioactivity Environmental Monitoring), created by the CEC following the accident at Chernobyl and producing the international exercise on Atmospheric Transport Model Evaluation Study (ATMES). He has belonged to the Atmospheric Physics unit at the Environment Institute, Italy, since 1990, in charge of atmospheric modeling activities for episodic releases.

Tony J. Jakeman is a Senior Fellow in Environmental Modeling at the Australian National University. His major research interests are in the development and application of system identification to quantitative environmental problems. He has worked in several resource areas including surface and subsurface hydrology, water quality and air quality. He is the author or co-author of 100 journal papers and several edited books. He is president of the Modeling and Simulation Society of Australia, Inc., and a director of the International Association for Mathematics and Computers in Simulation. Dr. Jakeman is also a member of the editorial boards for the journals of *Forecasting, Environmetrics, Ecological Modeling* and *Environmental Software.*

Aaron A. Jennings earned his undergraduate degree from the Rochester Institute of Technology, and his M.S. (Civil Engineering) and Ph.D. (Environmental Engineering) from the University of Massachusetts. Dr. Jennings has taught engineering at the University of Notre Dame and The University of Toledo and is currently a Professor of Civil Engineering at Case Western Reserve University. He teaches courses in Fluid Mechanics, Hydraulics, Water Supply, Solid and Hazardous Waste, Water Resources Engineering, Advanced Groundwater Analysis, and Environmental Engineering Modeling. Dr. Jennings is currently the Director of Environmental Engineering Programs at CWRU, and Editor-in-Chief of the International Journal *Environmental Software.*

Neil Lincoln has worked in his professional career for the Institute of Engineering Research at the University of California, Berkeley, as mathematician programmer supporting the Plasma Physics research; as Senior Consultant for URS Incorporated, Burlingame, California; as Senior Consultant for Control Data's Advanced Design Laboratory. He was the chief architect of the CDC CYBER-203 and CYBER-205 vector supercomputers. Prior to this he was the final architect of the CDC STAR-100 hardware system. He was vice-president and co-founder of ETA Systems Inc. In addition to being the Chief Architect of the ETA-10 computer. He is currently President and founder of Supercomputer Systems Engineering and Services Company (SSESCO), Minneapolis, Minnesota.

Ian G. Littlewood studied Environmental Sciences at Lancaster University then worked in the Water Data Unit of the Department of the Environment on a range of topics related to the UK national hydrometric network and the corresponding river flow database. This was followed by several years at the headquarters of the Welsh Water Authority dealing with a wide range of hydrological topics. From there he went to the University College Swansea where, as a member of a small team in the Geography Department collaborating with the Institute of Hydrology (IH), he participated in the multidisciplinary Llyn Brianne Acid Waters Project and gained a Ph.D. for a study of the dynamic behavior of streamflow acidity. He is currently employed at the National Water Archive at IH, Wallingford, UK.

Tom J. Lyons received his B.Sc. from the University of Melbourne and Ph.D. from the Flinders University of South Australia before joining Murdoch University where he is currently associate Professor of Atmospheric Science within the Environmental Science Program. He has contributed to over 100 publications dealing with mesoscale meteorology, air pollution, wind energy and urban transport, as well as acting as a consultant to Government and industry. He is the co-author of *Principles of Air Pollution Meteorology*. Dr. Lyons is currently an Associate Editor of the *Australian Meteorological Magazine* and on the editorial boards of *Atmospheric Environment (Urban Atmosphere)* and *Environmental Software*.

Stefano Marsili-Libelli received his Doctor Degree in Electronics Engineering from the University of Pisa in 1973. In the same year he joined the University of Florence on a postgraduate grant and has served in the Faculty of Engineering ever since. In 1983 he was appointed associate Professor of Process Control. His research interest is in the application of system theory and control engineering to environmental problems, in particular to the problems of water purification systems, aerobic wastewater treatment processes and river and ecosystems modeling. Since 1987 he has served as president of the Italian Research Group GRAGSA, an acronym for Research Group on Environmental Analysis and Management. In 1990 he joined Arno River Water Authority nominated by the Tuscany Regional Government.

Jana B. Milford received her B.S. in Engineering Sciences from Iowa State University, the M.S. in Civil Engineering and the Ph.D. in Engineering and Public Policy both from Carnegie Mellon University. She is currently assistant Professor of Civil Engineering and the Environmental Research Institute at the University of Connecticut. Her research interests are in air quality modeling and environmental policy. Much of her work has focused on analysis of control strategies for reducing photochemical air pollution, including the use of alternative fuels in motor vehicles. Prior to joining the University of Connecticut, she was a Congressional Fellow at the Office of Technology Assessment, a research arm of the United States Congress.

Ionel M. Navon received his B.Sc. in Mathematics and Physics and his M.Sc. in Atmospheric Science from the Hebrew University of Jerusalem in 1967 and 1971, respectively. He received a Ph.D. in Applied Mathematics from the University of Witwatersrand, Johannesburg, South Africa, 1979. He has been working at Florida State University since 1985, where he is Professor in the Department of Mathematics and Faculty Associate of the Supercomputer Computations Research Institute, and Program Director for Optimization and Optimal Control. His research includes computational fluid dynamics, finite-element modeling large scale optimization and optimal control of distribute parameters systems. He is Editor of the *Monthly Weather Review*, serves on the editorial board of several journals, and has authored about 120 papers as well as several book chapters.

Jacques C.J. Nihoul received his degree in Mechanical Engineering from Liège University in 1960, his M.S. in Mathematics from the Massachusetts Institute of Technology in 1961 and his Ph.D. in Applied Mathematics and Theoretical Physics from Cambridge University in 1965. In 1966, he joined Liège and Louvain Universities as Professor of Geophysical Fluid Dynamics. He is Editor of *Earth Science Reviews, Oceanography Section; Journal of Marine Systems, Physical Oceanography* and Proceedings International Liège Colloquia on Ocean Hydrodynamics, Elsevier Oceanography Series, as well as a member of the editorial board of *Applied Mathematical Modeling; Ecological Modeling; Environmental Software; Oceanologica Acta; Mathematical and Computer Modeling and Continental Shelf Research.*

Claudio Paniconi is currently a research scientist at CRS4 in Cagliari, Italy. He obtained a B.Math. degree in 1984 from the University of Waterloo, an M.A. in 1986 from Princeton University, and a Ph.D. in 1991 also from Princeton University. In 1991 and 1992 he conducted post-doctoral research at Princeton, in the Water Resources Program of the Department of Civil Engineering and Operations Research, and at the University of Padua, in the Department of Mathematical Models in Applied Sciences. During his undergraduate days Dr Paniconi accumulated programing and modeling experience working at the Atomic Energy of Canada in Chalk River, Joseph G. Debanne Company in Ottawa, and IBM Canada in Toronto. His current research activities include development of finite element hydrologic models, evaluation of conceptual models in catchment hydrology, and analysis of coupling, nonequilibrium and nonlinear phenomena in groundwater flow and contaminant transport processes.

Giorgio Pini received his Doctor Degree in Mathematics in 1970 at the University of Padua. In 1971 he was appointed as assistant Professor of Calculus, and then of Elementary Numerical Analysis. From 1985 he has served as associate Professor in the Department of Mathematical Models in Applied Sciences. His duties include teaching at undergraduate and graduate levels. His principal research activity is the development and implementation of finite element models of groundwater systems. In particular he has developed a 3-D flow model with automatic mesh generation. He has also contributed to the implementation of a quasi 3-D flow model for the analysis of the subsurface water resources of the Friuli Venezia Giulia area (northern Italy).

Owen Pitts received his B.Sc. in Physics and Ph.D. in Atmospheric Science from Murdoch University before joining Steedman Science and Engineering, a division of Oceanroutes Pty Ltd., as a consultant research scientist. He has specialized in atmospheric and oceanic modeling related to the preparation of environmental impact studies for a number of clients throughout Australia. Dr. Pitts has published papers in *Quarterly Journal of the Royal Meteorological Society, Monthly Weather Review, Atmospheric Environment* and the *Australian Meteorological Magazine.*

Mario Putti obtained his Doctor Degree in Civil Engineering in 1985 at the University of Padua, Italy, and his M.A. in 1987 and his Ph.D. in Civil Engineering in 1989 both from the University of California, Los Angeles, USA. Since 1989 he has been appointed as a research scientist in the Department of Mathematical Models for Applied Sciences at the University of Padua, and also works in the research group headed by Professor G. Gambolati. His current research interests include the development and application of numerical schemes for the solution of nonlinear problems in the transport of reactive contaminants in groundwater, and in the study and implementation of solution strategies in coupled flow and transport problems.

Varadhan Ravi received his Bachelor of Technology degree in Chemical Engineering from the Indian Institute of Technology, Madras, India, and a Doctoral degree in Civil and Environmental Engineering from the University of Toledo, Ohio in August 1993. He is currently employed at the Dynamac Corporation at the U.S. EPA's Robert S. Kerr Environmental Research Laboratory, Oklahoma. His primary research interests are in the area of mathematical modeling of contaminant transport in the unsaturated and groundwater regimes.

Armistead G. Russell received his B.S. from Washington State University, his M.S. (1980) and Ph.D. (1985) at Caltech in Mechanical Engineering. He is currently associate Professor of Mechanical Engineering and Public Policy at Carnegie Mellon University. His research interest is related to air pollution engineering, air pollution modeling, computational engineering and environmental fluid dynamics. Dr. Russell served on the National Research Council's Ozone Formation and Control Committee, and is currently on the NRC's Committee on Air Pollution Risk Assessment. He has been awarded the SAE Ralph Teetor Award and First Place in the Air Pollution Control Association Annual Paper Competition. At Carnegie Mellon, he is a member of the Environmental Institute's steering committee.

Christian Seigneur received an advanced degree in Chemistry from the Ecole Nationale Superieure de Chimie, Paris, and a Ph.D. in Chemical Engineering from the University of Minnesota. He is a Principal and General Manager of the San Francisco office of ENSR Consulting and Engineering. He is presently a member of the advisory committee for the Toxic Control and Risk Analysis research program of the University of California at Los Angeles (UCLA) and is on the Editorial Board of *Environmental Software*. His research interests include mathematical modeling, atmospheric science, environmental fate and transport of chemicals, health risk assessment, resource economics and environmental policy.

László Somlyódy received his M.Sc. and dr. tech. degree from the Budapest University of Technology in 1967 and 1973, respectively, and the C. Sc. and D. Sc. degree from the Hungarian Academy of Sciences in 1978 and 1985, respectively. He was elected to be a member of the above Academy in 1990. He is research Professor of the Water Resources Center, VITUKI, and professor of Water Quality Management at the Budapest University of Technology. Since 1991 he has been the leader of the Water Resources Project at the International Institute for Applied System Analysis, Laxenburg, Austria. His research interests include modeling of transport and water quality in recipient waters, as well as broader issues of environmental management.

Milton E. Teske received his B.S. degree from Iowa State University, his M.S. degree from California Institute of Technology, and his Ph.D. from the University of Michigan. All degrees were in Aerospace Engineering. He has concentrated his professional efforts in many areas of numerical analysis. From 1972 to 1979 he was employed at Aeronautical Research Associates of Princeton, Inc., where he was responsible for the development of their turbulence code. In the mid-1970s he developed the NASA-Langley WAKE code used to simulate vortex merging and decay behind jumbo jet aircraft. From 1979 to the present, at Continuum Dynamics, Inc., Dr. Teske continues to center his work around the development of numerical techniques, but with an emphasis on internal flows. He is also a consultant to the Spray Drift Task Force, and has written or co-authored over 300 publications.

Harold W. Thistle, Jr. received his Ph.D. in 1988 from the University of Connecticut. He specializes in forest meteorology, environmental instrumentations, air quality modeling and atmospheric boundary layer dispersion. He is currently employed in the Technology and Development branch of USDA-Forest Service Engineering. Primary responsibilities include the development of modeling techniques to accurately describe transport of pesticides in the atmospheric surface layer and the evaluation of meteorological instrument systems for use in environmental monitoring. Dr. Thistle is an AMS Certified Consulting Meteorologist and worked in private industry as an air pollution consultant before joining the Forest Service.

Renzo Tonin received his Ph.D. in Mechanical Engineering from the University of Adelaide in 1978. During the years 1976/1978 he was Associate Director and Sydney office manager at Vipac & Partners Pty Ltd. He is currently Director and Principal of Renzo Tonin & Associates Pty Ltd. His sponsored studies encompass such diverse activities as building, ships, motor vehicles, trains, industrial complexes, power stations, product development, material handling plants, machine health monitoring, coal washeries, public buildings and auditoria. His research interest is in the use of computer technology to solve problems in the field of acoustics and dynamics.

Marek Uliasz received his M.S. (1977) and Ph.D. (1986) in Environmental Engineering at Warsaw University of Technology, Poland, where he works as an assistant Professor at the Institute of Environmental Engineering Systems. During 1983-1989 he worked as a visiting scientist at the USSR Academy of Sciences, International Institute of Applied System Analysis, Austria, and Risø National Laboratory, Denmark. Since 1989 to date he has worked on extended leave of absence as a visiting scientist and then as a research associate in the Department of Atmospheric Science, Colorado State University, USA. He also works for Atmospheric Simulation Testing and Research (ASTeR), Inc. where he is the prime developer of the new Hybrid Particle and Concentration (HYPACT) model. He has written over 30 publications.

Olli Varis received the M.Sc., the Ph.D., and the Dr. Eng. degree in Civil Engineering from Helsinki University of Technology, in 1984, 1987, and 1991, respectively. Additionally, he received the M.Sc. degree in Limnology from the University of Helsinki in 1986. He has been working as The Academy of Finland Research Associate at Helsinki University of Technology, Finland, since 1987. He has been lecturing at the Asian Institute of Technology, Bangkok, Thailand, and has worked a number of times at the International Institute for Applied Systems Analysis, Laxenburg, Austria. His research includes mechanistic, statistical, and decision analytic modeling of the environment, with special interest in surface water quality problems and fisheries.

Eric G. Walther received the B.E.P. in Engineering Physics from Cornell University in 1964, and the M.S. and Ph.D. degrees in Atmospheric Science from the State University of New York in 1968 and 1970, respectively. Dr. Walther is Vice-President of Environmental Solutions, Inc. He is responsible for air quality, risk assessment, and environmental information management projects, and control technology engineering. He has developed technical specialization in air toxics, regulatory analysis, environmental information management, visibility, and soil gas investigations.

Paolo Zannetti received his Doctor Degree in Physics from the University of Padua in 1970. He has performed environmental studies for more than two decades, specializing in air pollution and environmental modeling and software. He has been associated with several private groups (IBM Scientific Centers of Venice, Palo Alto and Bergen; AeroVironment, Inc.; University of Stanford) and public organizations (Kuwait Institute for Scientific Research; University of Padua). He is currently Senior Managing Scientist at Failure Analysis Associates, Inc. in California. Among his several editorial activities, he is the Founding Editor of the Elsevier journal *Environmental Software*, Associate Editor of *Atmospheric Environment*, member of the Editorial Board of *Ecological Modeling* and the Director of the biennial conferences ENVIROSOFT and AIR POLLUTION. He has authored more than 100 technical publications.

Xiaolei Zou received her B.Sc. in Mathematics and her M.Sc. in Atmospheric Sciences from the Institute of Nanjing Meteorology, in 1982 and 1984, respectively. She received a Ph.D. degree in Atmospheric Sciences from the Institute of Atmospheric Physics, Academia Sinica, Beijing in 1988. She has been working at the Supercomputer Computations Research Institute of the Florida State University since 1989, where she is currently an associate research scientist. Her recent research centers on multiple applications of adjoint techniques – four-dimensional variational data assimilation, parameter estimate, general adjoint sensitivity analysis, and optimal perturbations – as applied to meteorological problems.

Chapter 1

Introduction to Environmental Modeling

P. Zannetti

Failure Analysis Associates, Inc., 149 Commonwealth Drive, Menlo Park, California 94025, USA

Abstract

This chapter provides an introduction to the topic of Environmental Modeling and a guide to the contents of the first two volumes of the book series.

Key words

Environmental Modeling, Pollution, Computer Simulation, Numerical Modeling, Computer Sciences, Research and Development

1. A Definition of Environmental Modeling

Definitions are always difficult. They can be too general and therefore mean nothing. Or they can be too specific and, consequently, restricting and incomplete. It is useful and necessary, however, to try to give a working definition to the subject of this book series.

Let us start with "environment" and "environmental sciences." Can we define the topics that are encompassed by the adjective "environmental?" The risk of giving a too general definition is very high. I have been working for more than two decades on environmental research and I can hardly imagine any subject that is more multidisciplinary than the environment. Understanding environmental issues requires a good knowledge of their physical, chemical, and biological aspects for both small and large geographical scales. Pollution control requires good engineering practice. Environmental simulations require sophisticated mathematical and numerical techniques and the use of state-of-the-art computational facilities. Environmental adverse effects cover the damage to human health, ecological impacts, and aesthetic damage, such as visibility degradation and the deterioration of art work. Environmental planning requires accurate cost-benefit analyses and econometrical assessments. Environmental issues often require a good understanding of laws and regulations, since these laws affect industrial and urban development, influence the environmental business, and drive environmental litigation. Last but not least, environmental problems pose new philosophical questions by challenging industrial managers–often under community pressure–to do what is ethically right and not just what is legally acceptable. In summary, physics, chemistry, biology, engineering, computing, medicine, psychology, art, economy, law, and philosophy all play an intricate and interconnecting role in the study of the environment.

But let us try to be more specific. Obviously, environmental sciences are the scientific disciplines that cover the *environment* in which we live. They therefore could be seen as synonymous with "earth sciences" or "planetary sciences" and include traditional and well established scientific fields such as geology, oceanography, and meteorology. However, in my opinion the label "environmental" gives a precise, new connotation. After all, to the best of my knowledge, environmental studies have been labeled "environmental" starting only in the 1960s even though "earth sciences" have been around for a much longer time.

The unique connotation that the label "environment" adds is the role of anthropogenic factors and manmade pollution. For example, oceanographic research aiming at a better understanding of ocean currents can be seen as a study in the field of oceanography; but if the same study aims, as a principal goal, at the understanding of transport and fate of oil spills of anthropogenic origin, the study is better characterized as environmental. Therefore, I would like to propose the following definition:

> *Environmental sciences are defined as the sciences that cover, as their principal subject, anthropogenic pollution: its generation, its transport and fate in different environmental media (air, water, soil, groundwater, and biota), and its adverse effects.*

We could use a more general characterization, if you prefer, and define environmental sciences as those concerned with pollution, population control, and resource conservation. But I would resist the temptation of "contaminating" environmental sciences with the inclusion of controversial and ideological topics that better belong to the political and economic sciences and to the agendas of modern-day environmentalism.

Now what about modeling? Let us first clarify that "modeling" may indicate two very different things: 1) small-scale physical or laboratory modeling, and 2) mathematical or numerical modeling. We are all very familiar with the first type of models since we all have played with *toys*. What are toys? Well, most toys are a small-scale representation of real things (e.g. cars, cranes, people, appliances, rifles, and tanks). And these toy "models" do possess some of the characteristics and physical properties of the original objects they try to imitate; for example, a toy boat does float on water and a toy crane is capable of lifting weights. A child discovers very soon, however, that the models are different from the real things. Even when the shapes and the colors of the models are very realistic, their structural behavior and response to external stimulation are different. For example, a toy car, even when built as an accurate small-scale representation of a real car, can withstand impacts and rollovers that would seriously damage a real automobile.

Physical models in environmental sciences are built to provide a tool for experimentation. For example, a physical model of a complex lagoon can simulate the expected behavior of the internal water currents if modifications are made to the lagoon structure. Similarly, a smog-chamber experiment in controlled laboratory conditions can lead to the identification and quantification of complex chemical reactions among atmospheric pollutants. Physical models are needed to understand some of the characteristics of a system that would be impossible to study otherwise. Moreover, physical models allow us to perform experimentation on small-scale replicas of large structures and entities, where the cost of experimenting directly on the real objects would be unaffordable.

Of course, physical models cannot replicate all the properties of the real system. For example, the laws that rule the transport of mass in the environment possess nonlinear terms (for example, viscosity). Therefore, any small-scale representation, no matter how sophisticated and detailed, will provide results that cannot perfectly replicate the real world. The scientific challenge is to understand, quantify, and minimize the inaccuracies of the physical models and use their simulation results with care.

This book series on environmental modeling does not exclude the topic of physical modeling. However, all the chapters so far deal with the second type of modeling approach: mathematical and numerical modeling. Historically, like many other sciences that originated from military needs, the science of mathematical modeling probably started with ballistic computations in the sixteenth century. The goal was to understand the trajectories of cannonballs and improve accuracy in hitting enemy targets. Under a set of simplifying conditions, equations were found that describe the trajectory of the cannonball. These equations were solved to provide an *analytical solution*, i.e. a *formula* that gives the point of impact as a function of the initial velocity and the shooting angle. These equations constituted a "mathematical model" of reality, i.e., the representation of physical properties solely with mathematical tools, pen, and paper.

In the last two centuries, scientists have been very successful in discovering new equations that describe reality. Differential calculus, in particular, has been the key mathematical tool in this endeavor. Equations representing reality became more and more complex and included fewer and fewer simplifying factors. Of course, finding new equations was only part of the problem. To be used for practical applications, equations needed to be solved. However, the more complex the equations representing reality, the more difficult it was to find analytical solutions. Scientists were capable of solving complex equations only in extremely simplified conditions, under steady-state, homogeneous assumptions, and in one or at most two dimensions.

Unfortunately, as far as environmental phenomena are concerned, the real world is three-dimensional, never at rest, and seldom homogeneous. However, in the last few decades, computers came to the rescue. The ability of electronic computers to perform hundreds, thousands, and today millions of operations per second has made it possible to solve the equations *numerically*. A numerical solution is not a general solution, such as an analytical solution. A numerical solution can only describe a specific set of conditions. However, we can solve any set of equations numerically, no matter how complex. This is the source of the success of numerical simulations and the use of computers in science. Computers have become indispensable tools to scientists in general and environmental scientists in particular. More precisely, computer techniques are more than just tools. Computer analysis is a method, an attitude, a research direction, a science in itself (the "third branch" of science), and, more importantly, an extension of the scientist's thinking power.

2. Models and Modelers versus the Rest of the World

As a modeler, my interactions with the rest of the world have been complex and sometimes frustrating. In most cases, I have received gratifying and competent appreciation of modeling work and a fair understanding of what models can do. But in other cases things have been complicated. Often, in certain circles, I have found it difficult to explain what computer modeling is. This is because computer illiteracy in certain segments of the population, including medium-high management levels, is still not uncommon today in spite of the clear progress made in the last decade by the personal-computer revolution.

In many cases, I faced a profound skepticism for the capabilities of environmental models. How many times I heard, even from scientists, the simplistic statement that "models do not work," or the other famous statement claiming that models are just "garbage in, garbage out!" In some cases, I have experienced profound diffidence toward the modelers themselves. They were seen as desk scientists working in isolation, living in a special world detached from the "real world," playing with computer toys, and unable to connect and interact with the rest of the environmental community, especially with the people who go into the field and understand what pollution really is from practical, direct experience.

Of course, any criticism has some fraction of truth. Nevertheless, I have never believed (and never will believe) that walking with boots on a polluted landfill or visiting a monitoring station can improve my ability to simulate groundwater pollution or perform a statistical analysis of the collected data. However, a few months ago, to better explore this issue, I mailed a questionnaire to a few dozen scientists. I selected these individuals because of their general environmental expertise. They are well-established environmental experts who are not modelers but know about environmental modeling and interact with computer modelers. The questionnaire asked the following three questions:

1) *What is your current opinion of environmental models? Has your opinion changed in the last 2-3 years? Which role do you see for environmental modeling in the future?*

2) *How do you characterize your interactions and communications with environmental modelers? Can you specify the positive and negative aspects of your current and past working relationship with modelers?*

3) *In your opinion, is there competition, contrast, or misunderstanding between computer modelers and other environmental professionals? If so, what are the roots of the problem and what can be done to improve communication and collaboration?*

I received few answers, but all were from very qualified people. A summary of the answers is presented below.

1) Most people who responded indicated that their opinions about models have risen. ("My opinion of current atmospheric models has improved in the last 2-3 years; so have the models.") They generally have a good opinion of models ("Models are more useful than ever"), especially after the arrival of low-cost, high-performance workstations that allow easier application of the most advanced models. Some people pointed out that models are sometimes grossly misused but clarified that this is not the fault of the modeler who developed the code. Some people criticized the "PC mentality," described as insistence on using personal computers instead of more powerful workstations. Almost anyone expected models to improve in their ability to represent reality and become more user-friendly. ("We will eventually come out of the present dark ages into a time when modeling is a proper tool that can be trusted to give believable answers.") Some people suggested that models should refuse to run when the input parameters are not appropriate and should optimize themselves given a few data points. (This particular feature is called "data assimilation" and is discussed at lengths in Chapter 9 of this volume).

2) The responses characterized the relationship with modelers as relatively good, but with some problems. The major criticism is that some modelers seem to believe more in their modeling simulations than in the actual measurements. Moreover, it was claimed that some modelers "seem to have rigged the input data" to obtain the results they wanted. Another criticism is that modelers "sometimes create model parameters that cannot be measured for calibration/validation." Regarding the last statement, I must add that it is apparently true that some model parameters are difficult or impossible to measure directly. Indirect measurements, however, are always possible. Also, I want to clarify that the identification of parameters that cannot be measured is the consequence of the application of the scientific method and not a trick invented by the modelers to escape accountability.

3) Some responses are conciliatory ("I do not believe there should be a competition"), but others are more realistic: "When resources for research are finite and dwindling, there is always competition, not only between modelers and experimentalists, but between local modelers and global modelers, and between atmospheric physicists and atmospheric chemists". Some people expressed strong criticism of the U.S. Environmental Protection Agency (EPA) for not providing better and more credible models: "They [the EPA] presently have a definite 'NIH' [not invented here] bias in what they recognize, and are using models known to be defective in reaching very expensive conclusions". One person identified the EPA as "the biggest problem by far, at least in the U.S.," because of the agency's "entrenched attitude" of not promoting more advanced modeling techniques: "Models which were 'approved' [by the EPA] more than a decade ago are still mandated in spite of their manifest inadequacies. Outdated models continue to be applied, at government insistence, to situations were accuracy is being sacrified in the name of 'consistency' or 'conservatism'."

I encourage readers to reflect seriously on this section and the statements above. I intend to collect more opinions on this topic and expand the analysis. Therefore, I would really appreciate it if many readers (modelers and nonmodelers) could send me their comments and frank opinions on this matter.

3. The Contents of *ENVIRONMENTAL MODELING,* Volume 2

Environmental Modeling, volume 2 is the second volume of an edited series of publications on computer methods for simulating environmental pollution and its adverse effects. It presents an organized collection of invited review papers covering environmental modeling topics. Each chapter was authored by a leading scientist in the field and was written to provide the reader with an organized and consistent approach to the field of mathematical and numerical simulation of environmental phenomena. In addition to the discussion of the mathematical, numerical, physical, chemical, biological and ecological aspects of environmental phenomena, *Environmental Modeling* provides a critical review of software available for environmental simulations.

This *Environmental Modeling* series represents an effort that is unique in its contents and especially in its long-term goals. Volumes I and II contain 17 technical chapters on different environmental subjects plus two introductory chapters. Subsequent volumes will expand the coverage of environmental topics by including new material and rearranging or rewriting previously published chapters. In other words, each new volume will change the structure of the entire publication series and provide the reader with an updated, expanded, and reorganized review. An introductory chapter in each new volume will assist the reader and guide the reading process through the different volumes.

In this volume (volume 2), Chapters 2 through 5 cover atmospheric topics at different scales, Chapter 6 deals with rainfall-runoff modeling and catchment hydrology, Chapters 7 and 8 discuss groundwater contamination issues, Chapter 9 covers variational data assimilation techniques, and Chapter 10 deals with expert-system support for environmental decisions. A brief description of each chapter is presented below.

Chapter 2 discusses atmospheric modeling at a short scale of 10-100 m. The chapter reviews **aerial spray drift modeling**, i.e. the assumptions, approaches, and techniques applied to the modeling of spray drift from aerial application of agricultural pesticides. In this review, the authors examine some of the traditional ways of modeling the trajectories of spray material that is released from aircraft, falls through the atmosphere, evaporates, and is deposited upon crops and the ground. It is important to note that aerial spray modeling has reached the point today where it is a tool running on personal computers and available to an application specialist seeking alternative spray strategies.

Chapter 3 discusses atmospheric modeling at the urban scale of 1-10 km. The chapter describes the development of a **metropolitan airshed pollution model** applied to the city of Perth, Australia. The simulation approach combines a numerical mesoscale meteorological model with a simple Langevin scheme to simulate the dispersion of pollutants emitted from both point and area sources. The method is applied to simulate the transport of automotive emissions in the city of Perth.

Chapter 4 discusses atmospheric modeling for mesoscale applications with a scale of 100-1000 km. The chapter focuses on **Lagrangian particle dispersion modeling** (or Monte-Carlo modeling) and describes two dispersion modeling systems: the Mesoscale Dispersion Modeling System (MDMS) and the Hybrid Particle Concentration Transport (HY-PAC) model. Two examples of applications of the Lagrangian particle dispersion model for complex terrain in the southwestern United States and eastern Europe demonstrate a design of computationally intensive air-quality studies with the aid of modern workstations.

Chapter 5 discusses atmospheric modeling at the long-range scale of 1000 km and more. The chapter presents an overview of state of the art in **long range dispersion models** of pollutants in the atmosphere. It includes the treatment of the most important mechanisms of pollutant removal from the atmosphere, such as dry and wet deposition and chemical reaction. The chapter also presents the results of a recent intercomparison study of long-range transport models applied to simulate the transport of the radioactive plume emitted during the Chernobyl accident on April 1986.

Chapter 6 discusses a new method of **rainfall-runoff modeling** and its applications in catchment hydrology. Rainfall-runoff models play a central role in catchment hydrology, assisting in a wide range of investigations such as assessment of the hydrological impacts of land use and possible climate changes, real-time flood forecasting and "design flood" estimation, assessment of the reliability of natural water resources, and investigations of river water quality. The chapter describes a new rainfall-runoff modeling method and presents several of its applications. A discussion is also included of the potential for information-transfer to ungauged catchments (regionalization) using statistical relationships linking Physical Catchment Descriptors (PCDs) and Dynamic Response Characteristics (DRCs) derived from Unit Hydrographs (UHs) parameters.

Chapter 7 discusses groundwater pollution modeling and focuses on **finite-element modeling of the transport of reactive contaminants in variably saturated soils**. The chapter presents the fundamental equation of nonlinear groundwater flow in variably saturated soils and develops a basic finite-element method for its solution under steady and transient conditions. The advantage of using finite-element techniques is that they provide high accuracy and can be used to simulate irregular three-dimensional domains with complex boundary conditions. The contaminant-transport model is expressed for both local equilibrium assumption (LEA) conditions (i.e. conditions in which contaminant sorption onto solid grains is simplified by assuming that chemical equilibrium is achieved instantaneously) and non-LEA conditions. The non-LEA model is solved by an integrodifferential approach involving a convolution integral, thus achieving a significant saving of computer storage and CPU time. The numerical models are successfully tested by comparing simulation results with solutions reported in the literature.

Chapter 8 discusses an important subtopic of groundwater pollution: the mechanisms and models for **aggressive-permeant interactions with soils**. This is a significant issue because aggressive permeants are capable of altering the hydraulic properties of the soils through which they flow. Consequently, they may alter the hydraulic conductivity of soil barriers used to contain them and escape into the environment. The chapter presents a comprehensive look at the subject of aggressive-permeant interactions with soils and discusses strategies for modeling the impact of aggressive permeants and for analyzing their preferential flow pathways. In addition, algorithms are presented for testing alternative aggressive-permeant interaction and for extracting the required model coefficients from typical permeameter observations.

Chapter 9 discusses theory and application of **variational data assimilation**. The objective of variational four-dimensional data assimilation is to find the solution to a numerical forecast that will best fit a series of observational fields distributed over some space and time interval. The basis of variational data assimilation is to use all the available information in order to produce the best possible initial state in a least-square norm optimal sense. Available information includes observations that are distributed more or less regularly in both time and space and vary greatly in their nature and accuracy, along with a numerical prediction model that describes physical conservation laws. This chapter presents the main aspects underlying a rigorous derivation of variational data assimilation, with special emphasis on meteorological applications.

Finally, chapter 10 discusses **expert-system support for environmental decisions**. Expert systems differ from traditional mathematical models in their ability to handle qualitative information that is very common in environmental decision problems. This chapter revises the architecture and the most important features of expert systems and shows how they can support environmental decision-makers. Two available environmental expert systems–OASIS and PigE–are described. Moreover, five roles for expert systems in environmental applications are identified: 1) to help in selecting appropriate models, 2) to help in calibrating the chosen models, 3) to predict adverse environmental effects, 4) to help users in understanding model outputs, and 5) to generalize model structures.

4. A Guide to the Reader of *ENVIRONMENTAL MODELING*, Volumes 1 and 2

This section provides some guidance to the reader of both volumes.

For a **general introduction to pollution modeling** in different environmental media, read Chapter 5 of volume 1. This chapter addresses multimedia modeling, i.e. the transport and fate of chemicals in the atmosphere, surface water, soil (including groundwater), and biota. The physico-chemical processes that govern the transport and fate of chemicals in each of these media are described, and the basic equations that represent these processes are discussed. The chapter includes a section on available software. Note that the topics in this chapter have recently been expanded into a series of technical articles in the quarterly journal *Environmental Software*, currently published by Elsevier. Additional useful information can also be found in Chapter 1 of each volume.

For the reader interested in **atmospheric modeling**, we suggest the following sequence of reading:

> Chapter 2 of volume 1 gives an introduction to atmospheric models and in particular the dynamics of atmospheric pollution. Different components and scales of the phenomenon are discussed: indoor air pollution; local-scale, urban, and regional scale pollution; and global air pollution. Different frameworks for air-quality modeling are presented, and several modeling applications are discussed. The chapter includes a section on available software.
>
> For a comprehensive description of air pollution modeling topics, the reader can also examine the textbook by Zannetti (1990).
>
> Chapters 2,3,4, and 5 of volume 2 provide expanded discussions of atmospheric modeling issues at different scales–from the local scale to the continental one. They cover, respectively, aerial spray drift modeling, the development of a metropolitan airshed pollution model, Lagrangian particle dispersion modeling, and long-range dispersion models. See the previous section for a brief outline of these four chapters.

For the reader interested in **surface hydrology and water pollution**, we suggest the following sequence of reading:

> Chapter 6 of volume 2 discusses a new method of rainfall-runoff modeling and its applications in catchment hydrology. See the previous section for a brief outline.
>
> Chapter 3 of volume 1 describes the application of mathematical modeling to the marine environment. In particular, the main characteristics of a general time-dependent three-dimensional model of the marine environment are presented, and the possibility of extracting smaller submodels from the general model is examined. Also, case studies with applications to the northwest European continental shelf and to the northern Bering Sea are described.
>
> Chapter 4 of volume 1 covers the modeling of the water quality of rivers and lakes. Strategies of model-building, governing equations, and case studies are presented. The case studies cover the Hungarian Danube, Lake Balaton, Lake Kuortaneenjarvi, and Lake Tuusulanjarvi. The chapter ends with a discussion of decision-support

systems, i.e. structures in which models are not used in an independent fashion but are part of a larger set of user-friendly interface tools.

For the reader interested in **groundwater hydrology and water pollution**, we suggest the following sequence of reading:

For a comprehensive description of groundwater modeling topics the reader can examine the textbook by Anderson and Woessner (1992).

Chapter 7 of volume 2 discusses groundwater pollution modeling and focuses on finite element modeling of the transport of reactive contaminants in variably saturated soils.

Chapter 8 of volume 2 discusses an important subtopic of groundwater pollution: the mechanisms and models for aggressive-permeant interactions with soils. See the previous section for a brief outline of these two chapters.

For a discussion of **ecological modeling**, read Chapter 6 of volume 1. This chapter presents a survey of some important features in mathematical ecology, with a special emphasis on the structural properties and behavior of ecosystems rather than just numerical aspects. Adverse conditions may induce structural perturbations in normal environmental processes, and the approach outlined in this chapter can be used to assess their extent and how they propagate throughout the entire ecosystem. The chapter includes a section on available software.

For a discussion of **environmental noise modeling**, read Chapter 7 of volume 1. This chapter covers the prediction of noise propagation from industrial plants, transportation noise models (for traffic, rail, and air transportation), modeling applications, a discussion on the accuracy of the predictions, and the modeling of long-term noise impact. The chapter includes a section on available software.

For a discussion of **environmental information management**, read Chapter 8 of volume 1. Since environmental regulations are imposing increased record keeping and reporting requirements, automating environmental information is becoming an important issue. This chapter provides a review of the different kinds of software available commercially, including on-line systems, databases, information services, data management systems, decision-support aids, expert systems, and training packages.

For a discussion of theory and application of **variational data assimilation**, read Chapter 9 of volume 2. See the previous section for a brief outline of this chapter.

For a discussion of **expert systems**, read Chapter 10 of volume 2. See the previous section for a brief outline of this chapter. Additional information on the use of expert systems for water-quality studies can also be found at the end of Chapter 4 of volume 1.

Finally, for a discussion of the **future of environmental modeling**, read Chapter 9 of volume 1. It presents a stimulating discussion on the growing role of environmental modeling. Nine facets of the evolution of environmental modeling over the next decade are examined from the perspective of a computer scientist: capability (performance, new science, and challenges), uniformity (databases, nomenclature, and interfaces), and accessibility (lower cost, user-friendliness, and networking). Additional useful information can also be found in Chapter 1 of each volume.

5. Conclusion

In conclusion, I want to thank all the authors of the 17 technical chapters in volumes 1 and 2 for their valuable efforts under challenging conditions and strict deadlines.

Hopefully, the *Environmental Modeling* series will not be just black ink on white paper. In fact, we plan to include "computerized" chapters(*) and user-friendly software in the series soon. By examining computerized chapters provided on disk, the readers will be able to access information in which text, data, equations, numerical solutions, and graphics are fully interconnected. The readers will then be able to work with all the material provided in the chapters and, with just a few keystrokes, adjust equations, modify data, and recalculate and replot the results under modified assumptions. Also, the inclusion of user-friendly versions of environmental models as part of the chapter material will allow readers to run environmental simulations to verify their understanding of modeling theories and numerical implementations.

Finally, I thank the readers for their support and I encourage the environmental scientific community to provide input, suggestions, contributions, and constructive criticism.

References

Anderson, M.P. and W. W. Woessner. *Applied Groundwater Modeling–Simulation of Flow and Advective Transport.* San Diego: Academic Press, 1992.

Zannetti, P. *Air Pollution Modeling–Theories, Computational Methods and Available Software.* New York: Van Nostrand Reinhold, 1990.

Acknowledgments

The author is grateful to Ms. Susan Ham of Failure Analysis Associates, Inc. for her editorial review of the manuscript.

Disclaimer

The opinions presented herein are those of the author alone and should not be interpreted as necessarily those of Failure Analysis Associates, Inc. (FaAA).

(*) Computerized chapters can be made today using special programs, such as *Mathcad, Maple V,* or *Mathematica*. The most striking feature of these software packages is their ability to create, on personal computers, interactive documents that mix text, animated graphics, and sound with active formulae. Using these packages, an author can write an article or a book in which the reader can modify and solve, analytically or numerically, a set of equations. These solutions generate figures, sound, animation. A simple change of a parameter or a variable in the text causes new solutions, new figures, new animation. In this way, the book or the article becomes a learning and testing environment for the reader.

Chapter 2

Aerial spray drift modeling

M.E. Teske,[a] **J.W. Barry,**[b] **H.W. Thistle, Jr.**[c]

[a] *Continuum Dynamics, Inc., P.O. Box 3073, Princeton, New Jersey 08543, USA*

[b] *USDA Forest Service, Forest Pest Management, 2121C Second Street, Suite 102, Davis, California 95616, USA*

[c] *USDA Forest Service, Missoula Technology and Development Center, Fort Missoula, Building 1, Missoula, Montana 59801, USA*

Abstract

A review of the assumptions, approaches and techniques applied to the modeling of spray drift from aerial applications is presented. Several topics of particular interest include: (1) the importance of an accurate description of the spray material issuing from the spray nozzle; (2) the differences between Lagrangian and Eulerian model approaches; (3) the need for an accurate representation of the flow field behind the aircraft; (4) the influence of meteorological properties and evaporative effects on the released spray material; and (5) the sensitivity of ground deposition results.

Key words

Modeling, Spray Behavior, Aerial Application, Computer, Pesticides.

Introduction

One of the most important professions in our society is that of ag pilot, a specialist who applies agricultural chemicals and biologicals by aircraft in an effort to ward off the destruction of crops vital to the world's health and well-being. With the banning of DDT (dichloro diphenyl trichloroethane) and 2,4-D (2,4-dichlorophenoxyacetic acid) in the 1960s, chemical companies have been moving into the development of biorational agents, while retaining their product lines through anticipated re-registration efforts with the U. S. Environmental Protection Agency. The Spray Drift Task Force was recently organized to coordinate a combined industry program to understand aerial spray drift, and respond to the environmental and social concerns present with the subjects of pest management and crop quality. The escalating costs of agricultural products, liability issues and our litigious society all trend to make aerial application a vulnerable -- yet necessary -- topic area.

These market factors suggest that the use of chemicals and biologicals now requires a higher degree of application precision than previously practiced. Candidate sprays demand more attention to application technology and timing, atmospheric conditions, and the target's physical and behavioral characteristics than ever before. Emphasis has shifted to developing field techniques that increase pesticide deposit on the target, and reduce spray drift, while also reducing the amount of spray material needed, yet depositing required levels to achieve all application objectives. Now, total accountancy of all released spray material from the aircraft is important: where does the material go after it leaves the spray nozzles (particularly downwind)?; how much spray evaporates (and what happens to the vapor and is it "important")?; where does the spray material deposit through the depth of the crop?; where do the smallest drops in the spray (the "fines") end up?; and what happens when the chemical enters the soil? These issues are complex, and will (unfortunately for the reader) not be resolved here. However, in this review article we will examine some of the traditional ways of modeling the trajectories of spray material released from aircraft, falling through the atmosphere, evaporating and depositing upon crops and the ground. Hopefully, we will provide some insight into the breadth of the problem, and the techniques employed to describe it today.

A spray aircraft flies over a crop, releasing spray material through a set of spray nozzles fitted to a spray boom that typically runs along the trailing edge of the aircraft wing, or beneath the fuselage of a helicopter. The spray material is mixed in a tank prior to flight, following label specifications with an eye toward the application parameters, particularly the ambient meteorological conditions. Small amounts of an adjuvant are sometimes added to inhibit evaporation, or otherwise achieve some other objective. The spray system pumps the material through the nozzles, where it is atomized into a drop size distribution, typically described by its volume median diameter $D_{v.5}$, the midpoint along the drop size distribution where half of the material volume is below, and half above, this point. Most agricultural spray materials are atomized such that their $D_{v.5}$ values are between 150 and 450 micrometers. A typical drop size distribution is shown in Figure 1. Plotted in this manner, distributions all show the inclusion of larger drop sizes (which deposit directly beneath the aircraft in its swath) and smaller drop sizes (the "fines").

These drops are sheared by the airflow and convected by local wind and turbulence effects around the aircraft, but are always moving under gravity toward the ground. Most aerial applications are made just after sunrise (with nearly quiescent atmospheric conditions), close to the top of the crop canopy (to avoid drift), around 5 to 15 feet. Evaporation effects act on individual drops to reduce their size further, and ambient atmospheric winds move the spray material downwind. Finally, the material deposits on the crop or ground.

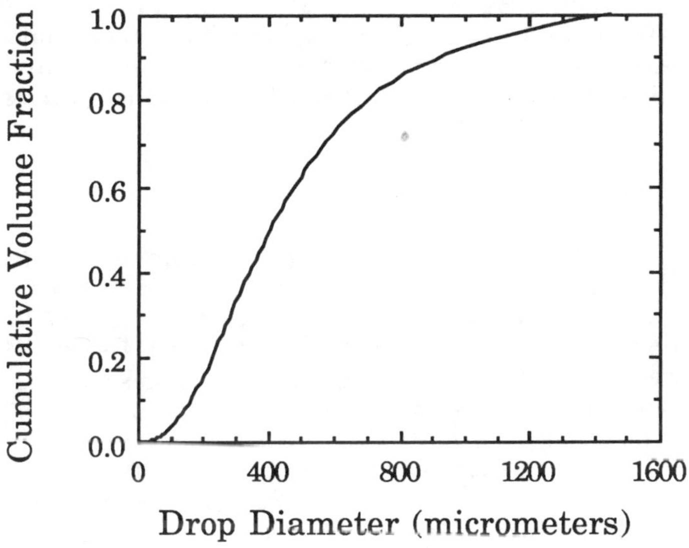

Figure 1: A typical drop size distribution from an agricultural spray nozzle. In this example $D_{v.5}$ = 408.0 micrometers.

Thus, drop size distribution, aircraft wake effects, and ambient atmospheric effects all influence the eventual deposition pattern. Within a canopy, target size and shape, density and type of foliage, and the velocity of the drops moving through the canopy, all influence deposition as well. Specific topics such as canopy penetration and deposition, target charge effects, foliage influence, and the like, will not be discussed here. These topics are less well-defined than the behavior of spray material above the canopy, to which we will restrict our remarks. Models have been constructed to approximate all of these effects, but we will confine our attention to what we know most about, namely the behavior of spray material from the nozzle to the top of the canopy or the ground.

The first known use of an aircraft to spray damaging insects in the United States occurred in 1921 and was reported by Neillie and Houser (1922). As they stated, "the authors began seeking an opportunity to conduct a practical test of the airplane as a distributor of insecticides. In a few instances the plan was received with favor; in others it was considered a theoretical, impractical and foolish undertaking. The outstanding feature of the application was the remarkable precision with which the poison could be placed at the point intended, thus dispelling the idea expressed by many before the test was made that the poison dust would be tossed willy-nilly by the air currents -- wholly beyond control. The success attending the initial effort to use the airplane in fighting destructive insect pests gives rise to the hope that the method may be extensively developed."

Since then, many aerial applications have occurred, and considerable effort has gone into understanding the engineering behind the process. The USDA Forest Service and the U. S. Army have both been major players in this development, but other organizations, and many individuals, have played significant roles as well. We will touch upon their contributions as we are able.

A Solution Primer

The physics of aerial spray modeling is contained in a designation of the spray "cloud" emitted from each nozzle on the aircraft (there could be 50 or more such source locations). The drop size distribution, usually obtained from a wind tunnel experiment (Skyler and Barry 1991, Esterly et al. 1993) is broken into convenient categories by the optical/laser scanning instrument taking the measurement. These categories are represented by mass fractions of released material (totaling 1.0) with an average drop size diameter -- typically the volume average of Herdan (1960), although other representations are possible (Sowa 1992). With the aircraft assumed in level flight, a continuous stream of material ejects from each nozzle, and the problem becomes two-dimensional in time (although the third dimension may be added for completeness). The transport of the released spray material within computational planes perpendicular to the aircraft flight path then becomes the focus of the analysis. Accuracy of prediction requires an accurate description of the aircraft wake and the local atmospheric effects (such as temperature and relative humidity), along with a description of the wind speed and direction.

Thus, the solution to the aerial spray problem encompasses two parts. First, background atmospheric and aircraft wake behavior must be resolved (or approximated). Then, the effects of the atmosphere and aircraft wake on the released spray material must be determined. These two distinct pieces to the solution (the structure of the background flow field and the behavior of the spray material within it) may be resolved in two ways:

1. Lagrangian. The behavior of each nozzle stream is tracked through the assumed flow field (modeling aircraft wake and atmospheric effects) by writing trajectory equations of motion. The nozzle streams are followed to canopy top or ground impact, and a model for dispersion is developed to recover deposition.

2. Eulerian. The Navier-Stokes equations of motion are solved in a two-dimensional, unsteady coordinate system, with spray material dispersion

determined through the solution of passive tracer equations of motion or Lagrangian trajectories. A subset of this approach is the Gaussian approach, where the form of the spray cloud is assumed, and integration across the cloud results in expressions for the behavior of the dispersion of the cloud. In this approach as well, simplifying assumptions must be made about the structure of the background flow fields.

These approaches will be reviewed in the text that follows. It is important to realize that none of these techniques would have received much play had it not been for the current focus on environmental concerns, as discussed above, and the advent of affordable personal computers. These approaches were first developed on mainframe computers, and, as personal computers developed and computing power grew exponentially, they adjusted to the changes, and are now available -- in one stage or another -- on faster personal computers and workstations. With their increased availability has come a need for simplifying inputs and "user-friendly" interfaces, ease of operation, and support (through contact persons and established user groups) in case of problems. Aerial spray modeling has reached the point today where it is a tool available to an application specialist seeking alternative spray strategies. Within the next five years, aerial spray models will become a part of expert systems, and may be utilized during litigation.

Lagrangian Approach

Reed (1953) first developed the equations of motion for material released from a nozzle on a spray aircraft. His insight was the realization that the wingtip vortices played a significant role in the subsequent behavior of the released spray. His model integrated the equations governing the dynamics of single drops in a vortex flow field modeled as two counter-rotating irrotational line vortices with separation distance equal to the span of the wing. Image vortices were used to simulate an inviscid ground plane. Resultant trajectories were determined for several drop sizes released at nozzle positions along the wing; then, with the drop size distribution known, an expression for the deposition on the ground was developed.

The strength of the Reed model rests in its simplicity. It has been the starting point for additional studies (Williamson and Threadgill 1974), additional models including propeller wash, crosswind and evaporation (Trayford and Welch 1977, Atias and Weihs 1985), and the addition of bound circulation on the wing, as well as some three-dimensionality in the aircraft wake (Bragg 1986). A consistent approach that includes the effects of wake turbulence on the growth of the cloud from each nozzle became known as the AGDISP model (Bilanin and Teske 1984, Bilanin et al. 1989). Since this model incorporates all of the features of the Lagrangian approach embodied in the original Reed model (and its derivative works), it will be reviewed here in detail.

Equations of Motion

The equations governing the behavior of a drop in motion are

$$\frac{d^2}{dt^2}(X_i + x_i) = [U_i + u_i - V_i - v_i]\left[\frac{1}{\tau_p}\right] + g_i \tag{1}$$

$$\frac{d}{dt}(X_i + x_i) = V_i + v_i \tag{2}$$

where the drag force on the drop is represented by the relaxation time

$$\tau_p = \frac{4}{3}\frac{D\,\rho}{C_D\,\rho_a\,|\,U_i - V_i\,|} \tag{3}$$

Here, the position and velocity vectors are written in tensor notation as the sum of an ensemble averaged mean (upper case) and fluctuating components about the mean (lower case), with

t	=	time
U_i	=	mean local velocity at the drop location
V_i	=	mean drop velocity
g_i	=	gravity = $(0,0,-g)$
D	=	drop diameter
ρ	=	drop density
ρ_a	=	air density

and C_D is the drag coefficient of the drop. C_D is evaluated empirically for spherical drops, and may be taken as

$$C_D = \frac{24}{Re}\left[1 + 0.197\,Re^{0.63} + 0.00026\,Re^{1.38}\right] \tag{4}$$

from Langmuir and Blodgett (1964), or other alternate expressions

$$C_D = \frac{24}{Re}\left[1 + 0.1\,Re^{0.75}\right] \tag{5}$$

from Ishii and Zuber (1979), or

$$C_D = \frac{24}{Re}\left[1 + Re^{0.687}\right] \tag{6}$$

from Reist (1984). In these expressions Re is the Reynolds number

$$Re = \frac{\rho_a\,D\,|\,U_i - V_i\,|}{\mu_a} \tag{7}$$

where μ_a is the air viscosity.

The relaxation time (defined by eqn 3) has physical significance with regard to dispersion in that it is the e-folding time required for a drop to catch up to its local ambient velocity. If eqns 1 and 2 are ensemble averaged, we obtain

$$\frac{d^2X_i}{dt^2} = [U_i - V_i]\left[\frac{1}{\tau_p}\right] + g_i \tag{8}$$

$$\frac{dX_i}{dt} = V_i \tag{9}$$

These equations were first examined by Reed (1953), and may be solved to obtain the mean trajectory paths for the spray material issuing from each nozzle. To close the problem, the local ambient velocity field U_i must be specified in eqn 8. Reed assumed that the ambient velocity field was generated by a counter-rotating pair of vortices, positioned at the aircraft wingtips. This field provides most of the ambient velocity effects close to the aircraft, and will be described later.

Substituting eqn 8 into eqn 1, and eqn 9 into eqn 2, results in the fluctuation equations, which may be pre-multiplied by x_i and v_i, ensemble averaged and manipulated to yield

$$\frac{d}{dt}\langle x_i x_i\rangle = 2\,\langle x_i v_i\rangle \tag{10}$$

$$\frac{d}{dt}\langle x_i v_i\rangle = \left[\langle x_i u_i\rangle - \langle x_i v_i\rangle\right]\left[\frac{1}{\tau_p}\right] + \langle v_i v_i\rangle \tag{11}$$

$$\frac{d}{dt}\langle v_i v_i\rangle = 2\left[\langle u_i v_i\rangle - \langle v_i v_i\rangle\right]\left[\frac{1}{\tau_p}\right] \tag{12}$$

where the i indices are not summed, and $\langle\ \rangle$ denotes a turbulent correlation. Equation 10 reflects the growth of the spray cloud, since it may be argued that $\sigma^2 = \langle x_i x_i\rangle$ is the square of the standard deviation of the growth of a Gaussian representation around the drop position. Equations 10 to 12 require the specification of $\langle x_i u_i\rangle$ and $\langle u_i v_i\rangle$ before solution is possible.

These correlations are developed by assuming that the fluctuating local ambient velocity is given by

$$u_i = \frac{1}{2\pi} \int_{-\infty}^{\infty} u_i(\omega)\, e^{i\omega t}\, d\omega \tag{13}$$

and solving the fluctuating equations

$$\frac{d^2 x_i}{dt^2} = (u_i - v_i)\left[\frac{1}{\tau_p}\right] \tag{14}$$

$$\frac{dx_i}{dt} = v_i \tag{15}$$

to obtain solutions for x_i and v_i

$$x_i = \frac{1}{2\pi} \int_{-\infty}^{\infty} u_i(\omega) \left[\frac{e^{i\omega t} - 1 + i\omega\tau_p\left(e^{-t/\tau_p} - 1\right)}{i\omega\left(1 + i\omega\tau_p\right)}\right] d\omega \tag{16}$$

$$v_i = \frac{1}{2\pi} \int_{-\infty}^{\infty} u_i(\omega) \left[\frac{e^{i\omega t} - e^{-t/\tau_p}}{1 + i\omega\tau_p}\right] d\omega \tag{17}$$

Upon multiplying these results by u_i and ensemble averaging, equations are then obtained for the needed correlations. Both equations contain the spectral density function

$$\Phi_u(\omega) = \langle u_i(\omega) u_i(\omega) \rangle \tag{18}$$

for transverse velocity fluctuations. For isotropic turbulence, von Karman and Howarth (1938) showed that fluctuations normal to the mean flow may be expressed as

$$\Phi_u(\omega) = \frac{1}{3\pi} \frac{\Lambda}{U} q^2 \frac{1 + 3\,(\omega\Lambda/U)^2}{\left[1 + (\omega\Lambda/U)^2\right]^2} \tag{19}$$

where Λ is the integral scale of the turbulence, q^2 is the mean square turbulence level, and U is the free stream velocity. Here we interpret U as the relative velocity $|U_i - V_i|$, thereby permitting us to integrate the equations analytically to find

$$\langle x_i u_i \rangle = \frac{q^2}{3}\left[-\tau_p K + \frac{\tau_t}{2}\right] \tag{20}$$

$$\langle u_i v_i \rangle = \frac{q^2}{3} K \tag{21}$$

where

$$K = \frac{1}{2}\frac{\left[3 - \left(\frac{\tau_p}{\tau_t}\right)^2\right]\left[1 - \frac{\tau_p}{\tau_t}\right] + \left(\frac{\tau_p}{\tau_t}\right)^2 - 1}{\left[1 - \left(\frac{\tau_p}{\tau_t}\right)^2\right]^2} \tag{22}$$

and τ_t is the travel time of the drop through a turbulent eddy of scale Λ

$$\tau_t = \frac{\Lambda}{|U_i - V_i| + \frac{3}{8}q} \tag{23}$$

Equations 10 to 12 are therefore governed by two time scales τ_p and τ_t. Their consistent limiting behavior is of interest here: for times t large compared to τ_p, these equations may be solved to obtain

$$\langle x_i x_i \rangle = 2\left[\langle x_i u_i \rangle + \langle u_i v_i \rangle\, \tau_p\right] t \tag{24}$$

As τ_p approaches zero, the drop will follow the fluctuations in the field and

$$\langle x_i x_i \rangle = \frac{q^2}{3}\, \tau_t\, t = \frac{8}{9}\, q\, \Lambda\, t \tag{25}$$

recovering the well-known large-time result that the position variance grows as the product of the turbulence level q and time (Monin and Yaglom 1973). In the limit of massive drops, the position variance may be shown to grow as

$$\langle x_i x_i \rangle = \frac{q^2}{3}\left(\frac{\tau_t}{\tau_p}\right)^3 t^2 \tag{26}$$

where the time-squared behavior is consistent with the short-time result (Monin and Yaglom 1973). As τ_p approaches infinity in this limit, the variance remains zero since the drops are too massive to be dispersed by turbulent fluctuations.

With the position and velocity information available for the drop at any time during the simulation, eqns 8 to 12 may be integrated exactly as an initial value problem for the solution at the next time step, with the assumption that the background conditions U_i , $\langle x_i u_i \rangle$ and $\langle u_i v_i \rangle$ are constant across a time step. For example, the mean equations become, for time step Δt

$$X_i = X_o - \tau_p \left[U_i - V_o + g_i \, \tau_p \right] \left[1 - e^{-\Delta t / \tau_p} \right] + \left[U_i + g_i \, \tau_p \right] \Delta t \tag{27}$$

$$V_i = - \tau_p \left[U_i - V_o + g_i \, \tau_p \right] e^{-\Delta t / \tau_p} + U_i + g_i \, \tau_p \tag{28}$$

where X_o and V_o are initial values at the beginning of the time step. Use of this solution restricts the size of the time step taken, to ensure consistency with evolving background conditions. Nonetheless, it is clear that the behavior of the released spray material is intimately related to the local ambient velocity field into which the spray drops are ejected by the nozzle. A discussion of the models for this ambient velocity field follows.

Flow Field Models

While many features may be modeled in the wake of an aircraft, such as vortex wake rollup, helicopter downwash, propeller swirl, and jet engine effects (Teske 1990), the dominant features are the wingtip vortical motion and crosswind.

When an aircraft flies at a constant altitude and speed, the aerodynamic lift generated by the lifting surfaces of the aircraft equals the aircraft weight. The majority of the lift is carried by the wings, and generates one or more pairs of swirling masses of air (vortices) downstream of the aircraft (Donaldson and Bilanin 1975). If the rollup of this trailing vorticity can be approximated as occurring immediately downstream of the wing, then the mean velocity field that results may be given by known aircraft characteristics and the wing load distribution. With a conventional rectangularly loaded wing, the strength of each wingtip vortex is computed from

$$\Gamma = \frac{W}{2 \, \rho_a \, s \, U_\infty} \tag{29}$$

with the local swirl velocity around each vortex approximated by

$$V_s = \frac{\Gamma}{2 \, \pi} \frac{r}{(r + r_c)^2} \tag{30}$$

where

W	=	aircraft weight
s	=	aircraft semispan
U_∞	=	aircraft speed
r	=	radius from vortex center
r_c	=	core radius

For a vortex pair the superimposed effects of four vortices are used to simulate the overall proximity to the ground, with image vortices maintaining the no-flow inviscid ground condition. The vortex strength Γ may itself decay with time because of atmospheric turbulence (Teske, Bilanin and Barry 1993).

The second most important effect in aerial sprays is the crosswind velocity profile, since the crosswind is responsible for drift off the target and downwind. Two typical profiles are selected, either the standard logarithmic profile

$$U = U_r \frac{\ln (z/z_o)}{\ln (z_r/z_o)} \tag{31}$$

where U_r is the known value of horizontal velocity at a given height z_r , and z_o is the surface roughness; or the power law

$$U = U_r \left(\frac{z}{z_r}\right)^{p} \tag{32}$$

where p is a specified exponent. A typical trajectory plot is shown in Figure 2, where the influence of the vortices and crosswind contribute to the predicted Lagrangian pattern from the spray nozzles.

Deposition

As the spray cloud approaches the ground, deposition begins, and continues until all non evaporated material deposits. Ground deposition in the original Reed model, updated by Bragg (1986), is computed by developing the concentration along the ground as

$$C = Q \frac{dS}{dD} \frac{dD}{dy} \tag{33}$$

where

Q	=	nozzle flow rate
S	=	drop size distribution (as in Figure 1)
y	=	distance along the ground

Alternately, we may assume that the concentration of material around the mean drop location (Y, Z) is Gaussian

$$C = \frac{1}{2\,\pi\,\sigma^2}\exp\left[-\frac{(y-Y)^2}{2\,\sigma^2}\right]\exp\left[-\frac{(z-Z)^2}{2\,\sigma^2}\right] \tag{34}$$

where z is the vertical distance. If the material deposits entirely at the point of surface contact, the normalized deposition pattern becomes

$$M = \frac{1}{\sqrt{2\,\pi}\,\sigma}\exp\left[-\frac{(y-Y)^2}{2\,\sigma^2}\right] \tag{35}$$

recovering the familiar Gaussian deposition about the impact point Y, with σ^2 obtained from the solution of eqn 10.

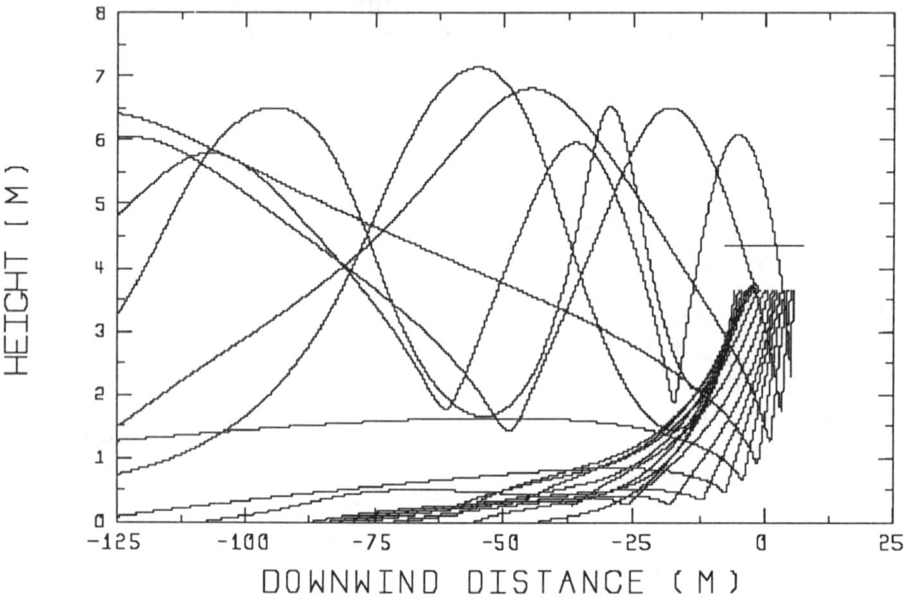

Figure 2: A trajectory plot for 70 micrometer drops emitted from nozzles at the trailing edge of a wing. The Lagrangian solution tracks each trajectory stream through its influence by the wingtip vortices and downwind to the ground.

A better result may be obtained by assuming the drop to deposit incrementally as it approaches the surface. This assumption results in the deposition

$$M = \frac{1}{2\sqrt{2\,\pi}\,\sigma}\exp\left[-\frac{(y-Y)^2}{2\,\sigma^2}\right]\mathrm{erfc}\left(\frac{Z}{\sqrt{2}\,\sigma}\right) \tag{36}$$

where erfc is the complementary error function. Ground deposition is obtained by summing all incremental contributions to M as the trajectory calculation proceeds. It may be noted that for material falling vertically toward the surface, the ground deposition pattern generated by eqn 36 will be identical to the Gaussian deposition of eqn 35. Equation 36, however, leads to more realistic ground deposition distributions.

A typical deposition pattern is shown in Figure 3. The in-swath deposition (near $y = 0$) falls off downwind (to the right) through a sharp drop (where buffer zones are positioned near sensitive areas), then a gradual reduction in deposition farther downwind. The lane separation between flight lines sets the level of deposition beneath the aircraft, and the degree of variability in the deposition beneath the aircraft (Teske, Twardus and Ekblad 1990), and all of the spray material, nozzle, aircraft and meteorological parameters contribute to the eventual downwind deposition pattern.

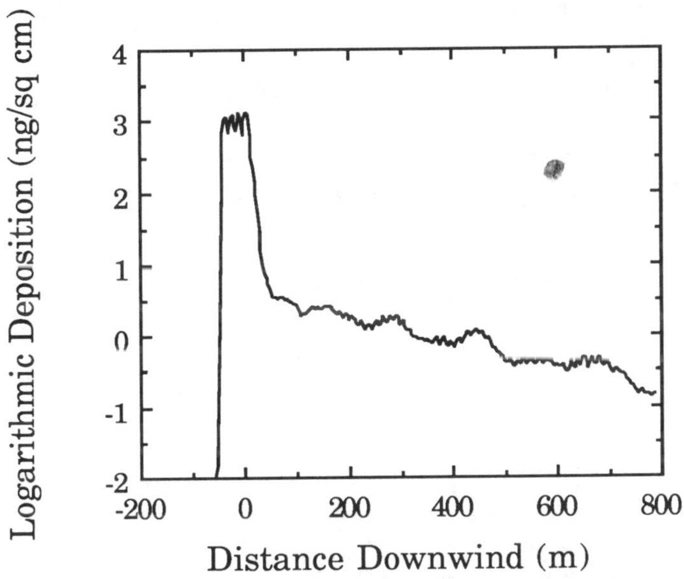

Figure 3: A deposition pattern predicting the movement of material 800 m downwind from several flight lines, through four orders of magnitude in deposition level.

Eulerian Approach

An alternate approach to modeling ambient background and aircraft wake behavior is to solve the Navier-Stokes equations to recover the flow field. This approach was first taken in the mid-1970s in response to the development of jumbo jet liners, and their potential impact on airport operations (because of the significant vortices generated). Papers on the subject have dealt with the behavior of aircraft vortices (Steger and Kutler 1976, Bilanin, Teske and Williamson 1977), particularly close to the ground (Bilanin, Teske and Hirsh 1978, Atias and Weihs 1984) where the phenomena of vortex "bouncing" occurs (Dee and Nicholas 1968, Harvey and Perry 1971,

Bradshears, Logan and Hallock 1975). Here, counter-sign vorticity generated by the approaching vortex pair actually "lifts" the pair off the ground. A concerted effort to develop computational tools to solve this problem led to the development of the WAKE code (Teske 1976) and its later enhancement UNIWAKE (Bilanin and Teske 1988, Teske 1988a, Teske, Quackenbush and Bilanin 1991), now used to analyze exhaust chemistry for the High Speed Civil Transport (Quackenbush, Teske and Bilanin 1993) and aircraft signature detection.

The development of the appropriate equations of motion is based on a description of the flowfield by invariant second-order closure of the turbulence (Donaldson 1973, Lewellen 1977). A comprehensive review of this subject may be found in Lewellen and Teske (1976), along with an examination of the differences between this model and similar approaches, particularly Mellor and Yamada (1974). The model has been extensively validated, including comparisons in the Monin-Obukhov surface layer (Lewellen and Teske 1973), axisymmetric jets and wakes (Lewellen, Teske and Donaldson 1974), and free convection (Lewellen, Teske and Donaldson 1976).

The Reynolds-averaged equations may be written

$$\frac{\partial U_i}{\partial x_i} = 0 \tag{37}$$

$$U_j \frac{\partial U_i}{\partial x_j} = - \frac{\partial}{\partial x_j} \langle u_i u_j \rangle + \frac{\partial}{\partial x_j}\left(\nu \frac{\partial U_i}{\partial x_j} \right) - \frac{1}{\rho_a} \frac{\partial P_a}{\partial x_i} \tag{38}$$

$$U_j \frac{\partial C}{\partial x_j} = - \frac{\partial}{\partial x_j} \langle u_i c \rangle + \frac{\partial}{\partial x_j}\left(\nu \frac{\partial C}{\partial x_j} \right) \tag{39}$$

$$U_k \frac{\partial}{\partial x_k} \langle u_i u_j \rangle = - \langle u_i u_k \rangle \frac{\partial U_j}{\partial x_k} - \langle u_j u_k \rangle \frac{\partial U_i}{\partial x_k} + 0.3 \frac{\partial}{\partial x_k}\left(q\Lambda \frac{\partial}{\partial x_k} \langle u_i u_j \rangle \right)$$
$$- \frac{q}{\Lambda}\left[\langle u_i u_j \rangle - \delta_{ij} \frac{q^2}{3} \right] + \frac{\partial}{\partial x_k}\left(\nu \frac{\partial}{\partial x_k} \langle u_i u_j \rangle \right) - \delta_{ij} \frac{q^3}{12\,\Lambda} \tag{40}$$

where index notation is implied, and

ν	=	absolute viscosity
P_a	=	ambient pressure
C	=	spray material passive tracer

The above equation set is solved in ground effect using the parabolic approximation. The problem is reformulated in streamfunction-vorticity variables, which results in a Poisson equation solved with the direct solver of

Swarztrauber and Sweet (1975). This approach removes the pressure as a variable in eqn 38. The Reynolds stresses $\langle u_i u_j \rangle$ and $\langle u_i c \rangle$ are approximated by the expressions

$$\langle u_i u_j \rangle = \delta_{ij} \frac{q^2}{3} - \frac{1}{4} q \Lambda \left[\frac{\partial U_i}{\partial x_j} + \frac{\partial U_j}{\partial x_i} \right] \quad (41)$$

$$\langle u_i c \rangle = - \frac{1}{3} q \Lambda \frac{\partial C}{\partial x_i} \quad (42)$$

from Sykes, Lewellen and Parker (1986). The UNIWAKE solution casts the equations into a fourth-order accurate scheme (solving for each variable and its first and second derivatives), using the approach suggested by Hirsh (1983). The solution is alternating-direction-implicit, with the streamfunction boundary condition taken as a series expansion of the Green's function of moments of vorticity within the computational domain.

This model of aerial spraying has been used successfully by Picot, Kristmanson and Basak-Brown (1987), and is now offered in a personal computer version of their PKBW code. The simplified wake models of the Lagrangian formulation are replaced by a more complex solution structure, with the effects of turbulence developed by the Navier-Stokes solution.

An illustration of the detail available from the Eulerian approach is shown in Figure 4, where spray material has been ejected into a simple aircraft wake. Spray material is seen to be entrained quickly by the vortex, even before ground effects begin.

Gaussian Approach

An alternate Eulerian approach is that of slant Gaussian plumes. In this technique, which can become algebraically complex, the traditional plume techniques of the 1960s have been applied to aerial spray dispersion. The form of the spray cloud is assumed, then substituted into the Navier-Stokes equations of motion, to obtain expressions for the downwind shape of the spray cloud. Background flow fields are modeled quite simply (to satisfy the equations of motion). One such successful model is FSCBG (Forest Service Cramer-Barry-Grim); a brief review of its history follows.

The USDA Forest Service's self-imposed ban on the use of DDT foreshadowed the importance of understanding where aerially sprayed material is actually deposited. The USDA Forest Service tapped into existing U. S. Army Gaussian plume models to account for the loss of material by gravitational settling of drops from elevated spray clouds, and to predict surface deposition (Cramer et al. 1972). Additional work by Dumbauld, Cramer and Barry (1975) and Dumbauld, Rafferty and Bjorklund (1977) added simple expressions for wake effects of the aircraft, namely, the approach that the released spray is subject to a downwash from the two wingtip vortices equal to the centerline downwash

$$U_w = \frac{4\,\Gamma}{\pi^2 s} \tag{43}$$

obtained from eqn 30 with the vortices positioned at $\pi s/4$ along the wing.

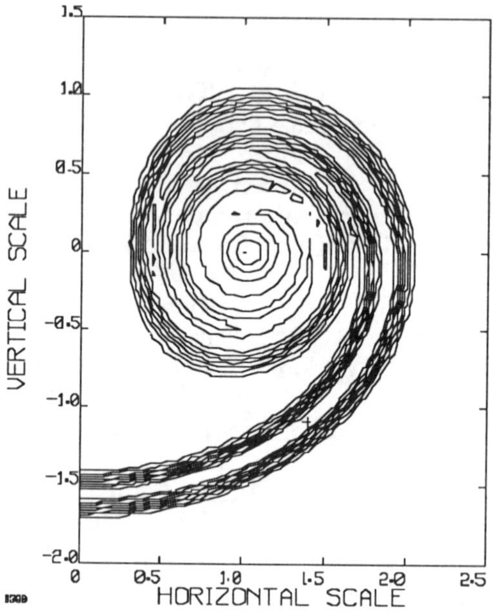

Figure 4: The entrainment of spray material downstream from two nozzles (for clarity) into an aircraft wake (right wing shown in figure). The contour lines give various levels of material concentration, and qualitatively illustrate the significant effect of the vortex on the initial movement of the spray material.

A first version of the FSCBG model was then introduced (Dumbauld, Bjorklund and Saterlie 1980). This model was used extensively for spray prediction and validation, beginning with an initial field study (Boyle et al. 1975). It was later applied to a pilot project in the Withlacoochee State Seed Orchard in Florida (Rafferty et al. 1982) that lead to wide acceptance of aerial application in forests in the Southeast.

Continued success led to the addition of a simplified version of AGDISP as its near-wake model (Bjorklund, Bowman and Dodd 1988), and the development of a revised user-interface and personal computer version (Teske and Curbishley 1991).

The FSCBG model calculates peak concentration, dosage and deposition downwind from nearly instantaneous elevated line sources oriented at an arbitrary angle with respect to the mean wind direction. The axis of the spray cloud is inclined from the horizontal plane by an angle that is proportional to the gravitational settling velocity for each drop size category. When evaporation is negligible, this inclination angle is invariant with distance from the line source.

The amount of spray material released from an instantaneous volume source that is deposited on the ground through gravitational settling is obtained from the expression

$$DEP_V = \frac{Q}{\sqrt{2\pi}\,\sigma_y} \exp\left[-\frac{y^2}{2\sigma_y^2} \right] \sum_{j=1}^{NJ} f_j \frac{d}{dx}\left[\frac{1}{\sqrt{2\pi}\,\sigma_z} \int_{-\infty}^{0} A(x,z)\, dz \right] \tag{44}$$

where

Q	=	volume source strength
σ_y	=	standard deviation of the crosswind spray distribution
σ_z	=	standard deviation of the vertical spray distribution
f_j	=	mass fraction of the jth drop size category (total = NJ)

and $A(x,z)$ is a complicated algebraic expression (Teske et al. 1993a).

The deposition from an elevated integrated line source may then be determined from

$$DEP_L = \int_0^L DEP_V\, d\delta \tag{45}$$

where L is the length of the spray line. The deposition is recovered from the assumption of a Gaussian plume structure.

An example of the Gaussian elevated line source technique is given in Figure 5. Here an aircraft flight line is flown 45 deg to the ambient wind, and spray material is deposited downwind in its familiar pattern.

Evaporation

Evaporation effects are significant in aerial application. As drops evaporate, they tend to drift more, missing the intended target area and becoming a potential liability. Thus, it is not surprising that, beyond the effects of wingtip vortices and crosswind, equations for the evaporation of spray material are a necessary part of any deposition model. Evaporation tends to complicate the Gaussian model considerably, and requires the Eulerian model to add a temperature equation and other features. In the Lagrangian model its application is straightforward.

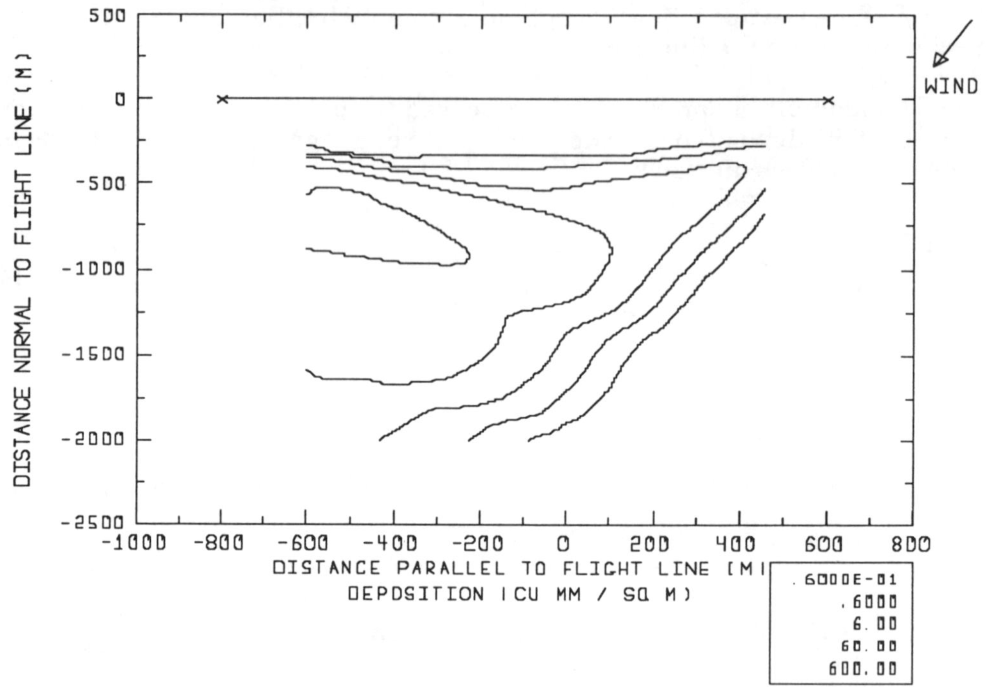

Figure 5: Isopleths of deposition downwind from a single flight line. The contour lines qualitatively illustrate the level of ground deposition.

The time rate of change of drop diameter is generally based on an extension of work by Fuchs (1959), alternately presented in Beard and Pruppacher (1971), Goering, Bode and Gebhardt (1972) and Williamson and Threadgill (1974) applied to the aerial application problem. The governing equation may be written as

$$\frac{dD}{dt} = -\frac{4\, M_p\, D_v\, \rho_a\, (e_s - e_\infty)}{M_a\, D\, \rho\, (P_a - e_s)} f_v \tag{46}$$

where

M_p = molecular weight of the evaporating vapor
M_a = mean molecular weight of the resulting vapor-air mixture (approximated by air)
D_v = molecular diffusivity of the evaporating vapor in air

e_s	=	partial pressure of evaporating vapor at the drop
e_∞	=	partial pressure of evaporating vapor far from the drop
f_v	=	ventilation factor

Drops are assumed to be spherical, with uniform temperature throughout. The variables on the right hand side of eqn 46 may be approximated by various formulae (Hall and Pruppacher 1976, Rasmussen 1978, Pruppacher and Rasmussen 1979), but the variable of most influence is the ventilation factor, which represents the effect of ambient wind moving over the drop. In Pruppacher and Rasmussen (1979) that factor is given by

$$f_v = 1.0 + 0.108\, Sc^{1/3}\, Re^{1/2} \qquad 0.0 \le Sc^{1/3}\, Re^{1/2} \le 1.4$$
$$f_v = 0.78 + 0.308\, Sc^{1/3}\, Re^{1/2} \qquad 1.4 \le Sc^{1/3}\, Re^{1/2} \le 51.4 \tag{47}$$

where Sc is the Schmidt number

$$Sc = \frac{\mu_a}{D_v\, \rho_a} \tag{48}$$

Alternate expressions for f_v are

$$f_v = 1.0 + 0.3\, Sc^{1/3}\, Re^{1/2} \tag{49}$$

from Goering, Bode and Gebhardt (1972), and

$$f_v = 1.0 + 0.276\, Sc^{1/3}\, Re^{1/2} \tag{50}$$

from Williamson and Threadgill (1974). These expressions all modify the evaporation rate slightly.

A simpler evaporation model was suggested by Trayford and Welch (1977), in which the time rate of change of drop diameter is taken as

$$\frac{dD}{dt} = -\frac{D}{2\,\tau_e\left(1 - \frac{t}{\tau_e}\right)} \tag{51}$$

where τ_e is the evaporation time scale of the drop, given by

$$\tau_e = \frac{D^2}{84.76\, \Delta\Theta\left(1 + 0.27\, Re^{1/2}\right)} \tag{52}$$

where the constant is in the units of m^2/sec - deg C and $\Delta\Theta$ is the wet bulb temperature depression in deg C. A typical comparison with data is shown in Figure 6.

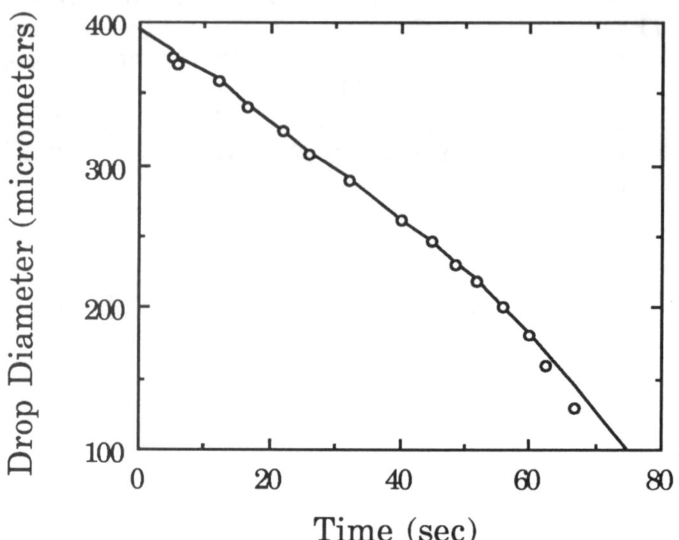

Figure 6: Comparison of the evaporation model (solid curve) with data (open circles) for water (Dennison and Wedding 1984).

Model Validation

Field evaluations play a critical role in the development of any aerial application model. Since the models approximate the complex processes occurring behind an aircraft and in the atmospheric boundary layer, a careful set of field measurements must be obtained before quantitative comparisons with a model can be made. The AGDISP and FSCBG models have benefited from a strong commitment by the USDA Forest Service and the U. S. Army for their validation. In this section we document these field studies and display several model comparisons.

The applicable field evaluations are summarized in Table 1. These evaluations were conducted in a wide range of settings from flat open terrain to forested mountain canyons, and have provided data to validate both AGDISP and FSCBG. Agreement between field measurements and model predictions is given in terms of the correlation coefficient, defined as

$$R^2 = \frac{\Sigma (x_p - \overline{x}_p)(x_d - \overline{x}_d)}{\sigma_p \, \sigma_d} \tag{53}$$

Table 1: Summary of field studies validating AGDISP and FSCBG

Location	Year	Site	Correlation	Reference
Rennic Creek MT	1974	Coniferous forest		Dumbauld, Rafferty and Bjorklund (1977)
Dugway UT	1974	Open desert	0.720	Boyle et al. (1975)
Withlacoochee FL	1980	Pine seed orchard		Rafferty et al. (1982)
Wallops Island VA	1981	Open field		Morris et al. (1984)
Dunphy New Brunswick	1984 1986	Forest	0.752	Mickle (1987)
Chico CA	1985	Almond orchard		Teske (1989a)
Red Bluff CA	1986	Mountain forest	0.762	Rafferty and Bowers (1993)
Sierra Nevada CA	1986	Open terrain	0.440	Teske (1989b)
State College PA	1988	Eastern oak forest	0.640	Anderson et al. (1992)
Heather OR	1989	Douglas-fir seed orchard	0.990	Teske et al. (1991)
Desert Ranch FL	1990	Open field	0.982	Teske & Barry (1993b)
Tauranga New Zealand	1991	Open field	0.644	Richardson et al. (1993)
Parley's Canyon UT	1991 1992	Mountain forest	0.630	Barry et al. (1993)
Plainview TX	1992	Open field	0.898	
Arbuckle CA	1993	Almond orchard	0.727	Teske et al. (1993a)

where x_p and x_d are model predictions and data, respectively, σ_p and σ_d are the standard deviations of model predictions and data, respectively, and an overbar denotes an average value. Since coefficients of 0.8 and above are considered highly correlated when comparing field data, it may be concluded (from Table 1) that the models do a good job of predicting the various data sets. Three comparisons are shown in Figures 7 to 9; others may be found in the references listed in Table 1, or in Teske, Barry and Grim (1993).

Model Application

Now that the equations have been displayed, the variables identified, and the models validated, it seems fair to ask some simple questions related to aerial application techniques: (1) out of all of the inputs that should be specified to these models, which ones are more influential in affecting model predictions (and hence field measurements); (2) since all three modeling approaches assume level flight, what variations are possible along the flight path of the aircraft (where zero variations are assumed by the models); and (3) because aerial applications are always done with multiple flight lines, how can the models be used to help optimize deposition pattern within the spray area.

Parameter Sensitivity

The importance of specific model parameters may be best seen in Table 2 (from Teske and Barry 1993b), where some of the inputs to FSCBG (with its

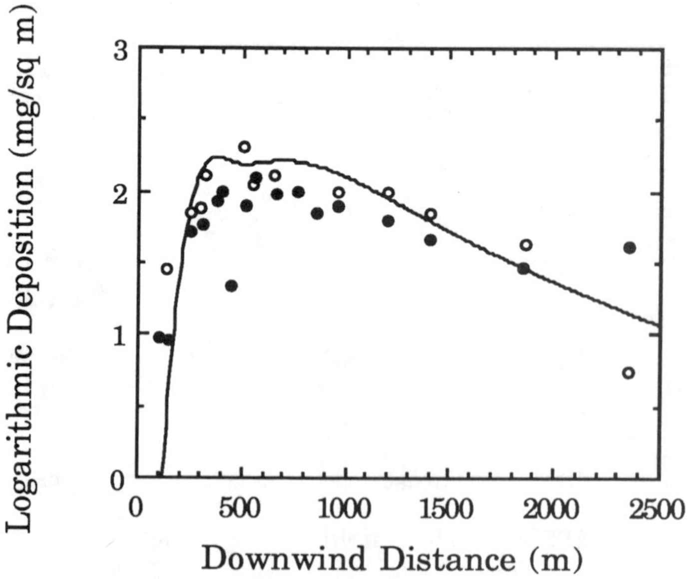

Figure 7: Comparison of AGDISP model predictions (solid curve) with Red Bluff spray trial A-4 (Teske and Barry 1993a from Rafferty and Bowers 1993). Open circles are dye drop counts on Mylar sheets; closed circles are manganese sulfate.

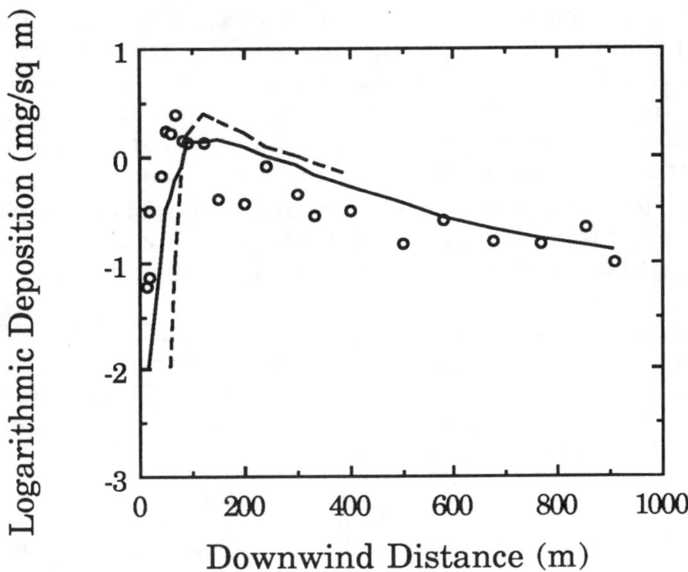

Figure 8: Comparison of FSCBG model predictions (solid curve) and PKBW model predictions (dashed curve) with the Dunphy spray trial 6 data (Mickle 1987). Ground deposition on cards is denoted by open circles.

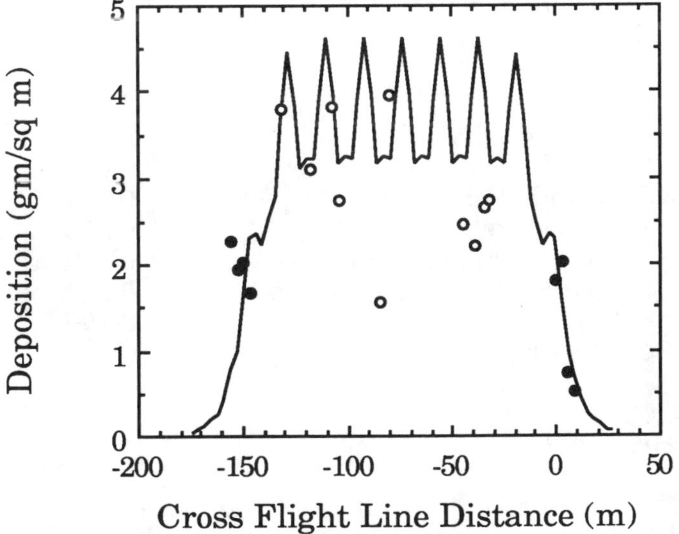

Figure 9: Comparison of FSCBG model predictions (solid curve) with the June application to Heather Seed Orchard (Teske et al. 1991). Ground deposition on cards set 5 m from ten sample trees are denoted by open circles; ground deposition on cards across the spray boundaries are the closed circles.

near-wake AGDISP model) are ranked by a sensitivity factor that reflects the importance of a change in the ground deposition pattern relative to a change in the input variable. The most sensitive variable is the position of the aircraft relative to the ground, in most cases a difficult input to quantify. It may be seen that meteorological data, particularly wind speed and direction, are also sensitive inputs to the model. And, even though $D_{v.5}$ is seventh on the list, spray drop size distribution remains an important input as well. A detailed representation of the entire distribution, down to 10 micrometers or so, may be needed to capture a good picture of downwind drift.

When assumptions must be made to complete the necessary model inputs, the subsequent predictions will only approximate the actual field measurements. Therefore, careful measurements of the meteorology, release height, and drop size distribution are most important when comparing predictions with actual deposition data.

Table 2: Sensitivity factors generated by AGDISP and FSCBG predictions

Variable	Sensitivity Factor
Release height	2.843
Specific gravity of tank mix	2.622
Spraying speed	2.118
Atmospheric pressure	1.853
Wind direction	1.716
Wind speed	1.701
$D_{v.5}$	1.154
Aircraft weight	0.832
Vortex decay effect	0.650
Solar radiation effect	0.404
Temperature	0.398
Volatile fraction of tank mix	0.319
Evaporation rate of tank mix	0.243
Relative humidity	0.150
Number of nozzles	0.108

Flight Line Sensitivity

A question often presents itself as to the variation of the deposition along the flight line, where the models assume that all planes perpendicular to the flight line exhibit the same downwind drift pattern. The presence of turbulence in the atmosphere does not cooperate with this assumption, however, and can result in distinct deposition profiles. One such example (with an AGDISP comparison) is shown in Figure 10.

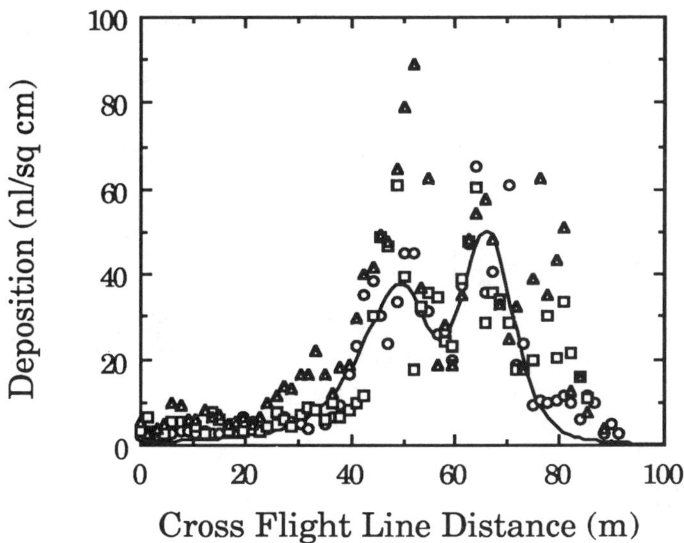

Figure 10: Comparison of AGDISP predictions (solid curve) with the three card lines of Mission run 17 (Teske 1988b). The three parallel card lines are denoted by open circles, squares and triangles.

A recent field study with two card variation lines along the aircraft flight path (Teske 1992) recovers the sensitivity results given in Table 3. The mean values along the card lines were computed from

$$\overline{D} = \frac{1}{N} \Sigma D_i \tag{54}$$

where N is the number of cards and D_i are the deposition levels on each card. Since the five tests were conducted in different meteorological conditions, the most effective comparison is with relative standard deviation RSD defined by

$$RSD = \left[\frac{1}{N} \Sigma \left(D_i \overline{D}\right)^2\right]^{1/2} \tag{55}$$

A set of data well-approximated by its mean (hence with little variation) might have a relative standard deviation no higher than 0.1 . All field data in these trials exhibit significantly larger RSD values, with the average RSD equal to 0.49 . It may be concluded, then, that variations down the flight line exist (as shown visually by Mickle 1990), and can be significant. Existing models will always provide an average value, obtained in the field by flying over several (at least three) card lines and averaging results. A detailed analysis of axial card variation awaits an anticipated study of the data in Taylor et al. (1972), where aircraft were flown over many card lines in each trial.

Table 3: Statistical analysis of axial flight line sensitivity

Trials	1	2	3	4	5
Number Deposition (drops/cm^2)					
Variation Line 1		9.44	2.98	9.42	3.94
Variation Line 2	6.05	12.64	5.79	5.97	1.53
Relative Standard Deviation					
Variation Line 1		0.36	0.54	0.33	0.49
Variation Line 2	0.40	0.44	0.39	0.66	0.65

Swath Width Sensitivity

Aerial applications almost always involve flying multiple flight lines, parallel to each other, each offset from the next by a distance called the lane separation. Aircraft are generally characterized before USDA Forest Service projects by flying a single spray path over a set of cards, to verify the swath width of the aircraft (a strong function of the nozzle type, location of the nozzles, and the drop size distribution). Swath width is the distance across which the deposition level will be effective in achieving the mission planned for the spray project. This level may be specified by a minimum number or volume per area criteria, or by a desire to minimize the standard deviation in the ground deposition pattern when multiple flight lines are overlapped.

The general measure of overlap is the coefficient of variation COV (Parkin and Wyatt 1982, Spillman 1983, Quantick 1985), which is effectively the re-interpretation of eqn 55 as COV across the width of the swath. These references suggest that $COV = 0.3$ is a desirable value. A typical deposition simulated by AGDISP, overlapped to construct COV , is shown in Figures 11 and 12. In this example a lane separation of 20.47 m would result in the preferred overlap effect. This information then helps determine the needed quantity of spray material, and its flow rate, to achieve the spray mission goals. The technique may be applied quite universally to recover swath widths of typical aerial application aircraft (Teske, Twardus and Ekblad 1990).

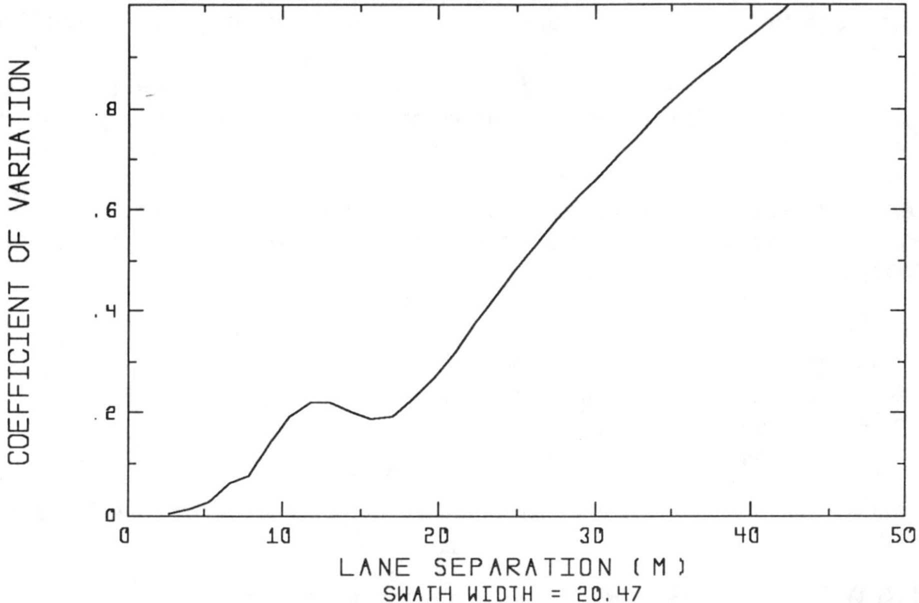

Figure 11: The quantification of swath width sensitivity with coefficient of variation for a typical agricultural aircraft configuration.

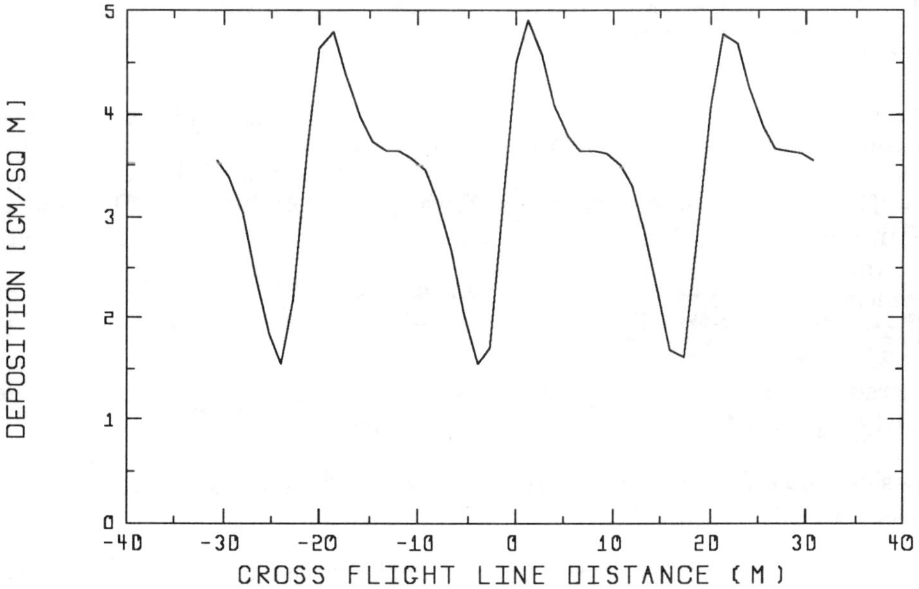

Figure 12: Deposition variation within the swath of multiple flight lines.

Summary

In this chapter we have summarized current models used to simulate aerial spray drift modeling of pesticides. We have argued that high accuracy in downwind drift prediction is becoming more and more important, and that models -- running on personal computers -- must be capable of accurate comparisons with field measurements. All three model solution techniques reviewed here accurately simulate the aerial spray drift problem, and are in use today.

References

Anderson, D. E., Miller, D. R., Wang, Y., Yendol, W. G., Mierzejewski, K. & McManus, M. L. (1992) Deposition of aerially applied Bt in an oak forest and its prediction with the FSCBG model, *J. Applied Meteorology*, **31**, 1457-1466.

Atias, M. & Weihs, D. (1984) Motion of aircraft trailing vortices near the ground, *J. Aircraft*, **21**, 783-786.

Atias, M. & Weihs, D. (1985) On the motion of spray drops in the wake of an agricultural aircraft, *Atomization and Spray Technology*, **1**, 21-36.

Barry, J. W., Skyler, P. J., Teske, M. E., Rafferty, J. E. & Grim, B. S. (1993) Predicting and measuring drift of *Bacillus thuringiensis* sprays, *Environmental Toxicology and Chemistry*, **12**, to appear.

Beard, K. V. & Pruppacher, H. R. (1971) A wind tunnel investigation of the rate of evaporation of small water drops falling at terminal velocity in air, *J. Atmospheric Sciences*, **28**, 1455-1464.

Bilanin, A. J. & Teske, M. E. (1984) Numerical studies of the deposition of material released from fixed and rotary wing aircraft, NASA Langley, VA, CR 3779.

Bilanin, A. J. & Teske, M. E. (1988) Unified wake analysis, Continuum Dynamics, Inc., Princeton, NJ, 88-06.

Bilanin, A. J., Teske, M. E., Barry, J. W. & Ekblad, R. B. (1989) AGDISP: the aircraft spray dispersion model, code development and experimental validation, *Transactions of the ASAE*, **32**, 327-334.

Bilanin, A. J., Teske, M. E. & Hirsh, J. E. (1978) Neutral atmospheric effects on the dissipation of aircraft vortex wakes, *AIAA J.*, **16**, 956-961.

Bilanin, A. J., Teske, M. E. & Williamson, G. G. (1977) Vortex interactions and decay in aircraft wakes, *AIAA J.*, **15**, 250-260.

Bjorklund, J. R., Bowman, C. R. & Dodd, G. C. (1988) User manual for the FSCBG aircraft spray and dispersion model version 2.0, USDA Forest Service, Davis, CA, FPM 88-5.

Boyle, D. G., Barry, J. W., Eckard, C. O., Taylor, W. T., McIntyre, W. C., Dumbauld, R. K. & Cramer, H. E. (1975) DC-7B aircraft spray system for large-area insect control, U. S. Army, Dugway, UT, DPG-DR-C980A.

Bradshears, M. R., Logan, N. A. & Hallock, J. N. (1975) Effect of wind shear and ground plane on aircraft wake vortices, *J. Aircraft*, **12**, 830-833.

Bragg, M. B. (1986) A numerical simulation of the dispersal of liquid from aircraft, *Transactions of the ASAE*, **29**, 10-15.

Cramer, H. E., Bjorklund, J. R., Record, F. A., Dumbauld, R. K., Swanson, R. N., Faulkner, J. E. & Tingle, A. G. (1972) Development of dosage models and concepts, U. S. Army, Dugway, UT, DTC-TR-72-609-I.

Dee, F. W. & Nicholas, O. P. (1968) Flight measurements of wing tip vortex motion near the ground, British Aeronautical Research Council, London, CP1065.

Dennison, R. S. & Wedding, J. B. (1984) Determination of evaporation rates of pesticide droplets, USDA Forest Service, Missoula, MT, 8434-2801.

Donaldson, C. duP. (1973) Atmospheric turbulence and the dispersal of atmospheric pollutants, *AMS Workshop on Micrometeorology* (D. A. Haugen, ed.), Science Press.

Donaldson, C. duP. & Bilanin, A. J. (1975) Vortex wakes of conventional aircraft, AGARDograph No. 204.

Dumbauld, R. K. Bjorklund, J. R. & Saterlie, S. F. (1980) Computer models for predicting aircraft spray dispersion and deposition above and within forest canopies: user's manual for the FSCBG computer program, USDA Forest Service. Davis, CA, 80-11.

Dumbauld, R. K., Cramer, H. E. & Barry, J. W. (1975) Application of meteorological prediction models to forest spray problems, U. S. Army, Dugway, UT, DPG-TR-M935P.

Dumbauld, R. K., Rafferty, J. E. & Bjorklund, J. R. (1977) Predictions of spray behavior above and within a forest canopy, H. E. Cramer Co., Salt Lake City, UT, TR-77-308-01.

Esterly, D., Hermansky, C., Valcore, D. & Young, B. (1993) The Spray Drift Task Force atomization program, Sixth Annual Conference on Liquid Atomization and Spray Systems, ILASS-Americas 93, Worcester, MA.

Fuchs, N. A. (1959) *Evaporation and Droplet Growth in Gaseous Media*, Pergamon, NY.

Goering, C. E., Bode, L. E. & Gebhardt, M. R. (1972) Mathematical modeling of spray droplet deceleration and evaporation, *Transactions of the ASAE*, **15**, 220-225.

Hall, W. D. & Pruppacher, H. R. (1976) The survival of ice particles falling from cirrus clouds in subsaturated air, *J. Atmospheric Sciences*, **33**, 1995-2006.

Harvey, J. K. & Perry, F. J. (1971) Flowfield produced by trailing vortices in the vicinity of the ground, *AIAA J.*, **9**, 1659-1660.

Herdan, G. (1960) *Small Particle Statistics*, Buttersworth, London.

Hirsh, R. S. (1983) Higher order approximations in fluid dynamics -- compact to spectral, von Karman Institute Lecture Series 1983-04.

Ishii, M. & Zuber, N. (1979) Drag coefficient and relative velocity in bubbly, droplet or particulate flows, *AIChE J.*, **25**, 843-55.

Langmuir, I. & Blodgett, K. B. (1964) A mathematical investigation of water droplet trajectories, Army Air Force, 5418.

Lewellen, W. S. (1977) Use of invariant modeling, *Handbook of Turbulence* (W. Frost and T. H. Moulden, eds.), Plenum.

Lewellen, W. S. & Teske, M. E. (1973) Prediction of the Monin-Obukhov similarity functions from an invariant model of turbulence, *J. Atmospheric Sciences*, **30**, 1340-1345.

Lewellen, W. S. & Teske, M. E. (1976) Second-order closure modeling of diffusion in the atmospheric boundary layer, *Boundary-Layer Meteorology*, **10**, 69-90.

Lewellen, W. S., Teske, M. E. & Donaldson, C. duP. (1974) Application of turbulent model equations to axisymmetric wakes, *AIAA J.*, **12**, 620-625.

Lewellen, W. S., Teske, M. E. & Donaldson, C. duP. (1976) Variable density flows computed by a second-order closure description of turbulence, *AIAA J.*, **14**, 382-387.

Mellor, G. L. & Yamada, T. (1974) A hierarchy of turbulent closure models for planetary boundary layers, *J. Atmospheric Sciences*, **31**, 1791-1806.

Mickle, R. E. (1987) A review of models for ULV spraying scenarios, Symposium on the Aerial Application of Pesticides in Forestry, National Research Council of Canada.

Mickle, R. E. (1990) Canadian laser mapping technique for aerial spraying, ASAE, Reno, NV, AA90-004.

Monin, A. S. & Yaglom, A. M. (1973) *The Structure of Atmospheric Turbulence*, Interscience Publishers.

Morris, D. J., Croom, C. C., van Dam, C. P. & Holmes, B. J. (1984) An experimental and theoretical investigation of deposition patterns from an agricultural airplane, NASA Langley, VA, 2348.

Neillie, C. R. & Houser, J. S. (1922) Fighting insects with airplanes: an account of the successful use of the flying machine in dusting tall trees infected with leaf-eating caterpillars, *National Geographic Magazine*, **41**, 333-339.

Parkin, C. S. & Wyatt, J. C. (1982) The determination of flight-lane separations for the aerial application of herbicides, *Crop Protection*, **1**, 309-321.

Picot, J. J. C., Kristmanson, D. D. & Basak-Brown, N. (1987) The PKBW model for prediction of aerial spraying deposition and drift, ACAFA Symposium on Aerial Application of Pesticides in Forestry, Ottawa.

Pruppacher, H. R. & Rasmussen, R. (1979) A wind tunnel investigation of the rate of evaporation of large water drops falling at terminal velocity in air, *J. Atmospheric Sciences*, **36**, 1255-1260.

Quackenbush, T. R., Teske, M. E. & Bilanin, A. J. (1993) Computation of wake/exhaust mixing downstream of advanced transport aircraft, AIAA, Orlando, FL, 93-2944.

Quantick, H. R. (1985) *Aviation in Crop Protection, Pollution and Insect Control*, Collins, London.

Rafferty, J. E. & Bowers, J. F. (1993) Comparison of FSCBG spray model predictions with field measurements, *Environmental Toxicology and Chemistry*, **12**, 465-480.

Rafferty, J. E., Dumbauld, R. K., Flake, H. W., Barry, J. W. & Wong, J. (1982) Comparison of modeled and measured deposition for the Withlacoochee spray trials, USDA Forest Service, Davis, CA, FPM 83-1.

Rasmussen, L. A. (1978) On the approximation of saturation vapor pressure, *J. Applied Meteorology*, **19**, 1564-1565.

Reed, W. H. (1953) An analytical study of the effect of airplane wake on the lateral dispersion of aerial sprays, NACA Langley, VA, 3032.

Reist, P. C. (1984) *Introduction to Aerosol Science*, Macmillan, New York.

Richardson, B., Ray, J., Miller, K., Vanner, A. & Davenhill, N. (1993) Measured and modelled spray drift from three nozzle types following aerial application of a herbicide simulant, *Transactions of the ASAE*, in review.

Skyler, P. J. & Barry, J. W. (1991) Compendium of drop size spectra compiled from wind tunnel tests, USDA Forest Service, Davis, CA, FPM 90-9.

Sowa, W. A. (1992) Interpreting mean drop diameters using distribution moments, *Atomization and Sprays*, **2**, 1-15.

Spillman, J. J. (1983) Lane separation determination, Lecture notes for short course on aerial application of pesticides, Cranfield Institute of Technology.

Steger, J. L. & Kutler, P. (1976) Implicit finite-difference procedures for the computation of vortex wakes, AIAA, San Diego, CA, 76-385.

Swarztrauber, P. & Sweet, R. (1975) Efficient FORTRAN subprograms for the solution of elliptic partial differential equations, NCAR Boulder, CO, TN/IA-109.

Sykes, R. I., Lewellen, W. S. & Parker, S. F. (1986) On the vorticity dynamics of a turbulent jet in a crossflow, *J. Fluid Mechanics*, **168**, 393-413.

Taylor, W. T., McIntyre, W. C., Barry, J. W., Sloane, H. S. and Sutton, G. L. (1972) Services development test PWU-5/A USAF modular internal spray system, U. S. Army, Dugway, UT, DTC-FR-73-317.

Teske, M. E. (1976) Vortex interactions and decay in aircraft wakes: user manual and program manual, Aeronautical Research Associates of Princeton, Inc., Princeton, NJ, 271.

Teske, M. E. (1988a) UNIWAKE user manual, Continuum Dynamics, Inc., Princeton, NJ, TN 88-04.

Teske, M. E. (1988b) AGDISP comparisons with the Mission swath width characterization studies, USDA Forest Service, Missoula, MT, MTDC-8834-2801.

Teske, M. E. (1989a) AGDISP comparisons with four field deposition studies, USDA Forest Service, Davis, CA, FPM 89-5.

Teske, M. E. (1989b) An examination of AGDISP helicopter model comparisons with data and detailed helicopter code predictions, USDA Forest Service, Missoula, MT, MTDC 89-29.

Teske, M. E. (1990) AGDISP user manual mod 6.0, Continuum Dynamics, Inc. Princeton, NJ, TN 90-16.

Teske. M. E. (1992) An interpretation of the DC-3 swath width field tests, Continuum Dynamics, Inc., Princeton, NJ, TN 92-06.

Teske, M. E. & Barry, J. W. (1993b) Parametric sensitivity in aerial application, *Transactions of the ASAE*, **36**, 27-33.

Teske. M. E. & Barry, J. W. (1993a) FSCBG spray drift predictions, ASAE-CSAE, Spokane, WA, 931101.

Teske, M. E., Barry, J. W. & Grim, B. S. (1993) Validation of the aerial spray dispersion model FSCBG, *Transactions of the ASAE*, in review.

Teske, M. E., Barry, J. W., Skyler, P. J. & Zalom, F. G. (1993b) FSCBG application to canopy spray penetration and deposition, ASAE-CSAE, Spokane, WA, 931062.

Teske, M. E., Bentson, K. P., Sandquist, R. E., Barry, J. W. & Ekblad, R. B. (1991) Comparison of FSCBG model predictions with Heather seed orchard deposition data, *J. Applied Meteorology*, **30**, 1366-1375.

Teske, M. E., Bilanin, A. J. & Barry, J. W. (1993) Decay of aircraft vortices near the ground, *AIAA J.*, **31**, 1531-1533.

Teske, M. E., Bowers, J. F., Rafferty, J. E. & Barry, J. W. (1993) FSCBG: an aerial spray dispersion model for predicting the fate of released material behind aircraft, *Environmental Toxicology and Chemistry*, **12**, 453-464.

Teske, M. E. & Curbishley, T. B. (1991) Forest Service aerial spray computer model FSCBG version 4.0 user manual, USDA Forest Service, Davis, CA, FPM 91-1.

Teske, M. E., Quackenbush, T. R. & Bilanin, A. J. (1991) Vortex roll-up, merging, and decay with the UNIWAKE computer program, FAA International Vortex Wake Symposium, Washington, DC.

Teske, M. E., Twardus, D. B. & Ekblad, R. B. (1990) Swath width evaluation, USDA Forest Service, Missoula, MT, 9034-2807.

Trayford, R. S. & Welch, L. W. (1977) Aerial spraying: a simulation of factors influencing the distribution and recovery of liquid droplets, *J. Agricultural Engineering Research*, **22**, 183-196.

von Karman, T. D. & Howarth, L. (1938) On the statistical theory of isotropic turbulence, *Proceedings of the Royal Society of London*, **164A**, 192-215.

Williamson, R. B. & Threadgill, E. D. (1974) A simulation for the dynamics of evaporating spray droplets in agricultural spraying, *Transactions of the ASAE*, **17**, 254-261.

Chapter 3

The development of a metropolitan airshed pollution model

T.J. Lyons,[a] **R.O. Pitts**[b]

[a] *Environmental Sciences, Murdoch University, Murdoch, WA 6150, Australia*

[b] *Steedman Science and Engineering, Jolimont, WA 6014, Australia*

Abstract

The evaluation of urban air quality requires an understanding of the spatial and temporal variations of emissions as well as the local meteorological fields. By combining a numerical mesoscale meteorological model with a knowledge of spatial driving patterns, an integrated airshed model is developed. Automotive emissions are derived from driving characteristics and vehicle emissions based on speed and acceleration to produce spatially and temporally varying emissions. A simple Langevin scheme is incorporated into a validated mesoscale meteorological model to simulate the dispersal of both point and area sources. With the incorporation of the predicted emissions, the model is shown to reproduce the measured concentrations for a selected case study and its potential as an urban airshed model is highlighted.

Key words

urban air pollution, vehicle emissions, Langevin equation

1 Introduction

Prediction of urban air quality within a metropolitan airshed requires a detailed knowledge of the spatial and temporal variation of the relevant meteorological fields as well as the emissions. So far most urban airshed models have used various mathematical procedures (Haugen, 1975; Landsberg, 1981; Hanna et al., 1982; Zannetti, 1990; Scheffe and Morris, 1993) to account for the transport and diffusion of pollutant based on a limited set of observations. Indeed, many authors have used the standard Gaussian approach based on surface observations of wind speed and direction from a network of stations and even more limited knowledge of the vertical structure of the atmosphere. Within a coastal region, this lack of resolution can be critical as the sea breeze - land breeze circulation leads to marked spatial inhomogeneties in the wind and turbulence fields (i.e. Pielke, 1974; Lyons, 1975).

Higher spatial and temporal resolution of the three dimensional meteorological fields, however, can be obtained by using a numerical mesoscale model to predict physically consistent flow fields (Pielke et al., 1983). Such a model, developed and improved by Pielke (1974), Mahrer and Pielke (1977), Segal and Pielke (1981) and McNider (1981) has been validated to simulate the mesoscale meteorology of the Perth, Western Australia, region (Lyons and Bell, 1990) and forms the basis of a metropolitan airshed model to describe the transport and diffusion of pollutant across the Perth urban area (Pitts and Lyons, 1992). Thus this chapter describes the development and application of the airshed model using Perth as an illustrative example.

Perth, with a population of one million inhabitants, lies at approximately latitude 32^0 S and longitude 116^0 E, on a coastal plain in the southwest corner of Australia (see Figure 1). The metropolitan area is adjacent to an almost straight north-south coastline and covers a relatively featureless sandy coastal plain. This plain has some low hills to the north of the city and the shallow estuarine valleys of the Swan and Canning Rivers which originate in the 300 m high Darling Scarp, along the eastern edge of the plain, about 25 km from the coast. The Darling Scarp, which descends to the plain in 3-5 kms, marks the edge of an extensive inland plateau and stretches for approximately 200 km in a north-south direction.

Within Perth, the emission of gaseous pollutants is confined to two major sources; an industrial area concentrated to the southwest of the metropolitan area and motor vehicles within the region (Lax et al., 1986; Lyons et al., 1990b). As emissions from the industrial area have been modelled extensively (Kamst and Lyons, 1982; KAMS, 1982), the vehicular contribution to the airshed needs to be ascertained.

Motor vehicles produce different emissions when driven under different conditions of speed, acceleration and idle. Thus, to spatially resolve these emissions requires an integration of data concerning traffic flow characteristics with data on vehicle numbers, vehicle types and vehicle kilometres travelled. Previous work in this area has emphasised the latter data on vehicles and has lacked any detailed input about how these vehicles are being driven (SATS, 1974; Kenworthy and Newman, 1984).

However, by incorporating a recent study on Perth's driving characteristics (Kenworthy et al., 1992), it is possible to spatially and temporally resolve driving patterns across the city and combine these with emission data to obtain fleet-representative vehicle emissions. Section 2 describes the production of the spatial and temporal automotive emission inventory, using the driving pattern data (Kenworthy et al., 1992; Newman et al., 1992) combined with vehicle emissions data for a representative Australian fleet (Post et al., 1985).

With both the temporal and spatial variations in meteorological fields and emissions resolved, a simple Langevin diffusion model is used to predict pollutant transport and diffusion. The Langevin model was chosen primarily because, unlike the advection diffusion equation, it can produce an accurate simulation of diffusion close to point sources, especially under convective situations. As large point sources exist within an urban area, the accurate prediction of both near source concentrations as well as overall regional concentrations are important considerations in the development of an

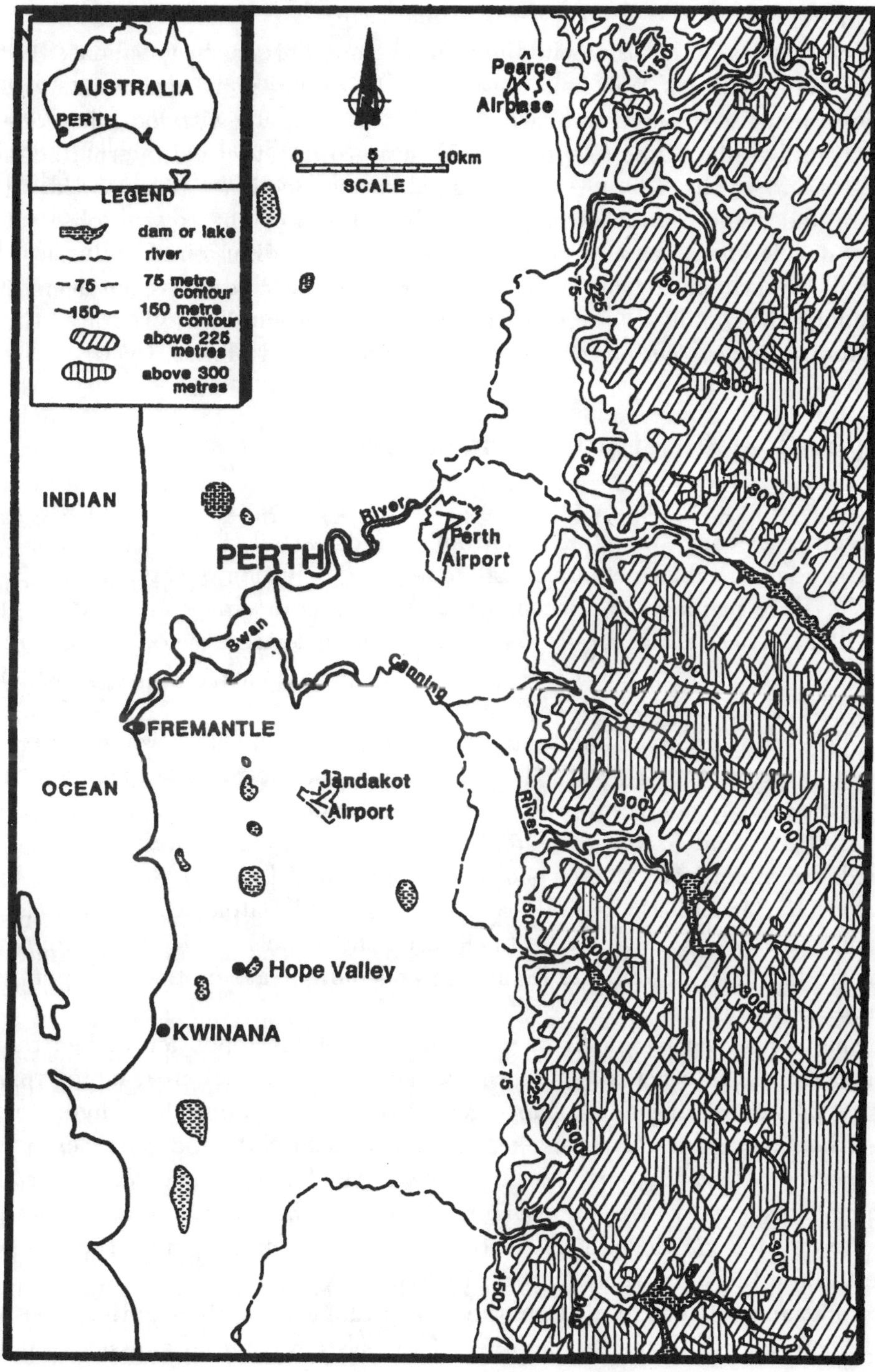

Figure 1: The Perth metropolitan area.

airshed model.

Section 3 presents the Langevin diffusion scheme. This has been validated against the water tank results of Willis and Deardorff (1978) and numerical results of Lamb (1978) for a point source and comparison has also been made with local field observations of a fumigating plume (Rayner, 1987) in a growing thermal internal boundary layer (Pitts and Lyons, 1992). Given this validation, Section 4 presents the airshed model run under selected meteorological conditions and compared with observations. As model development has addressed the transport and diffusion of pollutant across the coastal plain, chemical transformation and removal processes have not been considered although the significant chemical reactions could be simply incorporated through either a lumped or carbon bond approach (i.e. Scheffe and Morris, 1993).

2 Specification of pollutant sources

Air quality within Perth is for most days of the year exceptionally clean when compared to cities of comparable size (Bottomley and Cattell, 1974; Lax et al., 1986). However, under suitable meteorological conditions pollution events can occur that exceed the USA primary and secondary standards (Bottomley and Cattell, 1974; KAMS, 1982). These are generally associated with light stable synoptic pressure gradients (Bottomley and Cattell, 1974) or mesoscale phenomena, such as the sea breeze (KAMS, 1982).

The emission of gaseous pollutants within Perth is confined to two major sources: an industrial area concentrated to the southwest of the metropolitan area and motor vehicles within the region (Lax et al., 1986). The Kwinana industrial area emits large quantities of SO_2 and to a lesser extent NO_x (KAMS, 1982). These emissions have been modelled under both stable conditions (Kamst and Lyons, 1982) and sea breeze fumigation (KAMS, 1982), as well as in a standard climatological model based on Gaussian techniques (KAMS, 1982). In all cases, the models applied to the immediate region surrounding Kwinana (see Figure 1) and did not attempt to incorporate the broader airshed concept.

An analysis of the air quality across the broader metropolitan airshed requires the estimation of the other major source outside the industrial area, motor vehicles. These are the source of oxides of nitrogen (NO_x), brake lining dust, hydrocarbons (HC), carbon monoxide (CO), smoke, aldehydes, lead salts and particles, rubber, gaseous petrol and carbon particles (Lay, 1984). All of the CO and NO_x are emitted from the exhaust pipe whereas approximately 50% of the HC's from an uncontrolled vehicle are emitted via the exhaust, with the remainder coming from the crankcase, carburettor and fuel tank vents (SPCC, 1980).

Evaporative emissions result from the fuel system leaking HC's to the atmosphere at a rate determined by the temperature of the system (diurnal emissions) and hot soak emissions occurring after the vehicle has been driven some distance, through heating of the carburettor and fuel lines (Nelson, 1981). Hamilton et al. (1982) estimated evaporative emissions from typical early 1970s vehicles as 0.8 g km^{-1} and noted that these are generally constant throughout the life of the vehicle. Subsequent

	Date of manufacture	
	After 1 July 1976	After 1 January 1981
CO	24.2	18.6
HC	2.1	1.75
NO_x	1.9	1.9

Table 1: Australian emission standards, g km^{-1} (after SPCC, 1980)

to 1975, Australian emission standards (Table 1) have resulted in improved pollution control, as evidenced by Nelson (1981). He confirmed diurnal evaporative emission factors of 22.1 g vehicle^{-1} day^{-1} for uncontrolled (pre 1975), and 5.1 g vehicle^{-1} day^{-1} for controlled vehicles, respectively; with hot soak emissions of 12.5 g vehicle^{-1} for uncontrolled, and 4.2 g vehicle^{-1} for controlled vehicles. Consequently, evaporative emissions are a function of the age of the fleet whereas exhaust emissions are also dependent on vehicle driving characteristics. Hence the spatial and temporal variation of emission source strength is dependant on driving characteristics across the airshed as well as the age mix of the vehicle fleet.

Vehicle emissions in Sydney were estimated by Stewart et al. (1982) from the product of vehicle kilometres travelled (VKT), emissions per kilometre for each model year and travel fraction done by each model year. Although this gives a bulk estimate across the airshed it does not allow for the spatial resolution of the sources nor does it account for variations in driving conditions.

From a preliminary dynamometer test of 28 vehicles, Kent and Mudford (1979) found that typical emissions under Australian urban driving conditions, at that time, could be expressed as:

$$[CO] = 465s^{-0.97} \tag{1}$$

$$[HC] = 21.5s^{-0.73} \tag{2}$$

$$[NO_x] = 2.2 + 0.008s \tag{3}$$

where $[CO]$ is the carbon monoxide emission (g km^{-1}),$[HC]$ hydrocarbon emission (g km^{-1}),$[NO_x]$ oxides of nitrogen emission (g km^{-1}) and s the average vehicle speed (km h^{-1}). These are of a similar form to the estimates used by Iverach et al. (1976), based on US experience, and equivalent to the values employed by Taylor and Anderson (1982).

Such an estimation assumes emissions can be represented in terms of average speed alone. It neglects the influence of variations in driving conditions, particularly changes in acceleration, on emissions. For example, Kent and Mudford (1979) found that a three-dimensional plot of emission rates against speed and acceleration led to parabolic surfaces for CO and HC's, while NO_x showed a general increase in emission rates with speed and acceleration. In particular, they found that both CO and NO_x

show marked increases in emission rate with positive acceleration. This cannot be accounted for from an average speed model and highlights the need to incorporate a wide range of accelerations and speeds to produce reasonably representative emission inventories.

The spatial resolution of these emissions requires an integration of data concerning traffic flow characteristics with data on vehicle numbers, vehicle types and VKT. Previous work in this area has emphasised the latter data on vehicles and has lacked any detailed input about how those vehicles are being driven (SATS, 1974; Visalli, 1981). A very basic approach to incorporate driving patterns has been attempted for Melbourne but uses only the standard Los Angeles driving cycle for its traffic characteristics (Neylon and Collins, 1982).

2.1 Methodology

Newman et al. (1992) used an urban ecological approach (Kenworthy et al., 1992) in treating the city as an integrated system, to obtain representative driving cycles across Perth. They assumed that traffic conditions are determined by the macroscale of the city, its land use patterns, and private/public transport balance. Thus the conditions vehicles experience in driving on urban roads are influenced by a number of factors such as the density of the surrounding settlement, road type and availability, public transport factors, and other transport and land use characteristics. By defining homogeneous areas within a city based on these characteristics, routes within these areas should then provide a rational framework for sampling driving patterns and fuel consumption. Such data should enable the evaluation of spatial variations in vehicle driving patterns and ultimately emissions.

Kenworthy et al. (1992) divided the metropolitan area into six regions characterised in terms of socio-economic status and activity intensity (Table 2), ranging from high public transport availability and low private vehicle usage in the central core to very high dependence on private vehicle usage in the outer suburban areas. Using the chase car technique (Scott Research Laboratories, 1971), Newman et al. (1992) obtained detailed second by second speed time histories of typical urban driving in each of these regions. Although statistically representative driving cycles can be generated from such data, they suffer a considerable loss in speed resolution (Lyons et al., 1986), and emission estimates from modal synthetic cycles are affected by the loss of transient accelerations (Bulach, 1977). Hence, Newman et al. (1992) obtained representative driving cycles for each region by matching summary characteristics to the observed speed time traces, which is consistent with the methodology adopted by Kuhler and Karstens (1978). Accordingly, they obtained the representative urban driving speed time traces summarised in Table 3, for morning peak (MP), evening peak (EP) and off peak (OP) periods for each region.

These traces illustrate a graduation from the central business district (CBD) with higher speeds, longer cruise periods and shorter stops as you move further out. Average speed increases with distance from the CBD and is around 5 km h^{-1} higher in off peak driving except in the outer suburbs where driving is consistently free flowing.

Stops km^{-1} and root mean square acceleration are consistent with this and exhibit a general decreasing trend with distance from the CBD. Thus, driving characteristics are determined by location within the urban area and it is reasonable to expect that vehicle emissions will also be dependent on location.

The speed time traces, characterising driving in each region, were converted into speed acceleration probability matrices, where each matrix cell, of size 5 km h^{-1} by 1 km h^{-1} s^{-1}, contains the total number of one-second observations in the respective range from the representative driving cycle. Thus each matrix element represents the total time during the representative driving cycle that the vehicle was at that speed and acceleration.

Post et al. (1985) extended the analysis of Kent and Mudford (1979) to 177 Australian light duty vehicles in use, and obtained fleet averaged emission rates as a function of vehicle velocity and acceleration. Their results are presented at the same resolution as the speed acceleration probability matrices. Since cell averaged emission rates are independent of the velocity profile followed by the vehicle (Post et al., 1981a; 1981b), these can be used to estimate emissions for any driving pattern, assuming that the vehicles used by Post et al. (1985) are representative of the typical Australian urban fleet. Hence the total emissions over a representative driving cycle can be expressed as (Lyons et al., 1990)

$$[P] = \sum_{i=1}^{i=n} \sum_{j=1}^{j=n} e_{i,j} t_{i,j} \tag{4}$$

where $[P]$ is the emission (g) of pollutant species P, $e_{i,j}$ is the emission rate (g s^{-1}) of pollutant species P for the matrix element defined by velocity i and acceleration j (Post et al., 1985), $t_{i,j}$ total time (s) vehicle spent at that velocity and acceleration during the driving cycle, and the summation is over all possible speed acceleration cells. [P] is the total emission over the period of the driving cycle. Hence the characteristic emission factor (g km^{-1}) for that driving cycle can be represented as

$$[P]_k = \frac{[P]}{d_k} \tag{5}$$

where d_k is the distance covered during the representative driving cycle (see Table 3).

Within Australia, exhaust emission rates for CO and NO_x for heavy duty diesel powered vehicles remain uncontrolled and no locally validated data was readily available. Emission rates based on US experience and assumed independent of vehicle speed are listed in Table 4 (Stern, 1976; USEPA, 1977). These represent uncontrolled emissions averaged over a number of vehicles operating under a variety of conditions, and are consistent with the heavy duty emission rates used by Jakeman et al. (1984) for Australian conditions. Luria et al. (1984) obtained similar values for buses and expressed the emission factors for NO_x, HC and CO as a function of speed. They showed a marked decrease in CO and HC emissions with speed and an increase in

Table 2: A summary description of the six driving pattern areas in terms of the major factors used to derive them (after Kenworthy et al., 1992). Note all rankings of characteristics are made relative to Perth.

Characteristics	Social Economic status of residents (household income and vehicle ownership)	Activity Intensity of area, land use intensity, congestion, public transport availability	Dominant modal split split features of residents
Central core Area 1	Very low	Very high	*Peak periods* Low private vehicle usage, high public transport, walking, biking
Inner suburbs Area 2	Average to low	High	*All periods* Very high public transport
Middle Western suburbs Area 3	Average to high	Average to high	*Off peak* Very low public transport
Middle South, outer North and Eastern Suburbs Area 4	High	Low	*Peak periods* High private vehicle usage, low public transport, walking, biking
Outer South East and North East suburbs Area 5	Average	Very low	*Off peak* Very high private vehicle usage, very low walking, biking
Northern State housing suburbs Area 6	Low	Average to low	*Off peak* Very low private vehicle usage, very high walking, biking

Table 3: Characteristics of representative driving cycles for each area and time period, where MP is morning peak, OP off peak and EP evening peak (after Newman et al., 1992).

Area	Dist from CBD (km)	Average Speed (km h^{-1})	RMS Accel. (m s^{-2})	Stops per km. (km^{-1})	Idle Time (%)	Cruise Time (%)	Distance d_k (km)
1	2						
MP		30.0	0.89	1.60	20.6	13.8	11.9
OP		35.1	0.86	1.43	15.6	13.7	11.9
EP		30.6	0.87	1.60	18.4	12.5	11.9
2	5						
MP		36.4	0.80	1.39	17.9	27.5	14.5
OP		43.1	0.82	0.84	9.6	38.3	14.2
EP		38.8	0.85	0.95	17.9	30.3	16.8
3	9						
MP		40.6	0.78	0.79	11.4	30.8	17.7
OP		46.7	0.70	0.45	5.3	35.5	17.8
EP		45.3	0.71	0.56	8.2	38.5	17.7
4	13						
MP		37.6	0.77	1.08	14.3	30.6	19.4
OP		46.8	0.80	0.67	10.1	46.7	19.4
EP		41.1	0.76	0.82	18.3	33.7	19.4
5	19						
MP		52.9	0.69	0.33	3.4	59.4	18.4
OP		52.0	0.70	0.27	3.6	55.2	18.3
EP		50.0	0.78	0.30	4.6	54.9	19.8
6	11						
MP		42.4	0.74	0.72	13.3	36.7	19.4
OP		47.4	0.76	0.54	11.3	44.7	20.3
EP		45.2	0.76	0.64	14.4	49.5	20.3

Table 4: General emission factors (g km^{-1}) for heavy duty diesel powered vehicles (after [1] Stern, 1976; [2] USEPA, 1977) and those used in this study assuming an average speed of 42.3 km h^{-1} (after [3] Luria et al., 1984).

Pollutant	Emission Factor (g km^{-1}) (1,2)	(3)
Particulates	0.8	
CO	17.8	9.5
HC	2.9	1.5
NO_x	13.0	10.2
Aldehydes	0.2	
Organic acids	0.2	
SO_x	1.7	

NO_x emissions up to a speed of 40 km h^{-1}. In the absence of alternate emission factors these were used.

Trucks within the Perth metropolitan area are mostly able to maintain easy cruise conditions and appear to avoid built up areas and peak conditions (Lyons et al., 1987). Unlike automobiles, their driving cycle shows no dependence on location or time period. Consequently, as the heavy duty diesel emission factors are only expressed as a function of speed, the average speed from the Perth truck driving cycle of 43.2 km h^{-1} (Lyons et al., 1987) was assumed for all truck emissions, leading to the emission factors shown in Table 4.

The total emission in any period and any area of the city can be expressed as

$$E = \sum_{m=1}^{m=n} [P]_{k,m} VKT_{k,m} \tag{6}$$

where $[P]_{k,m}$ and $VKT_{k,m}$ are the emission factor and total VKT, respectively, for that time period and area and the summation is over vehicle type.

2.2 Vehicle emission inventory

Combining the characteristic driving patterns (Newman et al., 1992) and the fleet emissions (Post et al., 1985) leads to the automobile emission factors shown in Table 5 for each of the representative areas. As these factors are based on the same fleet data, the differences are directly attributable to the style of driving in each of the areas. This emphasises the contribution of variations in speed and acceleration patterns across a metropolitan area in determining the spatial variation of emissions.

The corresponding automobile emission factors based solely on average speed in each of the regions (Table 3) are shown in Table 6. With the exception of the NO_x emission factors, the average speed factors are lower, as would be expected, since the incorporation of acceleration leads to greater variability in the driving patterns and hence higher emissions. The NO_x emissions are higher because the average speed equation implies a speed independent emission of 2.2 g km^{-1} (Kent and Mudford

Table 5: Emission factors g km^{-1} for exhaust emissions for each time period and region based on the speed acceleration matrix (after Lyons et al., 1990b).

Area	1	2	3	6	4	5
NO_x						
MP	1.9	1.8	1.8	1.7	1.9	1.7
OP	1.9	1.9	1.7	1.8	2.0	1.7
EP	1.8	1.8	1.8	1.9	1.9	1.8
CO						
MP	21.8	18.4	18.1	16.9	19.2	14.8
OP	19.2	17.4	15.7	15.9	16.8	15.2
EP	21.3	18.5	16.7	16.9	17.5	16.0
HC						
MP	2.2	1.9	1.9	1.8	2.0	1.8
OP	2.0	1.8	1.7	1.8	1.8	1.7
EP	2.2	1.9	1.8	1.9	1.8	1.7

Table 6: Emission factors (g km^{-1}) for exhaust emissions for each time period and region based on average vehicle speed (after Lyons et al., 1990b).

Area	1	2	3	6	4	5
NO_x						
MP	2.4	2.5	2.5	2.5	2.5	2.6
OP	2.5	2.5	2.6	2.6	2.6	2.6
EP	2.4	2.5	2.6	2.6	2.5	2.6
CO						
MP	17.2	14.2	12.8	12.3	13.8	9.9
OP	14.7	12.1	11.2	11.0	11.2	10.1
EP	16.8	13.4	11.5	11.5	12.7	10.5
HC						
MP	1.8	1.6	1.4	1.4	1.5	1.2
OP	1.6	1.4	1.3	1.3	1.3	1.2
EP	1.8	1.5	1.3	1.3	1.4	1.2

1979) compared to the idle emission of 0.039 g min^{-1} of Post et al. (1985). Their results also suggest that emissions of the order of 2.2 g km^{-1} are only observed under high acceleration which is not maintained for any length of time in representative urban driving cycles (Lyons et al., 1986).

Within Perth, areas 1 and 5 illustrate the greatest differences in activity intensity ranging from the congested CBD, with its greater reliance on public transport, to the private vehicle dominated outer suburbs. Emission factors, based on the speed/acceleration distribution, show a decrease in emission factor between the CBD and the outer suburbs for NO_x corresponding to decreased accelerations characterised by high average speeds and maintained cruise conditions. Alternatively, emission factors based solely on average speed illustrate an increase as you move away from the congested CBD. Thus, a simple average speed model suggests higher emissions away from the congested CBD by not accounting for the marked accelerations changes induced by the congested stop start driving of the CBD.

The Perth metropolitan region was divided into grid squares of 1 km by 1 km and estimated daily VKT for each of these was obtained from traffic count information collected by the Main Roads Department (MRD, 1986). Automatic traffic counts, of 1 - 3 days duration, are carried out on all major roads in the region, as well as points on these at which a change in volume might be expected. They are expressed as annual average weekday traffic flow and represent the 24 hour traffic volume passing through a site on a typical weekday (MRD, 1986).

These individual grid values were summed to provided an overall measure of the recorded total daily VKT for Perth. Any shortfall between this figure and the estimated total VKT, listed in Table 7, can be attributed to subarterial roads. This was allocated across the region on the basis of the recorded traffic volumes.

The total truck VKT, given in Table 7, was allocated to high truck usage routes in the metropolitan area (Lyons et al., 1987) on the basis of the total grid VKT and subtracted from the individual grid totals. As the truck driving cycle is independent of peak periods, the truck VKT was divided by 24 to represent an average hourly truck VKT.

Twenty-four hour VKT weightings for Perth are 16.5% morning peak (0700 - 0900), 18.5% evening peak (1600 - 1800) and 65% for all off-peak times (Kenworthy et al., 1983). Consequently, the daily VKT for each grid square was corrected by these factors and divided by the length of the period to provide an hourly estimate of non-truck VKT.

The truck and non-truck VKT were then multiplied by the appropriate emission factors (Tables 4,5), to provide an estimate of the total vehicle exhaust emission across the metropolitan area. Additional evaporative emissions are accounted for from the distribution of registered vehicles and added to the total for HC. Figure 2 illustrates the spatial variation of the calculated morning and off peak NO_x emissions.The spatial distribution of CO and HC emissions are similar. The major arterial roads are clearly visible as well as the increased emission during the morning peak period. Given the relatively small variation in emission factors across the region, the major spatial variations are the direct result of variations in VKT.

Table 7: Estimated total VKT (Vehicle Kilometres Travelled) for Perth 1985. Note weekend VKT is estimated at 1.5 average weekday VKT (after ABS, 1985).

Vehicle Class	Total Annual VKT (10^9 km)	Equivalent average daily VKT (10^7 km)
Automobiles	6.996	2.064
Utilities/Panel vans	1.168	0.345
Total motor vehicles	8.164	2.409
Trucks	0.445	0.131
Motor cycles	0.138	0.041
Total	8.747	2.581

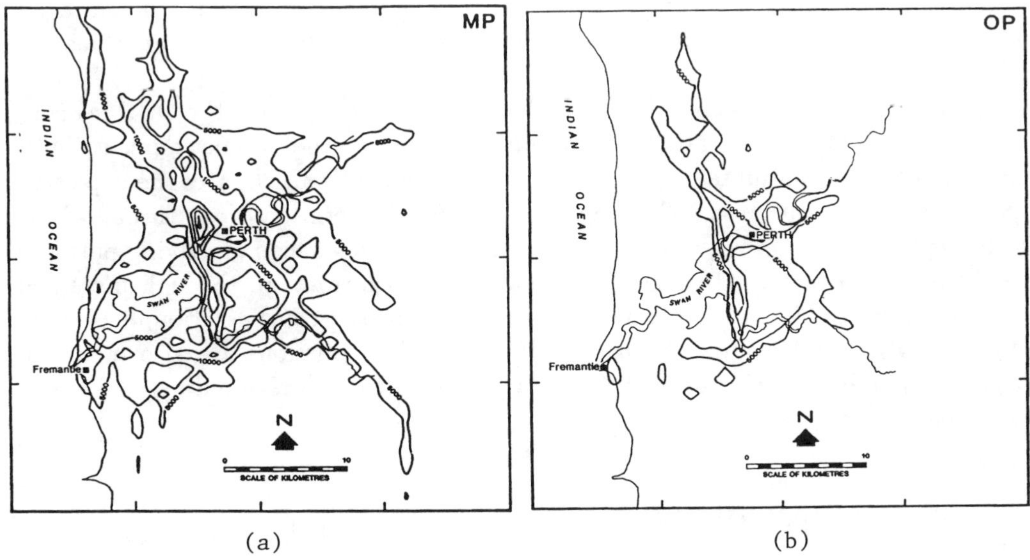

Figure 2: Predicted emission of NO_x (g hr^{-1} km^{-2}) across Perth for the (a) morning peak (MP), and (b) off peak (OP) periods (after Lyons et al., 1990b).

Table 8: Variation of CO/NO_x ratio across the Perth airshed resulting from temporal and spatial variations in driving patterns, where MP is morning peak, EP evening peak and OP off peak (after Lyons et al., 1990b).

Region	CO/NO_x Ratio		
	MP (0700-0900)	EP (1600-1800)	OP
1. Central Core	9.44	9.84	6.65
2. Inner suburbs	8.84	9.01	6.72
3. Middle western suburbs	9.06	8.45	7.17
4. Middle south, outer north and eastern suburbs	8.88	8.12	6.38
5. Outer south east and north east suburbs	8.07	8.34	7.47
6. Northern state housing suburbs	8.43	8.05	6.26

Emission totals presented in Figure 2 are not directly verifiable as they are based on average traffic conditions across the metropolitan area, which are not necessarily observed on any one day. However they do indicate the spatial variation in source strength and provide an indication of the relative magnitude of pollutant sources in different regions. An alternative statistic can be obtained from Tables 4 and 5 by computing the predicted CO/NO_x ratio on the basis of both time period and location (Table 8).

The greater congestion and higher accelerations as you approach the CBD leads to an increase in the CO/NO_x ratio of pollutants emitted from the exhaust pipe. As both smog-chamber and computed results suggest that added CO accelerates the depletion of NO and the generation of NO_2 as well as enhanced generation of O_3 through NO_2 photolysis (Demerjian et al., 1974; Drake et al., 1979), the change in driving patterns brought about by increased congestion enhances smog formation, through increased CO generation per kilometre of travel. Given the increased concentration of vehicles using the CBD this becomes significant. If evaporative emissions are also included, the greater number of vehicles in the CBD would lead to a corresponding increase in HC emissions (Lyons et al., 1990a).

3 Dispersion modelling

The analysis of air quality across metropolitan Perth requires the incorporation of these motor vehicle emissions with the spatial and temporal variation of meteorological fields. These have been derived from a numerical mesoscale model that has been validated for the region (Lyons and Bell, 1990).

In simulating regional pollutant dispersion, Kimura (1983), Kitada et al. (1984) and Kitada and Kitagawa (1990), amongst others, used numerically derived wind and turbulence fields and an advection diffusion model of pollutant transport. However, such models, based on the advection diffusion equation, cannot accurately simulate the near source growth of plumes and resultant ground level concentrations. This is primarily the result of numerical diffusion when the plume width is less than twice the grid resolution. Furthermore, under convective situations, they cannot simulate the plume knockdown and subsequent lift off (Willis and Deardorff, 1978; Lamb, 1978). As large point sources exist within the urban area, particularly in the vicinity of the Kwinana industrial area (Kamst and Lyons, 1982), an accurate prediction of their near source concentrations as well as the overall regional concentrations requires an alternative dispersion scheme.

Recent modelling using Lagrangian particle schemes have demonstrated a fair degree of success for convective dispersion of plumes in the near field (Baerentsen and Berkowicz, 1984; De Bass et al., 1986; Hurley and Physick, 1991), under stable conditions (Yamada et al., 1989) and through to mesoscale (Uliasz, 1993) and long range transport (McNider et al., 1988). Such a scheme does not suffer from numerical problems associated with the grid size resolution in the advection diffusion equation, though its major drawback has been the large computational time required for simulations. This problem has been addressed by Yamada et al. (1989) who used a kernel density function to reduce the number of particles needed. However, in the following simulations this technique was not incorporated as the largest simulations involving up to 50,000 particles over a period of 18 hours took only 30 minutes CPU on a *Sun Sparc 1*.

Therefore a diffusion model using a Lagrangian particle scheme that utilises wind and turbulent fields from the numerical mesoscale model was developed and validated (Pitts and Lyons, 1992). Such an approach addresses both area and large point sources and is particularly appropriate to an urban area which often consists of a combination of different pollutant sources.

3.1 Urban diffusion model

The spread of pollutant is simulated by a simple Langevin diffusion model derived from McNider et al. (1988). This uses a stationary, homogeneous, Gaussian solution to the Langevin equation. As such, it is not strictly valid under all conditions, such as under convective conditions when the vertical velocity distribution is skewed. However, research by Hurley and Physick (1991) for a fumigating plume, as well as that of Baerentsen and Berkowicz (1984) and Brusaca et al. (1989) under convective releases, have shown that this form with additional approximations can give very satisfactory results. In particular, under convective conditions, an additional velocity derived from a simple probability density function needs to be added to the particle velocities.

Following McNider et al. (1988) and Hurley and Physick (1991), the movement of particles in the x direction can be modelled after

$$x(t+\Delta t) = x(t) + \Delta[\overline{u}(t) + u'(t)] \quad (7)$$

where $\overline{u}$ is the mean wind speed, u' the turbulent wind speed component, t the time and Δt the timestep. Using the solution to the Langevin equation under homogeneous, stationary, Gaussian turbulence, u' can be determined as

$$u'(t+\Delta t) = R_u u'(t) + [1 - R_u^2]^{\frac{1}{2}} \sigma_u r_u \quad (8)$$

The first term in equation 8 is correlated with the previous velocity using the autocorrelation function. This is defined from the Lagrangian time scale, T_L^u as

$$R_u = \exp(-\frac{\Delta t}{T_L^u}) \quad (9)$$

whereas the second term in equation 8 is a random component with r_u a random number from a Gaussian distribution with zero mean and unit variance and σ_u the standard deviation of the Eulerian velocity in the x direction. Similar equations are used for diffusion in the y and z direction .

To account for particle accumulation in zones of low turbulent kinetic energy, an additional drift correction term as proposed by Legg and Raupach (1982) is added to the w form of equation 8 and is expressed as:

$$Drift\ correction = 2(1 - R_w) T_L^w \sigma W \frac{d\sigma_w}{dz} \quad (10)$$

Hurley and Physick (1991) have shown that this form is only appropriate under near homogeneous conditions. However, with an additional random convective velocity, Pitts and Lyons (1992) suggested that the more rigorous form is not necessary.

To close the turbulent diffusion scheme, Lagrangian timescales and variances of the velocity fluctuations in each of the three component directions need to be determined. McNider (1981), McNider et al. (1988) and Segal et al. (1988) based their turbulent statistics upon a number of schemes utilising predicted eddy diffusivities, Richardson and critical Richardson numbers, planetary boundary layer (PBL) depth and surface parameters. This scheme worked well under convective conditions but appears to under predict the turbulent vertical velocities under stable conditions (McNider, 1981). Therefore, to provide estimates of the standard deviations of the velocities and the Lagrangian time scales, Pitts and Lyons (1992) used equations suggested by Hanna (1982). These were derived from the Minnesota PBL experiment and are based on the depth of the boundary layer and surface parameters.

Under unstable conditions the vertical velocity distribution is not Gaussian but rather characterized by convective cells (updrafts) with large vertical velocities and local downdrafts occupying a larger area but with smaller velocities. Brusasca et al. (1989), Baerentsen and Berkowicz (1984) and Pitts and Lyons (1992) accounted for this organised motion by adding a convective velocity constructed from two Gaussian distributions to the particles (Weil, 1988). As the stability tends towards neutral conditions, this additional velocity decreases to zero and the distribution asymptotes to the Gaussian (Pitts and Lyons, 1992).

3.2 Validation

Pitts and Lyons (1992) validated this model by verifying that it satisfied the well mixed criterion as well as being able to reproduce the water tank dispersion results of Willis and Deardorff (1978) and Lamb (1978) representing dispersion from a single point source. These validations showed that the simple approximation was able to simulate reasonably well the dispersion in the convective boundary layer.

For coastal cities, fumigation under a thermal internal boundary layer is an important consideration. This phenomenon has been studied (KAMS, 1982; Rayner, 1987) for the industries at Kwinana to the south west of the city (Figure 1) and the model was compared directly to these observations. This illustrated good agreement in both the overall highest concentrations and spread (Pitts and Lyons, 1992).

4 Perth Automotive Emission

As a comparison of the models ability to predict the dispersion of the Perth automotive plume, an individual episode was selected from the Western Australian Environmental Protection Authority (EPA) records. Up until 1989, the EPA had only several stations monitoring CO and NO_x and these were located at either Kwinana (Figure 1) or in the central city. Since the observed concentrations in the central city would be seriously effected by the ventilation in street canyons, the monitors at Kwinana (Hope Valley) were chosen for a comparison with the simulation. Also, as the model does not include any chemistry, a night time event was chosen to minimise the effects of chemical transformations.

On the night of 19-20 May, 1989 the observed wind speed and direction at Kwinana are presented in Figure 3. These show essentially northerly winds with a weak north westerly sea breeze developing in the afternoon, returning to a north easterly land breeze overnight. Later on, the flow swings north westerly and increases in speed with the passage of a weak cold front. The mesoscale meteorological model predicted wind speed and wind direction plotted in Figure 3 show good agreement with these observations.

The observed concentrations of CO and NO_2, NO and NO_x (Figures 4 and 5) indicate that a plume traversed the monitor from 1700 LST on May 19 to 0400 LST on the 20th. Within this plume, there appears to be two distinct sections, a central section from 2000 - 2300 LST on the 19th where NO comprises more than 60% of the NO_x and the surrounding region where the NO_x is comprised mainly of NO_2. This difference in the relative composition of the plume is probably due to its age, with the outside edges of the plume being from older emissions whilst the 'central' plume is from more recent emissions and therefore has less NO converted to NO_2. For the central peak of the plume, the CO/NO_x ratio is 10.2 which is slightly above the value predicted for the Perth region in Table 8.

Pollutant emissions were simulated by the number of particles in a 2 km square being calculated each time step and released from a height of 2 m. These emissions were further increased by 19.1% to account for the estimated annual increase in total

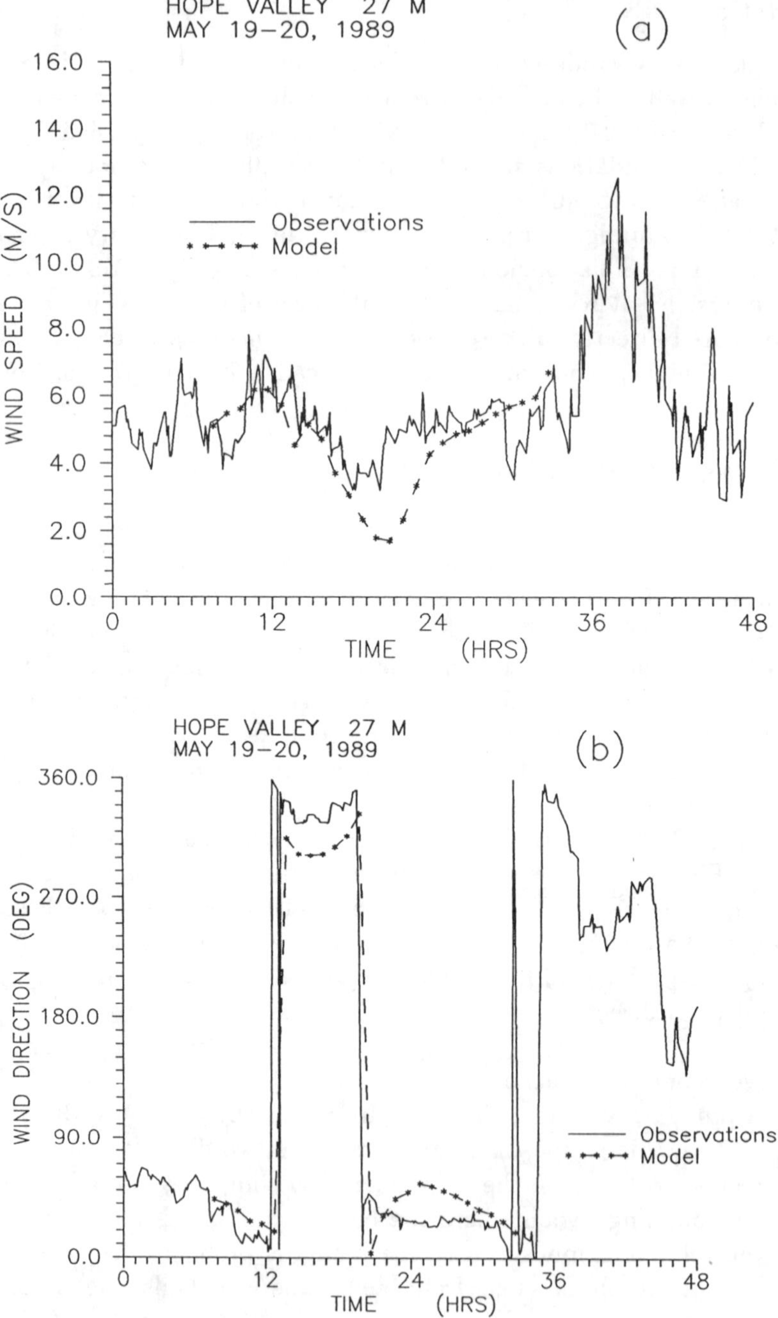

Figure 3: Observed and modelled wind speed and direction at 30 m at Hope Valley for May 19-20, 1989 (after Pitts and Lyons, 1992)

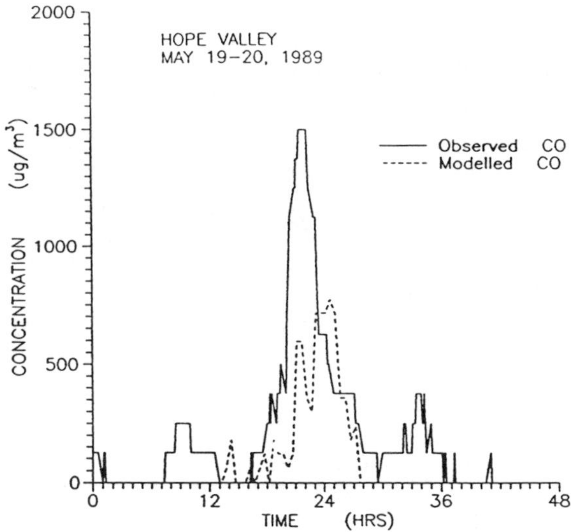

Figure 4: Observed and modelled CO concentrations at Hope Valley for May 19-20, 1989 (after Pitts and Lyons, 1992)

VKT of 5.1% (ABS, 1988), since the initial traffic survey data was collected. As noted there, 16.5 and 18.5%, respectively, of emissions occur in the morning and off peak rush hours, with a further 45% of the emissions occurring during the day time and the remaining 20% over the rest. To further resolve the temporal change in emission strength throughout the off peak period, source strengths were varied according to measured traffic characteristics.

The results of these simulations (Figures 4 and 5) indicate reasonable agreement in the timing of the episode with the predicted plume impacting Hope Valley from 1900 to 0400 LST, compared to the observed times of 1700 to 0400 LST and with the predicted peak concentration at 0030 compared to 2200 LST. This difference in time can be explained by noting that the model predicted northerly change occurred later and was weaker than that observed. The magnitude of the predicted concentrations however, in particular the CO, reproduces the observations less well. This relatively lower value of CO may be partly the result of under-predicting the CO emissions. This is supported by the fact that the observed CO/NO_x ratio is higher than any predicted in Table 8. Further reasons for the lower values may be due to the later time of the peak episode and therefore correspondingly lower emission source strength.

As the model simulates NO_x and not NO and NO_2, a direct comparison with the observed NO/NO_2 ratio and suggested relative age of the central and outer plume cannot be made. However, the good agreement with the time of impact of the Perth plume and that the highest NO/NO_2 occurs when the centre of the Perth plume is at its closest to Hope Valley between 2000 and 2300 LST (Figures 6 and 7) suggests that this is the case.

Figures 6 and 7 also illustrate the movement of the plume around Perth and the corresponding temporal and spatial variation in the maximum concentration. It is seen

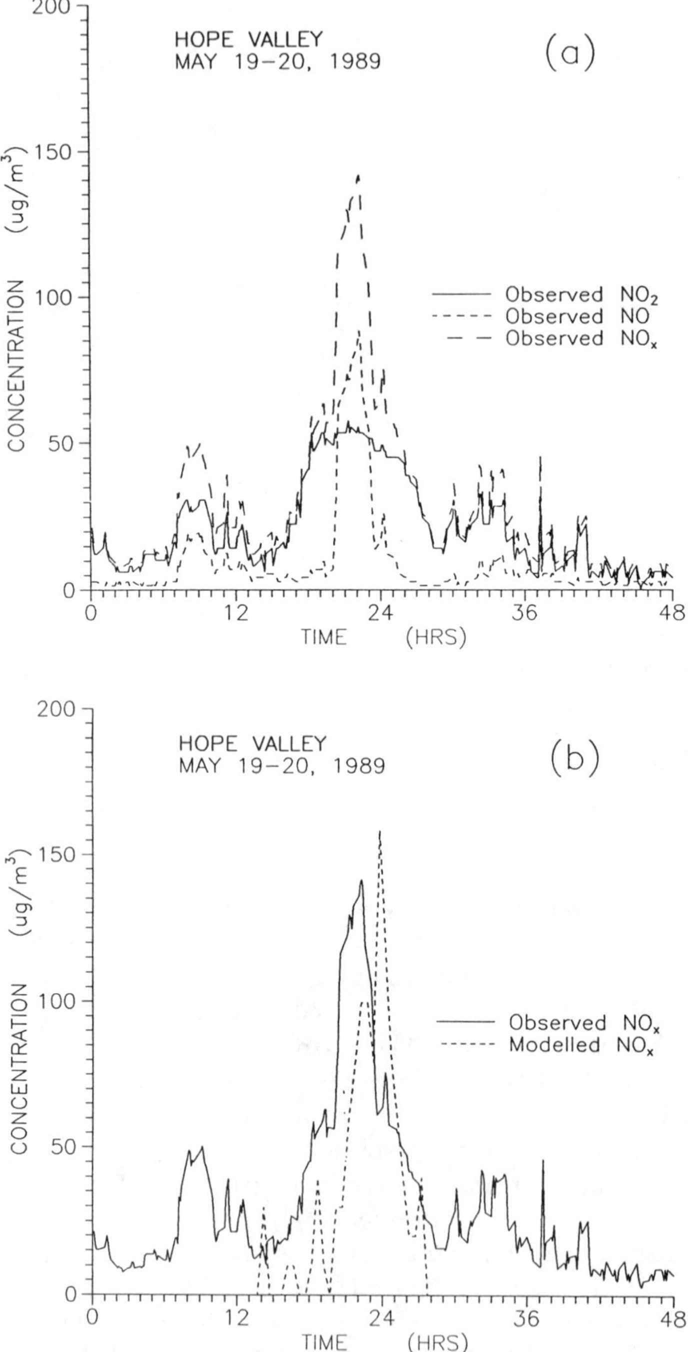

Figure 5: (a) Observed NO, NO_2 and NO_x, and (b) observed and modelled NO_x concentrations at Hope Valley for May 19-20, 1989 (after Pitts and Lyons, 1992).

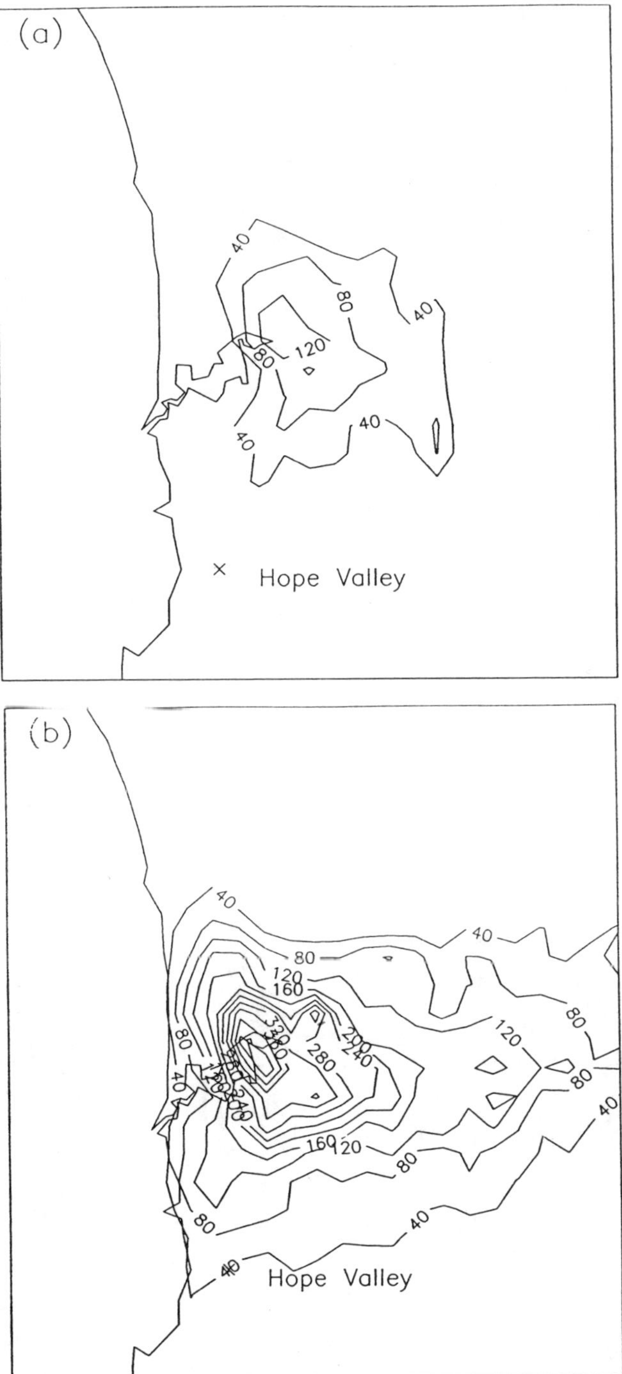

Figure 6: Modelled NO_x concentrations in $\mu g\ m^{-3}$ at (a) 1745 LST, and 2030 LST on 19 May 1989. The contour interval is 40 $\mu g\ m^{-3}$ (after Pitts and Lyons, 1992).

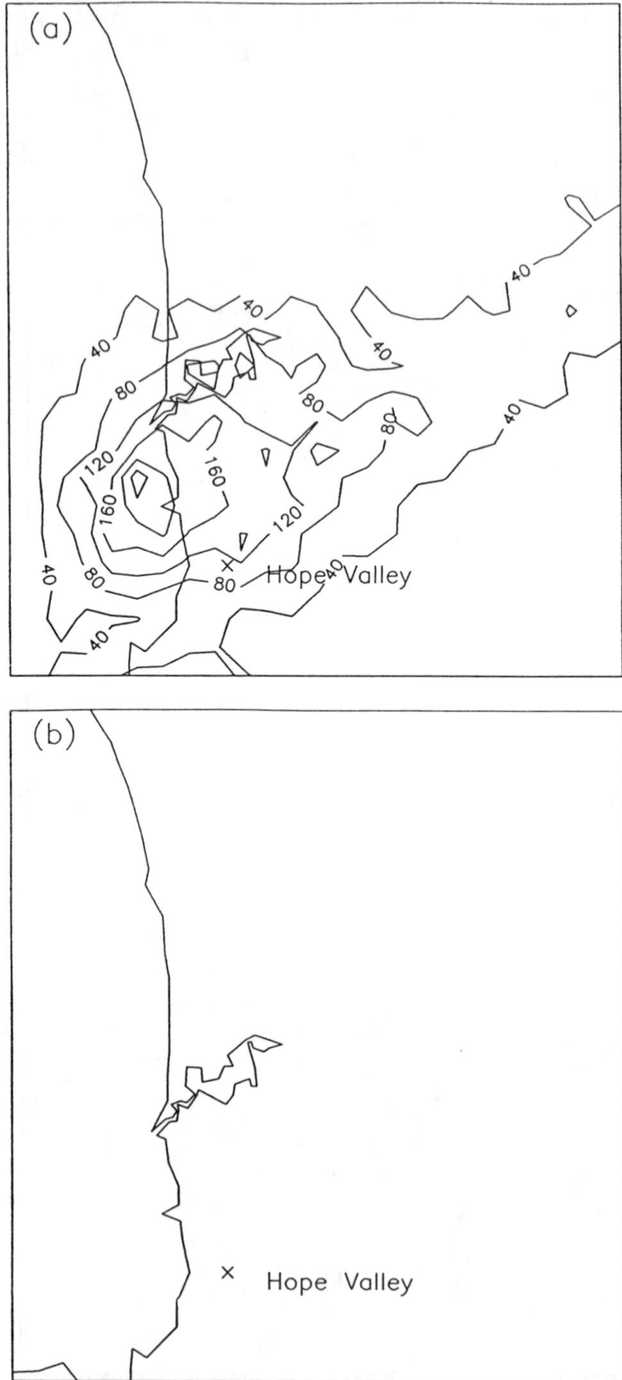

Figure 7: Modelled NO_x concentrations in μg m^{-3} at (a) 2345 LST on 19 May 1989, and (b) 0245 LST on the 20 May 1989 (after Pitts and Lyons, 1992).

that the maximum concentration at 1745 LST occurs just to the east of the city centre. At 2045 LST, with the sea breeze dying and a nocturnal inversion developing, the concentration of NO_x reaches a maximum of 570 $\mu g\ m^{-3}$. In Western Australia there is as yet no standards for NO_x to compare this level to, though the National Health and Medical Research Council have proposed a guideline for NO_2 of 320 $\mu g\ m^{-3}$ for a 1 hour average (NHMRC, 1986). As Figure 5 indicates that the majority of the NO_x would be in the form of NO it would appear that the guideline was not exceeded. With the re-establishment of the north easterlies and decrease in the emission strength, the plume is advected towards the south west with the concentrations steadily decreasing. The above example, though neglecting chemical transformations and removal processes, indicates that the model simulates reasonably well the advection and diffusion of the Perth plume. It also indicates the importance of resolving both the spatial and temporal emission strengths.

5 Conclusion

A motor vehicle emission inventory was developed for the Perth metropolitan airshed by integrating data on traffic conditions with emission factors that incorporate the effects of both speed and acceleration. This highlights the impact of varying driving conditions on the spatial and temporal resolution of vehicle emissions, and illustrates that traffic congestion enhances pollutant production through increased variations in vehicle accelerations.

In particular, the greater congestion and corresponding variations in acceleration within the CBD, increases the production of pollutants and the potential for photochemical smog through enhanced CO production. As the central core facilitates air quality emission stress through its greater congestion, this suggests that urban planning can influence air pollution through planning for reduced congestion and more free flowing traffic. Such a conclusion assumes a linear relationship between traffic flow and emissions which Lyons et al. (1990a) and Newman et al. (1992) dispute in arguing that the extra road building required to reduce congestion would only further extend urban sprawl and create greater vehicle usage. It is a prescription for increasing total emissions whereas increased intensity of urban activity can be used to reduce automobile dependence leading not necessarily to a reduction in congestion but a reduction in overall emissions (Newman et al., 1988; 1992; Kenworthy et al., 1989).

The mesoscale model and emissions inventory were combined to form a regional air pollution model that can simulate accurately the dispersal of both point and area sources within a coastal environment. The model is based on the solution of the Langevin equation for homogeneous turbulence, with the special case of convective conditions approximated by adding a convective velocity. The scheme has been shown to satisfy the well mixed criterion and showed good agreement with the dispersion results of Willis and Deardorff (1978) and Lamb (1978) for an elevated source in a convective boundary layer. In comparison with observations of sea breeze fumigation (Rayner, 1987), it demonstrates good agreement, reproducing well the location, shape and magnitude of the ground level concentrations. With the incorporation of the predicted emissions, the model for a selected event was able to reproduce well the measured concentrations. This example highlights the importance of correctly simulating the meteorological fields as well as determining the spatial and temporal variations in emissions.

Acknowledgements

The development of the Perth airshed model was supported by the Australian Research Council and Murdoch University's Special Research Grant, whereas the driving cycle data was collected under funding from the State Energy Commission of Western Australia and the National Energy Research, Development and Demonstration Programme, which is administered by the Commonwealth Department of Resources and Energy. Throughout it Owen Pitts was in receipt of a Commonwealth Postgraduate Research Award. All of this assistance is gratefully acknowledged.

References

ABS, 1985: *Survey Motor Vehicle Usage for 12 months ended 30 September 1985.* Australian Bureau of Statistics.

ABS, 1988: *Survey Motor Vehicle Usage for 12 months ended 30 September 1988.* Australian Bureau of Statistics.

Baerentsen, J.H. and Berkowicz, R., 1984: Monte Carlo simulation of plume dispersion in the convective boundary layer. *Atmos. Environ.*, **18**, 701-712.

Bottomley, G.A. and Cattell, F.C., 1974: Nitrogen oxide levels in the suburbs of Perth, Western Australia. *J. R. Soc. W. Aust.*, **58**, 65-74.

Brusasca, G., Tinarelli, G. and Anfossi, D., 1989: Comparison between the results of a Monte Carlo atmospheric diffusion model and tracer experiments. *Atmos. Environ.*, **23**, 1263-1280.

Bulach, V., 1977: *Prediction of line source of emissions along urban roads.* Dept. Mechanical Engineering Report T26/77, University of Melbourne, Parkville, Victoria, Australia.

De Baas, A.F., Van Dop, H. and Nieuwstadt, F.T.M., 1986: An application of the Langevin equation for inhomogeneous conditions to dispersion in a convective boundary layer. *Q. J. R. Meteorol. Soc.*, **112**, 165-180.

Demerjian, K.L., Kerr, J.A. and Calvert, J.G., 1974: The mechanism of photochemical smog formation. *Adv. Env. Sci. Tech.*, **4**, 1-262.

Drake, R.L., Hales, J.M., Mishima, J. and Drewes, D.R., 1979: *Mathematical models for atmospheric pollutants.* Appendix B: Chemical and Physical Properties of Gases and Aerosols. EA-1131, Appendix B, Research Project 805, Electric Power Research Institute, 3412 Hillview Avenue, Palo Alto, CA 94304, U S A.

Hamilton, R.B., Cass, M.R, Angus, G.A. and Watson, H.C., 1982: Forecasting hydrocarbon emissions from motor vehicles in the Sydney air shed. In Carras, J.N. and Johnson, G.M. (eds), *The urban atmosphere - Sydney, a case study.* CSIRO, Division of Fossil Fuels, 525-570.

Hanna, S.R., 1982: Applications in modeling. In Nieuwstadt, F.T.M. and Van Dop, H., (eds), *Atmospheric turbulence and air pollution modelling.* D. Reidel Publishing Company, Dordrecht, Holland, 275-310.

Hanna, S.R, Briggs, G.A. and Hosker, R.P., 1982: *Handbook on atmospheric diffusion.* Technical Information Centre, U.S. Department of Energy, DE82002045 (DOE/TIC- 11223), 102pp.

Haugen, D.A., (editor) 1975: *Lectures on air pollution and environmental impact analyses.* American Meteorological Society, Boston, 296pp.

Hurley, P. and Physick, W., 1991: A lagrangian particle model of fumigation by breakdown of the nocturnal inversion. *Atmos. Environ.*, **25A**, 1313-1325.

Iverach, D., Mongan, T.R., Nielsen, N.J. and Formby, J.R., 1976: Vehicle related air pollution in Sydney. *J. Air. Poll. Control Assoc.*, **26**, 39-44.

Jakeman, A.J., Simpson, R.W. and Taylor, J.A., 1984: *A simulation approach to assess air pollution from road transport.* Discussion Paper, Centre for Resource and Environmental Studies, Australian National University, Canberra.

KAMS, 1982: *The Kwinana air modelling study.* Western Australian Department of Conservation and Environment, Report 10, 96pp.

Kamst, F.H. and Lyons, T.J., 1982: A regional air quality model for the Kwinana industrial area of Western Australia. *Atmos. Environ.*, **16**, 401-412.

Kent, J.H. and Mudford, N.R., 1979: Motor vehicle emissions and fuel consumption modelling. *Transpn. Res.*, **13A**, 395-406.

Kenworthy, J.R., Newman, P.W.G. and Lyons, T.J., 1983: *A driving cycle for Perth.* National Energy Research Development and Demonstration Council, Dept. Resources and Energy, Report 79/9252, 326pp.

Kenworthy, J.R. and Newman, P.W.G., 1984: *Motor vehicle emission inventories: a review of methodology, applications and potential for Australian cities.* Transport Research Paper 4/84, Murdoch University, 44pp. [Available through Institute for Science and Technology Policy, Murdoch University, Murdoch, WA 6150, Australia]

Kenworthy, J.R., Newman, P.W.G. and Lyons, T.J., 1989: Urban planning and traffic congestion. *Urban Policy and Research*, **7**, 67-80.

Kenworthy, J.R., Newman, P.W.G. and Lyons, T.J., 1992: The ecology of urban driving 1 - Methodology. *Transpn. Res.-A*, **26A**, 263-272.

Kimura, F., 1983: A numerical simulation of local winds and photochemical air pollution (1): Two-dimensional land and sea breeze. *J. Met. Soc. Japan.*, **61**, 862-878.

Kitada, T., Carmichael, G.R. and Peters, L.K., 1984: Numerical simulation of the transport of chemically reactive species under land- and sea-breeze circulations. *J. Climate Appl. Meteor.*, **25**, 767-784.

Kitada, T. and Kitagawa, E., 1990: Numerical analysis of the role of sea breeze fronts on air quality in coastal and inland polluted areas. *Atmos. Environ.*, **24A**, 1545-1559.

Kuhler, M. and Karstens, D., 1978: Improved driving cycle for testing automotive emissions. Presented *Passenger Car Meeting*, Troy Hilton, Troy, Michigan, June 5-9, SAE Paper 780650.

Lamb, R.G., 1978: A numerical simulation of dispersion from an elevated point source in the convective boundary layer. *Atmos. Environ.*, **12**, 1297-1304.

Landsberg, H.E., 1981: *The urban climate.* Academic Press, New York, 275pp.

Lax, F., Robertson, W.A. and Garkaklis, B.V., 1986: Air pollution components in Perth. *J. R. Soc. W. Aust.*, **69**, 19-27.

Lay, M.G., 1984: *Source book for Australian roads.* 2nd edition. Australian Road Research Board, 551pp.

Legg, B.J. and Raupach, M.R., 1982: Markov-chain simulation of particle dispersion in inhomogeneous flows: The mean drift velocity induced by a gradient in eulerian velocity variance. *Bound.-layer Meteorol.*, **24**, 3-13.

Luria, M., Vinig, Z. and Peleg, M., 1984: The contribution of city buses to urban air pollution in Jerusalem, Israel. *J. Air. Poll. Control Assoc.*, **34**, 828-831.

Lyons, T.J. and Bell, M.J., 1990: Mesoscale contributions to available wind power. *Solar Energy*, **42**, 483-485.

Lyons, T.J., Alimoradian, B. and Newman, P.W.G.,1987: *A truck driving cycle for Perth.* Transport Research Paper 5/87, Murdoch University, 1-29. [Available through Institute for Science and Technology Policy, Murdoch University, Murdoch, WA 6150, Australia]

Lyons, T.J., Kenworthy, J.R., Austin, P.I. and Newman, P.W.G.,1986: The development of a driving cycle for fuel consumption and emissions evaluation. *Transpn. Res.*, **20A**, 447-462.

Lyons, T.J., Kenworthy, J.R. and Newman, P.W.G., 1990a: Urban structure and air pollution. *Atmos. Environ.*, **24B**, 43-48.

Lyons, T.J., Pitts, R.O., Blockley, J.A., Kenworthy, J.R. and Newman, P.W.G., 1990b: Motor vehicle emission inventory for the Perth airshed. *J. Roy. Soc. West. Aust.*, **72**, 67- 74.

Lyons, W.A., 1975: Turbulent diffusion and pollutant transport in shoreline environments. In Haugen, D.,(ed), *Lectures on Air Pollution and Environmental Impact Analyses*, Workshop Proceedings, American Meteorological Society, Boston, 136-208.

Mahrer, Y. and Pielke, R.A., 1977: A numerical study of the airflow over irregular terrain. *Contrib. Atmos. Phys.*, **50**, 98-113.

McNider, R.T., 1981: *Investigation of the impact of topographic circulations on the transport and dispersion of air pollutants.* Ph.D. dissertation, University of Virginia. 210pp.

McNider, R.T, Moran, M.D. and Pielke, R.A., 1988: Influence of diurnal and inertial boundary layer oscillations on long-range dispersion. *Atmos. Environ.*, **22**, 2445-2462.

MRD, 1986: *Average weekday traffic flows 1980/1 - 1985/86 Perth metropolitan region.* Main Roads Department Western Australia, 69pp.

NHMRC, 1986: *National guidelines for control of emission of air pollutants from new stationary sources. Recommended methods for monitoring air pollutants in the environment.* National Health and Medical Research Council, Australian Government Publishing Service, Canberra, 19pp.

Nelson, P.F., 1981: Evaporative hydrocarbon emissions from a large vehicle population. *J. Air. Poll. Control Assoc.*, **31**, 1191-1193.

Newman, P.W.G., Kenworthy, J.R. and Lyons, T.J., 1988: Does free-flowing traffic save energy and lower emissions in cities? *Search*, **19**, 267-272.

Newman, P.W.G., Kenworthy, J.R. and Lyons, T.J., 1992: The ecology of urban driving II - Driving cycles across a city: their validation and implications. *Transpn. Res.-A*, **26A**, 273-290.

Neylon, M. and Collins, B., 1982: Mobile source emissions inventory for the Melbourne airshed study. In *Proc. Joint SAE-A/ARRB 2nd Conference on Traffic, Energy and Emissions*, Melbourne, May 19-21, (SAE 82144).

Pielke, R.A., 1974: A three-dimensional numerical model of the sea breeze over south Florida. *Mon. Weather Rev.*, **102**, 115-139.

Pielke, R.A., McNider, R.T., Segal, M. and Mahrer, Y., 1983: The use of a mesoscale numerical model for evaluations of pollutant transport and diffusion in coastal regions and over irregular terrain.

Bull. Amer. Met. Soc., **64**, 243-249.

Pitts, R.O. and Lyons, T.J., 1992: A coupled mesoscale/particle model applied to an urban area. *Atmos. Environ.*, **26B**, 279-289.

Post, K., Gibson, T., Maunder, A., Tomlin, J., Carruthers, N., Pitt, D., Kent, J.H. and Bilger, R.W., 1981a: *Motor Vehicle Fuel Economy Report to NERDDC by the University of Sydney, 1979-1981.* Charles Kolling Research Laboratory Tech. Note ER-37, University of Sydney, N.S.W.

Post, K., Tomlin, J., Pitt, D., Carruthers, N., Maunder, A., Gibson, T., Kent, J.H. and Bilger, R.W., 1981b: *Fuel Economy and Emissions Research Annual Report by the University of Sydney for 1980-81.* Charles Kolling Research Laboratory Tech. Note ER-36, University of Sydney, N.S.W.

Post, K., Kent, J.H, Tomlin, J. and Carruthers, N., 1985: Vehicle characterisation and fuel consumption prediction using maps and power demand models. *Int. J. Vehicle Design*, **6**, 72-92.

Rayner, K.N., 1987: *Dispersion of atmospheric pollutants from point sources in a coastal environment.* Ph.D dissertation, Murdoch University, 249pp.

SATS, 1974: *Sydney area transportation study. Volume 2: Travel model development and forecasts.* N.S.W. Ministry of Transport, Sydney, N.S.W.

Scheffe, R.D. and Morris, R.E., 1993: A review of the development and application of the urban airshed model. *Atmos. Environ.*, **27B**, 23-39.

Scott Research Laboratories., 1971: *Vehicle operations Survey I and II.* CRC APRAC Project No. CAPE-10-68(1-70), PO Box 2416, San Bernardino, CA 92406, U.S.A.

Segal, M. and Pielke, R.A., 1981: Numerical model simulation of biometeorological heat load conditions - summer day case study for the Chesapeake Bay area. *J. Appl. Meteorol.*, **20**, 735-749.

Segal, M., Pielke, R.A., Arritt, R.W., Moran, M.D., Yu, C.-H. and Henderson, D., 1988: Application of a mesoscale atmospheric dispersion modeling system to the estimation of SO2 concentrations from major elevated sources in southern Florida. *Atmos. Environ.*, **22**, 1319-1334.

SPCC, 1980: *Control of pollution from motor vehicles.* State Pollution Control Commission, Publication MV-3, N S W Ministry for Planning and Environment, Sydney, N.S.W., 1-24.

Stern, A.C., 1976: *Air Pollution.* Volume 4, Academic Press, New York.

Stewart, A.C., Pengilley, M.R., Brain, R., Haley, J.J. and Mowle, M.G., 1982: Motor vehicle emissions into the Sydney air basin. In Carras, J.N. and Johnson, G.M., (eds), *The urban atmosphere - Sydney, a case study.*, CSIRO, Division of Fossil Fuels, 485-502.

Taylor, M.A.P. and Anderson, B.E., 1982: Modelling pollution and energy use in urban road networks. *Proc. 11th ARRB Conference*, **11(6)**, 1-17.

Uliasz, M., 1993: Atmospheric mesoscale dispersion modeling system - MDMS. *J. Appl. Meteorol.*, (in press).

USEPA, 1977: *Compilation of air pollutant emission factors.* 3rd edition, USEPA, Office of Air Programs, Research Triangle Park, NC, U.S.A., AP-42.

Visalli, J.R., 1981: Effects of variable vehicular age and classification distributions in mobile source modeling. *J. Air Poll. Control Assoc.*, **31**, 68-71.

Weil, J.C., 1988: Dispersion in the convective boundary layer. In Venkatram, A. and Wyngaard, J.C., (eds), *Lectures on air pollution modeling*, American Meteorological Society, Boston, 167-227.

Willis, G.E. and Deardorff, J.W., 1978: A laboratory study of dispersion from an elevated source within a modeled convective boundary layer. *Atmos. Environ.*, **12**, 1305- 1311.

Yamada, T., Kao, C-Y.J. and Bunker, S., 1989: Airflow and air quality simulations over the western mountainous region with a four-dimensional data assimilation technique. *Atmos. Environ.*, **23**, 539-554.

Zannetti, P., 1990: *Air Pollution Modeling.* Van Nostrand Reinhold, New York, 444pp.

Chapter 4

Lagrangian particle dispersion modeling in mesoscale applications

M. Uliasz

Department of Atmospheric Sciences, Colorado State University, Fort Collins, Colorado 80523, USA
(also at Warsaw University of Technology, Poland)

Abstract

This chapter presents some experiences resulting from development and applications of two dispersion modeling systems based on the Lagrangian particle modeling: (1) Mesoscale Dispersion Modeling System (MDMS), and (2) Hybrid Particle Concentration Transport (HYPACT) model. A special attention is paid to the development of an efficient modeling tool to perform intensive calculations of air pollution dispersion on mesoscale and regional scales. These efforts go in several directions: (1) evaluation of simplifications of the Lagrangian particle models acceptable in mesoscale applications; (2) combining different modeling techniques in hybrid dispersion models; (3) using an alternative receptor-oriented approach in dispersion modeling. Methods for concentration calculations and physical parameterizations in the particle models are also shortly reviewed. Two examples of applications of the Lagrangian particle dispersion model for complex terrain in the southwestern United States and eastern Europe demonstrate a design of computationally intensive air quality studies with the aid of the modern workstations.

Key words

air pollution modeling, Lagrangian particle models, mesoscale models, receptor modeling

1 Introduction

Lagrangian particle models have recently become a very important tool for studying air pollution dispersion (e.g. Zannetti, 1992). They are based on assumption that atmospheric diffusion can be modelled by a Markov chain process first proposed by Obukhov (Obukhov, 1959) and Smith (Smith, 1968). Although the concept of particle dispersion modeling is not new, its widespread application has been limited by (1) a lack of available 3-D meteorological input data,and (2) considerable computational requirements. It was especially true in the case of mesoscale dispersion of pollutants in the 20 to 2000 km range where influence of the landscape variability and complex terrain on atmospheric transport must be taken into account. Recent advances in computer technology make it possible to link Lagrangian particle dispersion models to numerical mesoscale meteorological models suitable to simulate atmospheric circulations in complex terrain (Pielke et al., 1991; Lyons et al., 1993). A real revolution in mesoscale dispersion applications has been introduced by powerful and affordable workstations (Grubb and Borchers, 1991) which can be dedicated to specific tasks. The workstations allow us to fully utilize visualization capabilities of the particle modeling technique.

This chapter presents some experiences resulting from development and applications of two dispersion modeling systems based on the Lagrangian particle modeling:

- Mesoscale Dispersion Modeling System (MDMS)
- Hybrid Particle Concentration Transport (HYPACT) model

The MDMS was originally developed on a personal computer in Warsaw University of Technology, Poland (Uliasz, 1990a; Uliasz, 1990b; Uliasz, 1993) and then used in different applications on Unix workstations at Colorado State University (CSU). This system includes a 3-D hydrostatic meteorological mesoscale model (MESO), a Lagrangian Particle Dispersion (LPD) model and an Eulerian Grid Dispersion (EGD) model. The LPD model is used not only with the MESO model but also with other meteorological models including the CSU Regional Atmospheric Modeling System (RAMS) (Pielke et al., 1992). The HYPACT is a new dispersion code being developed at ASTER, Inc. It is designed to be used with the newest version of the CSU RAMS and includes more advanced physical parameterizations in the particle dispersion model as well as the concept of a hybrid Lagrangian-Eulerian dispersion modeling.

Despite of advances in computer technology, the particle models are still computationally expensive if it is necessary to track a large number of particles for long distances and for a long time. This problem appears in mesoscale applications, especially, when multiple pollution sources with continuous emissions are considered. The goal of the presented research is the development of an efficient modeling tool to perform intensive calculations of air pollution dispersion on mesoscale and regional scales. These efforts go in several directions: (1) evaluation of simplifications of the Lagrangian particle models acceptable in mesoscale applications; (2) combining different modeling techniques in hybrid dispersion models; (3) using an alternative receptor-oriented approach in dispersion modeling. Our experiences in concentration calculations and physical parameterizations in the particle models are also shortly reviewed. Finally, two examples of applications of the LPD model for complex terrain in the southwestern United States and eastern Europe demonstrate a design of computationally intensive air quality studies with the aid of the modern workstation. These projects emphasize the application of the receptor-oriented dispersion modeling.

2 Model equations

Pollution dispersion in the LPD model is simulated by tracking a large set of particles. Subsequent positions of each particle, representing a discrete element of pollutant mass, are computed from the following relations

$$X(t+\Delta t) = X(t) + (u+u')\Delta t \quad (1)$$
$$Y(t+\Delta t) = Y(t) + (v+v')\Delta t \quad (2)$$
$$Z(t+\Delta t) = Z(t) + (w+w'+w_p)\Delta t \quad (3)$$

The resolvable scale components of wind velocity u, v and w are obtained directly from the meteorological model. Three Markov chain schemes (LPD2a, LPD2b, LPD2c) and two fully random walking schemes (LPD1b, LPD1c) are considered to create the turbulent wind components u', v', and w' (Uliasz and Pielke, 1993). The most advanced model, LPD2a, is used as a reference and all other model versions are obtained by its subsequent simplifications. In addition, an option without turbulent diffusion (LPD0) where particles are moved explicitly by the resolved wind can be used to calculate trajectories and streaklines. An additional vertical velocity component, w_p, in (3) takes into account gravitational settlement of heavy particles and buoyancy phenomena.

Model LPD2a. The Markov process including wind velocity covariances is defined by the scheme proposed by Zannetti (1986):

$$u'(t) = \phi_1 u'(t-\Delta t) + \sigma_{ru}\eta_u \quad (4)$$
$$v'(t) = \phi_2 v'(t-\Delta t) + \phi_3 u'(t) + \sigma_{rv}\eta_v \quad (5)$$
$$w'(t) = \phi_4 w'(t-\Delta t) + \phi_5 v'(t) + \phi_6 u'(t) + (\sigma_{rw}\eta_w + w_d) \quad (6)$$

The coefficients $\phi_1,...,\phi_6$ are expressed by the wind velocity variances, σ_u^2, σ_v^2, σ_w^2, covariances, $\overline{u'w'}$, $\overline{v'w'}$, $\overline{u'v'}$, and Lagrangian autocorrelations R_u, R_v, R_w. The last terms are random normally-distributed components; η_u, η_v, and η_w are random numbers from a standard Gaussian distribution. The random component of vertical velocity has a nonzero mean value w_d, called a drift velocity, to prevent the spurious accumulation of particles in regions of low turbulence (Legg and Raupach, 1982). The Lagrangian velocity autocorrelations are expressed by Lagrangian time scales, e.g. $R_w(\Delta t) = \exp(-\Delta t/T_{Lw})$.

The time step Δt used to move particles in the Markov chain model versions is variable in inhomogeneous turbulence and depends on the Lagrangian time scale, T_{Lw}: $\Delta t = max\,(0.1T_{Lw},\ \Delta t_{min})$. The minimum time step Δt_{min} is arbitrary prescribed to avoid a zero time step near the ground surface.

Model LPD2b. If the covariances of wind velocity components are neglected, the model LPD2a is simplified to the form:

$$u'(t) = R_u u'(t-\Delta t) + (1-R_u^2)\sigma_u \eta_u \quad (7)$$
$$v'(t) = R_v v'(t-\Delta t) + (1-R_v^2)\sigma_v \eta_v \quad (8)$$
$$w'(t) = R_w w'(t-\Delta t) + (1-R_w^2)\sigma_w \eta_w + w_d \quad (9)$$

This model version can be further simplified to the model LPD2c by neglecting the horizontal turbulent wind components:

$$u'(t) = 0 \quad (10)$$
$$v'(t) = 0 \quad (11)$$
$$w'(t) = R_w w'(t-\Delta t) + (1-R_w^2)\sigma_w \eta_w + w_d \quad (12)$$

Model LPD1b. This model version is obtained from the model LPD2b by increasing the time step Δt used to move particles. Assuming that Δt is much larger than the Lagrangian time scales, the scheme is derived where particles have no memory and move fully randomly at each time step:

$$u'(t) = \sigma_u \eta_u \quad (13)$$
$$v'(t) = \sigma_v \eta_v \quad (14)$$
$$w'(t) = \sigma_w \eta_w + w_d \quad (15)$$

The time step Δt is kept constant (typically $\Delta t = 180$ s). After neglecting the horizontal turbulent velocity components in the above equations the simplest variant, model LPD1c, is obtained. The random walk model versions do not require the knowledge of Lagrangian time scales.

The wind velocity variances and covariances required by the LPD model are calculated diagnostically from available meteorological information using a simplified second-order closure technique developed by Mellor and Yamada (Mellor and Yamada, 1982; Helfand and Labraga, 1988; Andrén, 1990). A so-called level 2.5 scheme modified for a case of growing turbulence (Helfand and Labraga, 1988) is applied if the fields of wind, potential temperature, and turbulent kinetic energy are provided by the meteorological model which uses the same turbulence parameterization. This scheme is based on the prognostic equation for the turbulent kinetic energy solved in the meteorological model. A simpler level 2.0 scheme is used if only wind and potential temperature fields are available from the meteorological model or observations. This scheme assumes an exact balance between production of turbulent energy and dissipation. The Lagrangian time scales are calculated from turbulent length scale, l, and wind velocity variances

$$T_{Lu} = c_T l/\sigma_u, \quad T_{Lv} = c_T l/\sigma_v, \quad T_{Lw} = c_T l/\sigma_w \quad (16)$$

The constant c_T is assumed to be 1 which provides a good agreement between the Lagrangian time scale, T_{Lw}, simulated in the model and given by empirical formulae for an idealized case of a horizontally homogeneous atmospheric boundary layer (Hanna, 1982). Equations (16) indicate that the Markov chain particle models may be quite sensitive to the turbulent length scale, l. Formulation of a proper turbulent length scale in mesoscale meteorological models applied for complex terrain simulations still requires further research. It should be also pointed out that the Lagrangian time scale, T_{Lw} plays an important role in a plume rise parameterization discussed in Section 4.4.

3 Concentration calculations

The pollution concentration, c, at a given time and location may be determined by counting the number of particles in an imaginary sampling volume centered at (x,y,z):

$$c(x,y,z,t) = \frac{1}{\Delta x_s \Delta y_s \Delta z_s} \sum_{i=1}^{N} m_{pi} I \tag{17}$$

where

$$I = \begin{cases} 1 & for\ |X_i - x| < \Delta x_s/2\ and\ |Y_i - y| < \Delta y_s/2\ and\ |Z_i - z| < \Delta z_s/2 \\ 0 & otherwise \end{cases},$$

and m_{pi}, X_i, Y_i, Z_i are the mass and coordinates of the ith particle at time t. The computed concentrations depend on the size of the sampling volume $\Delta x_s \Delta y_s \Delta z_s$ and on the number of particles N used in the computations. Confidence in this statistical estimate increases with the increase of a number of particles found in the sampling volume and can be achieved by increasing N or by expanding the sampling volume dimensions. However, an increase in N results in a growth of computer time, while an increase of Δx_s, Δy_s, or Δz_s reduces the resolution of the dispersion model. To obtain a smooth and statistically steady concentration field it is usually necessary to release a large number of particles on the order of several thousands. In a rigorous concentration calculation, the contribution of each particle mass should be weighted by the total time spent by the particle inside the sampling volume during each time step (Lamb et al., 1979). However, this requirement seems to be unnecessary in mesoscale applications where sizes of sampling volumes are usually much larger than distances traveled by the particle in one time step.

Computational efficiency of the particle model can be significantly improved by application of a kernel density estimator to calculate the concentration field from particle locations (Lorimer, 1986; Boughton et al., 1987; Yamada and Bunker, 1988; Grossmann, 1989; Uliasz, 1990b; Zannetti, 1992). The kernel method requires no imaginary sampling volumes and produces a smooth concentration distribution with a much smaller number of particles. The concentration at a given point is calculated as the sum of contributions from all particles taking into account a reflection of particles from the ground surface:

$$c(x,y,z) = \sum_{i=1}^{N} \frac{m_{pi}}{h_{xi} h_{yi} h_{zi}} \left[K(r_x, r_y, r_z) + K(r_x, r_y, r_z') \right] \tag{18}$$

where the kernel K satisfies the condition

$$\frac{1}{h_{xi}h_{yi}h_{zi}}\int_{-\infty}^{\infty}\int_{-\infty}^{\infty}\int_{-\infty}^{\infty}K dx dy dz = 1 \quad (19)$$

and $r_x = (X_i - x)/h_{xi}$, $r_y = (Y_i - y)/h_{yi}$, $r_z = (Z_i - z)/h_{zi}$, $r'_z = (Z_i + z)/h_{zi}$. The parameters h_{xi}, h_{yi}, h_{zi} are the bandwidths which determine the degree of smoothing in each coordinate direction. The bandwidths should not be kept constant, as is done in many applications, but they should be particle dependent and change in relation to a natural length scale. Various functional forms can be used for the kernel K. The concentration estimations are not very sensitive to the functional form of the kernel. However, they depend fairly critically on the bandwidths.

Yamada and Bunker (1988) used a Gaussian kernel

$$K(r_x, r_y, r_z) = \frac{1}{(2\pi)^{3/2}}\exp\left(-\frac{r_x^2}{2}\right)\exp\left(-\frac{r_y^2}{2}\right)\exp\left(-\frac{r_z^2}{2}\right) \quad (20)$$

with particle dependent bandwidths related to σ_x, σ_y, and σ_z calculated for each particle with the aid of Taylor diffusion theory. However, this choice of the bandwidths leads to oversmoothing of concentration fields and values of these bandwidths must be limited, especially, in unstable conditions. A more computationally efficient three-dimensional parabolic kernel was discussed by Grossmann (1989):

$$K(r_x, r_y, r_z) = \frac{15}{8\pi}(1 - r^2)I, \quad I = \begin{cases} 1 & for\ r^2 = r_x^2 + r_y^2 + r_z^2 < 1 \\ 0 & otherwise \end{cases} \quad (21)$$

The application of the kernel density estimation to the Lagrangian particle model presented by Lorimer (1986) and Grossmann (1989) is a purely mathematical technique for estimating a probability density (concentration) from a sample (particle positions). It can not be interpreted physically in the context of the atmospheric dispersion problem. The bandwidths are dependent on the number of particles and are related to the standard deviations of the particle distribution. The same bandwidths are applied for all particles at a given time. The proposed method for selecting bandwidths is derived under the assumption that the concentration has a trivariate Gaussian distribution. It seriously limits applications of their approach in practice.

Since the problem of bandwidth selection for a kernel technique is not generally solved, we calculate grid-volume average concentration in most of our mesoscale applications using a very simple uniform kernel:

$$K(r_x, r_y, r_z) = \frac{1}{8}I_x I_y I_z, \quad I_\alpha = \begin{cases} 1 & for\ r_\alpha^2 < 1 \\ 0 & otherwise \end{cases}, \qquad \alpha = x, y, z \quad (22)$$

with constant bandwidths proportional to the increments of the grid used in the calculations: $h_x = \alpha\Delta x$, $h_y = \alpha\Delta y$, $h_z = \alpha\Delta z$. Typically, $\alpha = 0.5$ is selected. Larger bandwidths can be prescribed if stronger smoothing of concentration field is desired. A given particle can contribute to the concentration in several cells of the concentration grid. For $h_x << \Delta x$, $h_y << \Delta y$, and $h_z << \Delta z$, the uniform kernel is practically equivalent to counting particles in a grid cell (17).

The importance of the technique used to calculate concentration fields from particle distributions is illustrated by an example taken from the study performed for the Shenandoah National Park in the eastern United States (Uliasz, 1993). The meteorological simulation was performed with the MESO model for idealized summer conditions when a synoptic flow interacts with sea-breeze and mountain mesoscale circulations. The modeling domain (550 × 550 × 5 km; 67 × 67 × 30 gridpoints) covers the state of Virginia and includes the Chesapeake Bay in the eastern part and the Appalachian and Blue Ridge mountains in the northwestern part of the domain. A 60-hour 3-D meteorological simulation (from 0400 LST on day 1 until 1600 LST on day 3) with the model MESO was performed for cloudless June conditions with a steady synoptic wind of 5 m/s from the northeast. In the discussed example five point emission sources with SO_2 emission rates greater than 100 g/s were taken into account. Continuous emission was simulated by releasing 240 particles per hour from the effective stack height calculated for each source. Figure 1 shows the particle distribution at 1200 LST (day 3) and 3-hour (0900-1200 LST) average surface concentration fields calculated with the aid of the uniform kernel (22) using different grid spacing ($\Delta x = \Delta y = 2.5$, 5, 10, and 20 km, $\Delta z = 0.1$ km) and different smoothing ($\alpha = 0.5$, and 1.0). Maximum surface concentrations vary more than order of magnitude: 73.3, 43.4, 38.9, 22.7, 17.1, 10.2, 7.5, and 4.8 $\mu g m^{-3}$ in the cases from A to H, respectively. Based on the discontinuities in the concentration fields in cases A, B, and C, it is obvious that the grid spacing and smoothing parameters are not adequate for the number of particles that were used in the simulation.

Unfortunately, the technique used in concentration calculations and its implication for obtained concentration values is rarely discussed in applications of particle modeling presented in literature. When designing particle simulations for given applications, it is necessary to answer a difficult question: what number of particles should be released to calculate concentrations with the required resolution, or what concentration resolution can be obtained from the released number of particles. This problem can be approached by studying sensitivity of concentration estimations in respect to the number of particles. Another approach is to use source- and receptor-oriented techniques described later in this chapter. Both techniques should provide the same values of concentration calculated at a receptor. In the discussed study, a difference between concentrations calculated at selected receptors with $\Delta x = 10$ km and $\alpha = 0.5$ (case E) using source- and receptor-oriented techniques were smaller than 5 %.

Since particle simulations in mesoscale usually require a lot of computer time, it is convenient to store distributions of particles during the simulation and then to calculate the concentrations in a postprocessing module. It allows us not only to vary parameters of the kernel estimator but also to calculate different time averages of the concentration.

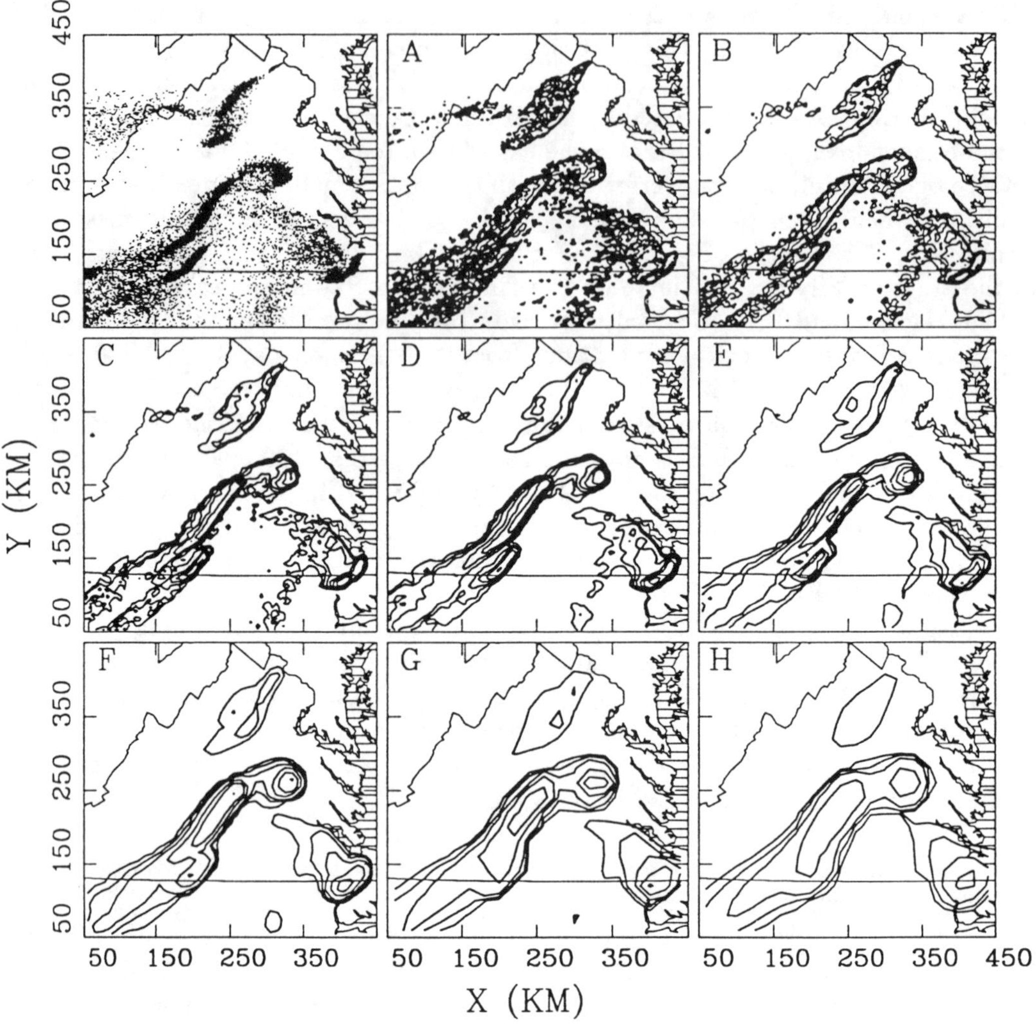

Figure 1: Distribution of particles in the Shenandoah National Park simulation and corresponding 3-hour average surface concentration fields calculated with different grid spacing and smoothing: A - $\Delta x = 2.5$ km, $h_x = 0.5\Delta x$, B - $\Delta x = 2.5$ km, $h_x = \Delta x$, C - $\Delta x = 5$ km, $h_x = 0.5\Delta x$, D - $\Delta x = 5$ km, $h_x = \Delta x$, E - $\Delta x = 10$ km, $h_x = 0.5\Delta x$, F - $\Delta x = 10$ km, $h_x = \Delta x$, G - $\Delta x = 20$ km, $h_x = 0.5\Delta x$, H - $\Delta x = 20$ km, $h_x = \Delta x$, (contours of concentration: $0.25, 0.5, 1, 2.5, 5, 10, 25, 50\ \mu g\, m^{-3}$).

4 Physical parameterizations

4.1 Chemical transformation and radiological decay

Since each particle represents a certain mass of one or more pollutants, linear chemical transformations or radiological decay can be easily included in the model. Let us consider a case of two species with concentrations, c_1 and c_2 respectively, which transformations are given by a simple chemistry mechanism (e.g. $SO_2 \rightarrow$ particulate sulfate):

$$\frac{dc_1}{dt} = -\alpha_1 c_1, \quad \frac{dc_2}{dt} = \alpha_1 c_1 - \alpha_2 c_2 \tag{23}$$

With the aid of the analytical solution of the above equation set at a given time step the mass of each particle is expressed as a sum of masses of two species $m_p = m_{p1} + m_{p2}$ which vary in time:

$$\begin{aligned} m_{p1}(t+\Delta t) &= m_{p1}(t)\exp(-\alpha_1 \Delta t) \qquad (24) \\ m_{p2}(t+\Delta t) &= m_{p1}(t)\frac{\alpha_1}{\alpha_1 - \alpha_2}[\exp(-\alpha_2 \Delta t) - \exp(-\alpha_1 \Delta t)] + m_{p2}(t)\exp(-\alpha_2 \Delta t) \end{aligned}$$

The transformation coefficients, α_1 and α_2, may depend on local meteorological variables, e.g. temperature, humidity, solar radiation. A proper treatment of nonlinear chemistry may be very difficult or impossible within a Lagrangian framework (Zannetti, 1992).

4.2 Dispersion of heavy particles

Dispersion of gaseous species or fine particles with diameter $D_p \leq 1\ \mu m$ can be represented by the dispersion of the passive tracer as described in Section 2. For larger particles the assumption that the turbulence characteristics of the tracer fluid or particle are similar to those of the surrounding fluid is no longer valid. The effect of gravitational forces on the mean particle motion should be included in the dispersion model by the addition of the mean particle settling velocity, v_t in equations (3). The particle attains its terminal settling velocity $v_t = T_p g$ for $t >> T_p$. The relaxation time, T_p, which is extremely short, can be determined for spherical aerosol particles as

$$T_p = \frac{D_p^2}{18\nu}\frac{\rho_p}{\rho} \tag{25}$$

where ρ_p and ρ - densities of particle and air, g - gravitational acceleration, ν - kinematic molecular viscosity. For larger particles which do not satisfy the Stokes law (Reynolds number $Re = D_p v_t/\nu > 0.1$), an empirical correction factor $f(Re)$ is required in (25).

Since the dynamic response of a heavy particle to turbulent flow is different from that of a fluid element, two additional effects must be taken into account (Yudine, 1959; Csanady, 1963):

- Inertia effect. A particle with inertia does not respond to all frequencies of atmospheric turbulence but only to those with a period much greater than its relaxation time, T_p. This selective response means that there will always be some relative motion between the particle and the surrounding air. The effect of inertia can be neglected for small particles up to 400-500 μm diameter.

- Crossing trajectories effect. As a consequence that a particle has a mean downward velocity the particle does not remain within a particular turbulent eddy but continuously drops out from the influence of eddies. The result is that a particle with a large enough terminal velocity crosses the trajectories of many eddies and loses correlation more rapidly than would a passive tracer. The crossing trajectories effect is important down to even quite small diameters.

Several authors proposed modifications to the particle dispersion model which take into account these two effects (Hunt and Nalpanis, 1985; Walklate, 1986; Walklate, 1987; Hashem and Parkin, 1991; Sawford and Guest, 1991). The inertia effect reduces the particle variance σ_{pi}^2 and increases the particle autocorrelation $R_{pi}(\Delta t)$ in comparison to the corresponding values of air velocity σ_i^2 and $R_i(\Delta t)$ while the cross trajectory effect reduces the particle autocorrelation only. It means that heavy particles may disperse faster than fluid elements if the inertia effect controls the dispersion. On the other hand, if the inertia effect is negligible, the dispersion of heavy particles is less than the dispersion of a passive tracer following fluid elements. E.g., if the inertia effect can be neglected, the autocorrelation of particle vertical velocity is expressed as

$$R_{pw}(\Delta t) = R_w(\Delta t) \exp\left(-C\frac{v_t}{\sigma_w}\right) \tag{26}$$

where C is a constant in the range 1 to 2 (Walklate, 1987). A comprehensive analysis of heavy particle dispersion was recently presented by Wang and Stock (1993).

4.3 Dry deposition

The interaction of the particle with the ground surface is parameterized following Boughton et al.(1987). Above a certain height h, the probability of particle deposition is negligible. If the particle comes below the height $z < h$, the probability that the particle is absorbed during time Δt is computed from the transition probability density given by Monin (1959)

$$\begin{aligned} P(z,\Delta t) &= \phi\left[-\frac{z - v_t\Delta t}{\sqrt{2K\Delta t}}\right] + \frac{v_d}{v_d - v_t}\exp\left(\frac{v_t z}{K}\right)\phi\left[-\frac{z - v_t\Delta t}{\sqrt{2K\Delta t}}\right] - \\ & \frac{2v_d - v_t}{v_d - v_t}\exp\left[\frac{v_d z}{K} + \frac{v_d(v_d - v_t)\Delta t}{K}\right]\phi\left[-\frac{z + (2v_d - v_t)\Delta t}{\sqrt{2K\Delta t}}\right] \end{aligned} \tag{27}$$

where v_d is the deposition velocity, $\phi(x)$ is the standard Gaussian distribution, K is the eddy diffusivity, and z is the height of the particle above the ground surface. For particles with a zero settling velocity, v_t, this equation reduces to:

$$P(z,\Delta t) = 2\left(\phi\left[-\frac{z}{\sqrt{2K\Delta t}}\right] - \exp\left[\frac{v_d(z+v_d\Delta t)}{K}\right]\phi\left[-\frac{z+2v_d\Delta t}{\sqrt{2K\Delta t}}\right]\right) \qquad (28)$$

The probability of absorption given by the above equations allows one to simulate a partial deposition of the pollutant, as well as the extreme cases of perfect reflection and perfect absorption. For a later case, the probability $P(z,\Delta t)$ is derived by taking the limit of equation (27) as $v_d \rightarrow \infty$. Two treatments of ground-level particles are possible:

- The particle which appears at $z < h$ is first moved as if no boundary were present. Then, if a random number from the uniform distribution on $[0,1]$ is less than $P(z,\Delta z)$, the particle is absorbed. Otherwise, the particle is moved as if the boundary were perfectly reflecting. This option is designed for heavy particles with a non-zero settling velocity.

- The ground-level particle is perfectly reflected from the ground surface and the particle mass is reduced by a fraction equal to the absorption probability $P(z,\Delta t)$. This option is applied for simulations with linear chemistry where each particle represents multiple species which may have different deposition velocities. For each species, the probability P is calculated separately and then a proper mass reduction is applied.

Both options can be used in the case of the passive tracer. Specification of the height h is not critical because the absorption probability falls off very rapidly as the particle moves away from the boundary.

The deposition velocities are computed with the aid of the resistance-model approach (Walcek et al., 1986):

$$v_d = (r_a + r_b + r_c)^{-1} \qquad (29)$$

The atmospheric resistance r_a is parameterized using surface layer similarity theory, the boundary resistance r_b is related to molecular transfer across a thin laminar sublayer immediately adjacent to the surface, and the surface resistance r_c is a function of land use and vegetation physiology.

This parameterization of dry deposition in the LPD model was compared against prediction of the high resolution K-theory dispersion model. A very good agreement was found for deposition fluxes calculated by the Markov-chain versions of the particle model and the K-theory model. The simplified random-walk versions of the LPD model did not perform as well. This is due to the time step used to move particles being too long for a proper treatment of the interaction of particles with the ground surface.

4.4 Buoyancy phenomena

Several modelers suggested different approaches to simulate buoyant plumes with Lagrangian particle models (Zannetti and Al-Madani, 1984; Cogan, 1985; Gaffen et al., 1987; Shimanuki and Nomura, 1991; Van Dop, 1992; Luhar and Britter, 1992; Anfossi et al., 1993; Hurley and Physick, 1993). The most promising seems to be a model proposed by Van Dop (1992) where the additional stochastic equation describes the buoyancy evolution for each particle. However, even simpler approaches which derive an additional vertical velocity for each particle from an analytical plume rise model may still be acceptable in mesoscale applications. In our implementation, two options for treatment of buoyant plumes are considered:

- particles are released at an effective stack height, z_{eff};
- particles are released at stack top, z_s, and an additional vertical velocity, $w_p = w_b$, due to buoyancy is applied to each particle until the particle reaches the effective stack height, z_{eff}.

Plume rise calculations are based on an analytical model developed by Netterville (1990) for point emission sources. This model is based on the momentum and buoyancy conservation equations written as:

$$\frac{dM}{dt} = F - fM, \quad \frac{dF}{dt} = -N^2M - fM \tag{30}$$

where $M = UR^2w_b$ and $F = UR^2g(T_s - T_a)/T_a$ are the downwind fluxes of vertical momentum and buoyancy, U is wind velocity, T_s and T_a are temperature of exit gases and ambient atmosphere, R is a plume radius, $f = T_{Lw}^{-1}$ is a frequency of atmospheric turbulence, and N is the Brunt-Väisälä frequency defined as $N^2 = g/T_a\, d\vartheta/dz$. The above equations may be solved analytically for atmospheric layers with constant U, N^2, and f in order to obtain height of the plume centerline above the stack, $Z(t)$, and its vertical velocity, $w_b(t)$, in each layer. The model takes into account the vertical structure of the atmosphere (vertical changes of wind, thermal stratification and turbulence) but still assumes that all meteorological fields are horizontally homogeneous in the the area where plume rise takes place. The advantage of Netterville's model is that the plume rise is terminated by the effect of ambient turbulence in neutral or unstable conditions. Vertical profiles of wind, potential temperature and Lagrangian time scale, T_{Lw}, required by this algorithm are provided by the meteorological model.

The effective stack height and/or plume vertical velocity are calculated at each time when new meteorological fields are available. Then, the buoyant vertical velocity component can be picked from the calculated table at each time step of particle motion. Optionally, it is possible to take into account the effect of the plume internal turbulence following Anfossi et al. (1993). A spectrum of z_{eff} and/or w_b profiles is calculated for a range of F values. An initial buoyancy flux is assigned randomly for each released particle from a normal distribution with mean value F_s calculated from provided stack parameters and the standard deviation equal to $F/3$. At each time step the buoyant vertical velocity component is selected for each particle from a precomputed set of values according to its initial buoyancy and current height. If the first option of plume rise calculation is selected, the particle is simply released at the height, z_{eff}, corresponding to its initial buoyancy.

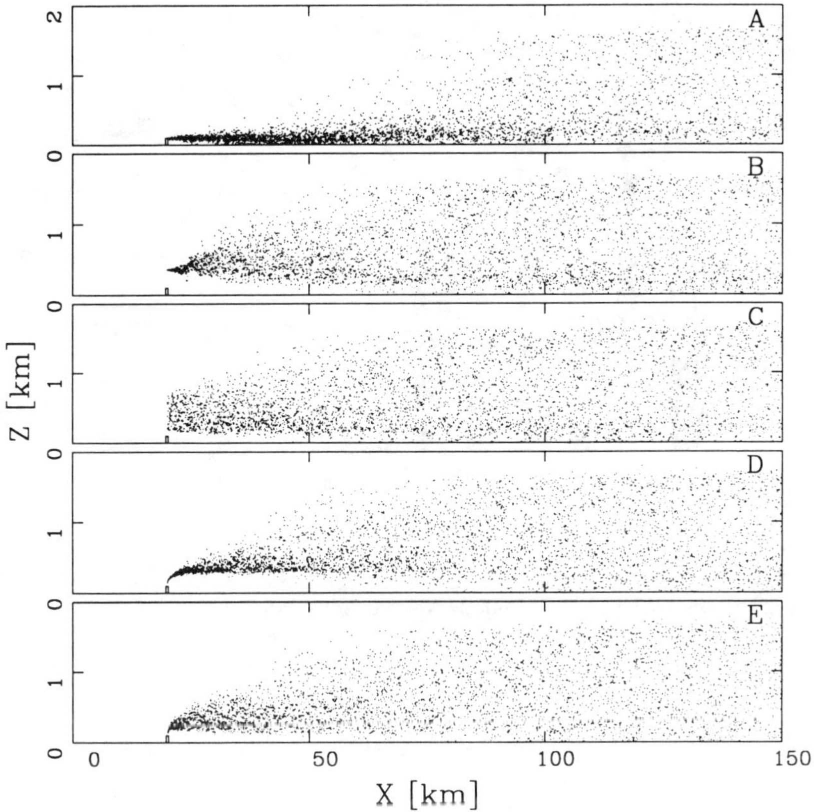

Figure 2: Comparison of plume rise parameterizations in the LPD model: A - no plume rise, B - particles released at the effective stack height, C - same as B but with initial buoyancy flux perturbations, D - additional vertical velocity due to buoyancy, E - same as D but with with initial buoyancy flux perturbations.

Figure 2 illustrates the discussed variants of buoyant plume simulation with the LPD model. Meteorological profiles are obtained from a simulation of atmospheric boundary layer over homogeneous terrain. Particles are released at a rate of 720 particles per hour from a 100 m stack into a near-neutral residual layer above the stable nocturnal boundary layer. The buoyancy flux, F, for this stack is about 950 m^4/s^2.

5 Evaluation of model simplifications

Computation time for the LPD model strongly depends on the number of particles to be tracked as well as on meteorological conditions and the variant of turbulent diffusion parameterization used. Only particles traveling within a specified area are considered. Figure 3 compares the computer time required by an IBM RISC 6000/550 workstation to trace 10000 particles during 1 hour using different meteorological input:

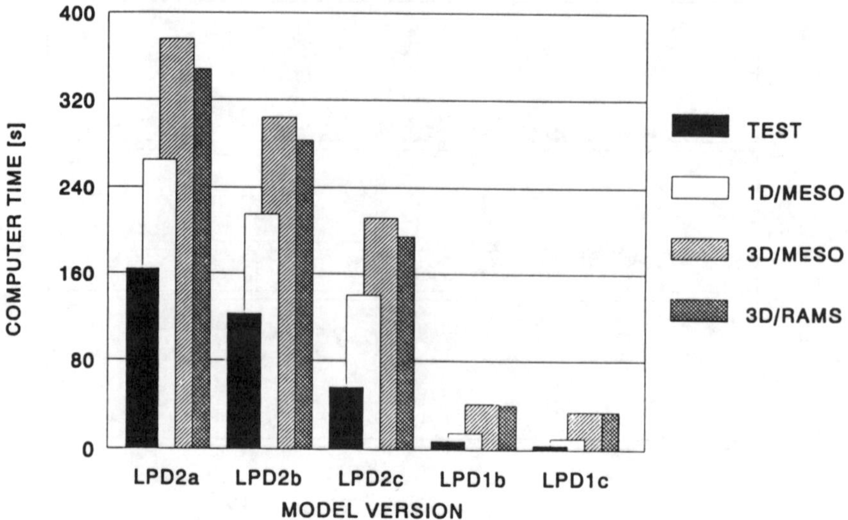

Figure 3: Computer time required by different versions of the LPD model to trace 10000 particles for 1 hour on an IBM RISC-6000/550 workstation.

- a test homogeneous meteorology;
- 1-D simulation of a horizontally homogeneous boundary layer using the MESO model (60 levels up to 3 km);
- 3-D simulation of a mesoscale circulation using the MESO model (67 $\times$ 67 horizontal grid points, $\Delta x = 6$ km, 30 levels up to 5 km);
- 3-D simulation of a mesoscale circulation using the CSU RAMS (2 nested grids, particles were traced on the internal grid: 73 $\times$ 53 horizontal grid points, $\Delta x =$ 12.5 km, 13 levels up to 3 km).

The particles were released randomly within a 200 $\times$ 200 $\times$ 1 km volume at the start of the test simulations. It should be pointed out that additional computer time is needed to read meteorological fields created by the meteorological model, to perform diagnostic calculations of turbulent variables, and to interpolate all variables in space for the position of each particle at each time step of the model. The interpolation may require more computer time than the advection of particles, especially, in the case of 3-D high resolution meteorological fields. The most advanced model version (LPD2a) requires interpolation of nine turbulent variables in addition to three wind components, while the simplest version (LPD1c) uses only one turbulent variable (σ_w).

The different versions of the LPD model were examined with the aid of the meteorological simulation performed for a region of complex terrain in the eastern United States (Uliasz, 1993) used already in the example presented in Section 3. SO_2 concentration fields from the VA Power-Cumberland power station with an SO_2 emission rate of 489.8 gs^{-1} were simulated by each version of the LPD model. Particles

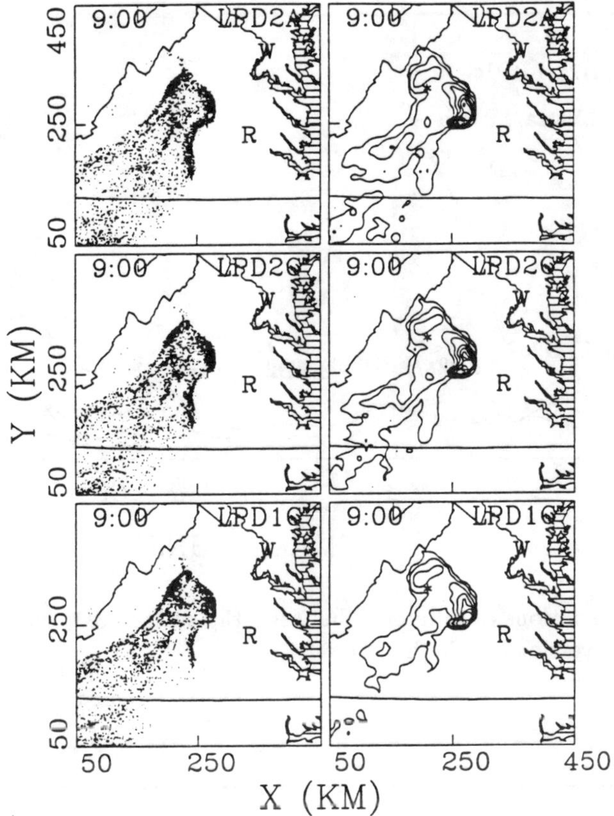

Figure 4: Particle distribution at 0000 LST (2nd day of meteorological simulation) and SO_2 surface concentration fields simulated by models LPD2a, LPD2c, and LPD1c (contours of concentration: 0.1, 0.5, 2, ..., 8, 10, 20, ... $\mu g\, m^{-3}$, W - Washington, R - Richmond, $*$ - the receptor in Shenandoah National Park).

were released continuously at a rate of 1440 particles per hour starting at 2400 LST at the effective stack height. A time series of 3-hour average surface concentration fields were calculated from each particle simulation in $10 \times 10 \times 0.1$ km boxes with the aid of the uniform kernel (22) with $\alpha = 0.5$. Figure 4 demonstrates examples of particle distributions and surface concentrations simulated by LPD2a, LPD2c, and LPD1c models. The differences between simulations were analyzed in terms of the root mean square difference (RMS). The RMS was calculated using concentration fields from each couplet of simulations and normalized by the maximum concentration taken as a mean from these two simulations (Figure 5). The differences between all three Markov chain simulations are very small and also differences between the two random walk simulations are negligible. Differences between the most advanced model (LPD2a) and each of the random walk version are evident, however, as expressed by the RMS, they do not exceed 15% of the maximum concentrations. The RMS was also calculated for concentration fields sorted from maximum to minimum values. This operation does not significantly change the RMS which means that the random walk version

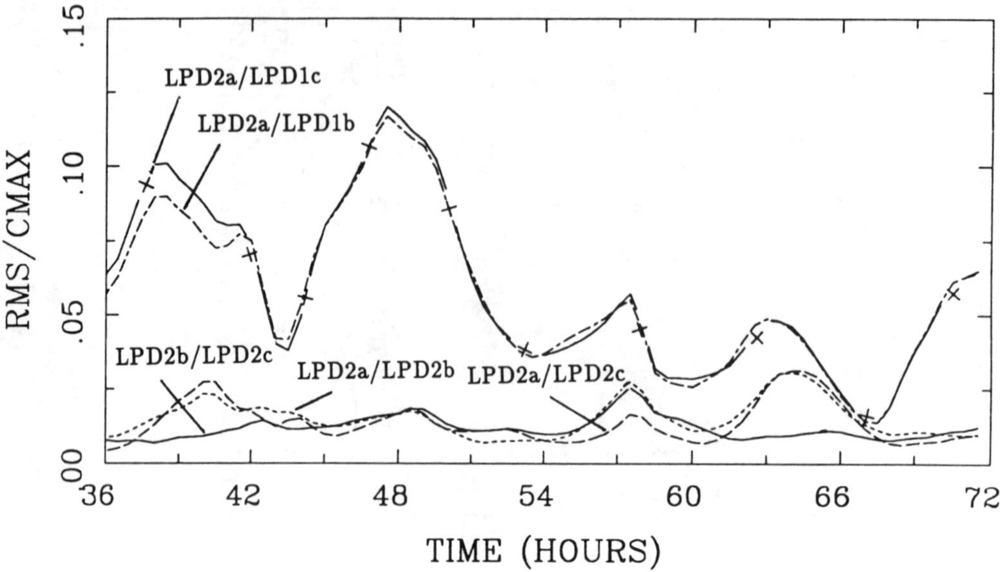

Figure 5: Time variations of difference between the surface SO_2 concentration fields predicted by different versions of the LPD model.

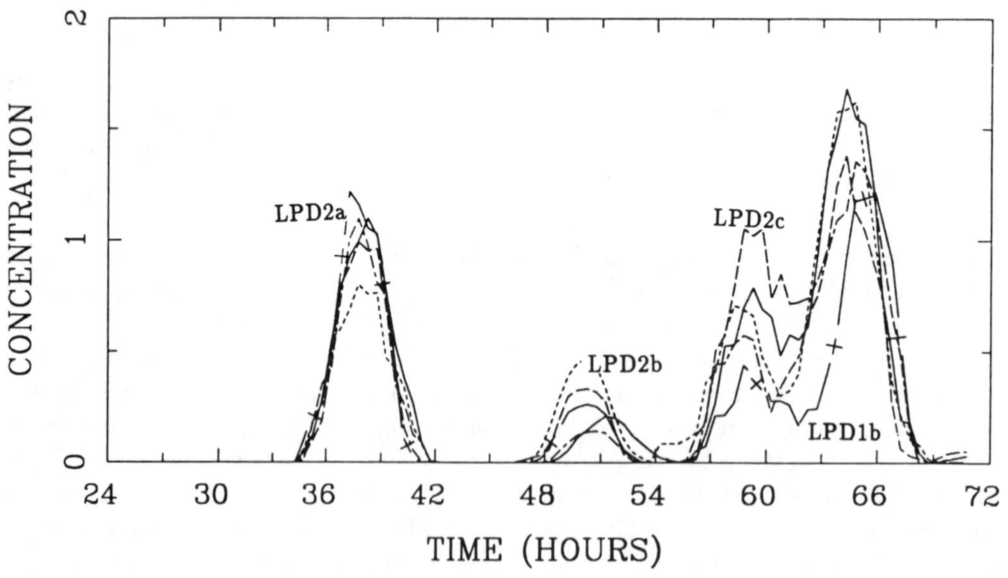

Figure 6: Time series of the surface SO_2 concentration [$\mu g\, m^{-3}$] at Shenandoah National Park predicted by different versions of the LPD model.

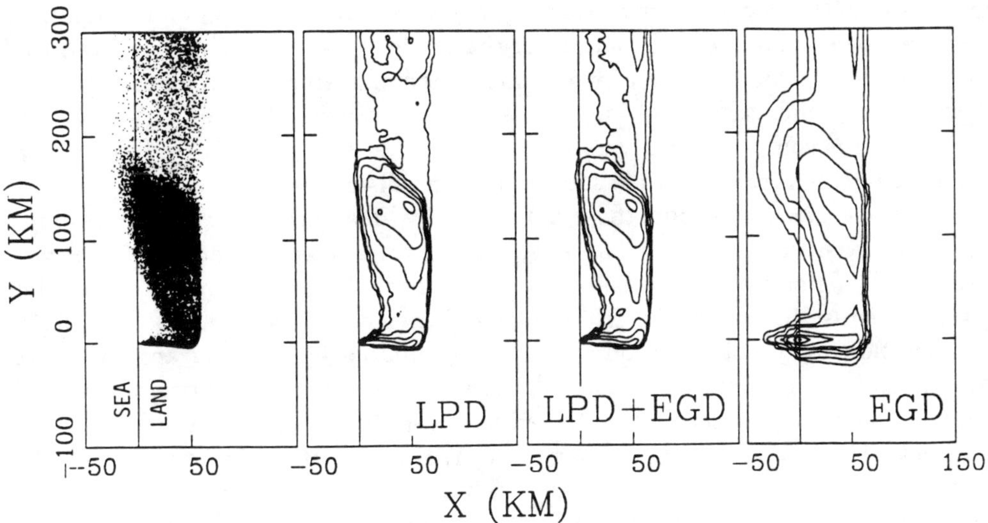

Figure 7: Particle distribution and surface concentration fields calculated by the LPD model, the hybrid dispersion model (LPD+EGD), and the EGD model at 1800 LST (logarithmic contours: -2, -1.5, -1,...).

predicts similar concentration patterns as the Markov chain version but they may under- or overestimate maximum concentrations. This conclusion is confirmed by the time series of SO_2 concentration calculated for a receptor located in Shenandoah National Park (Figure 6).

The performed simulations indicate that in mesoscale and regional applications horizontal turbulent diffusion of pollutants is not important and may be neglected. There is also no need for more refined particle models (e.g. models including wind covariances). The random walk particle models predict the same concentration pattern as the more advanced particle models, although some differences in maximum concentrations are evident. These simplified particle models are a very attractive tool for mesoscale studies since they can be an order of magnitude faster than the Markov chain models. It should be noted, however, that the random walk particle models with long time steps Δt may not be suitable for use with some physical parameterization, e.g. dry deposition or plume rise. The presented analysis of the particle models is being extended for a larger set of mesoscale dispersion simulations.

6 Hybrid dispersion modeling

Computer efficiency of the LPD model in mesoscale applications may be improved by combining this modeling technique with an Eulerian grid dispersion model based on numerical solution of diffusion equations. Due to assumption of the K-theory, the EGD model is applicable to problems whenever plume sizes are much larger than a scale of turbulent eddies in the atmosphere. On the other hand, small sources (e.g.

a single grid cell) associated with strong gradients of concentration cause significant perturbations in the numerical solution of the model equations. Therefore, the EGD model can be used for large area or volume emission sources only. The LPD model is more general and can be applied for arbitrary emission sources. However, the particle model is computationally expensive if it is necessary to track a large number of particles released from multiple sources for long distances.

It is possible to overcome these difficulties by linking the LPD and EGD models together to constitute a hybrid Lagrangian-Eulerian dispersion model (Uliasz and Pielke, 1990). After sufficiently long travel time particles are assumed to contribute to a volume emission field in the grid model and then disappear. Therefore, pollution dispersion close to point emission sources is simulated by the Lagrangian particle technique and as time increased and plume sizes are large in comparison with the grid steps, the numerical solution of the advection-diffusion equations is applied. The condition that determines when the particle contributes to the volume emission field is related to its horizontal sigma coefficients $\max(\sigma_x, \sigma_y) > \delta$ since the plume is usually resolved much faster by a vertical grid than by horizontal grid in the EGD model. The coefficients σ_x and σ_y are determined for each particle using Taylor diffusion theory by time integration of the wind velocity variances encountered during the history of the particle. The value of δ depends on the resolution of the grid model and should be assumed large enough to ensure that at least two grid cells are under the influence of each particle in the kernel calculations. The pollution concentration is determined as a sum of the concentration given by the EGD model and the concentration calculated from particles remaining in the LPD model. The simulation with the grid model can be started with a certain delay in relation to the beginning of the particle simulation when the nonzero emission field is created.

Figure 7 demonstrates an example of results from the hybrid LPD-EGD model (Uliasz and Pielke, 1991a). A two-dimensional simulation of a sea breeze over an idealized coastal area was used as meteorological input for the dispersion models. Simulations were performed for a continuous emission from a point source located at $(0, 0, 0.1)$ km. 1000 particles per hour were released in the LPD1c model. The EGD model was run with 5 km horizontal grid spacing and 20 levels in the vertical. The number of particles was reduced by 50% (with $\delta = 1500$) in comparison with the pure particle model run at the 42-nd hour of the simulation. Concentrations were calculated from particles using the uniform kernel (22) with $\Delta x = \Delta y = 5$ km, $\Delta z = 0.1$ km, and $\alpha = 0.5$. The surface concentration fields predicted by the LPD model and the hybrid model are nearly identical while results from the pure EGD model show oversmoothing of concentrations, especially, close to the emission source.

Another possibility of hybrid modeling consists in combining two particle models with a different level of sophistication. An example of this would be the use of the Markov-chain version of the LPD model right after the release of particles from a source, and then after a prescribed travel time, the particles would be switched to the random-walk model which runs with much longer time step, Δt.

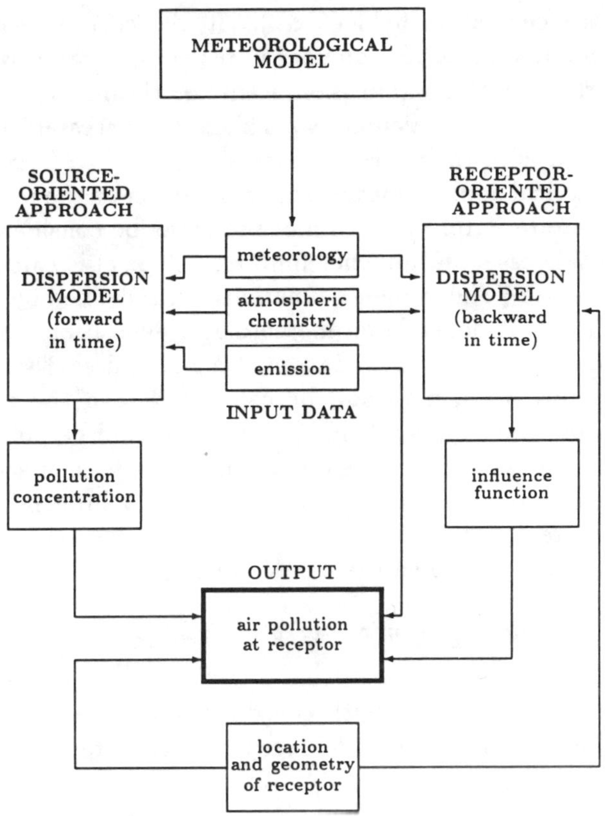

Figure 8: Source- and receptor-oriented techniques in air pollution dispersion modeling.

7 Receptor-oriented dispersion modeling

The source-oriented and receptor-oriented dispersion modeling techniques can be used as complementary tools in air quality studies (Uliasz and Pielke, 1991b; Uliasz, 1993). These two alternative approaches illustrated in Figure 8 can be defined as follows:

$$\Phi[C] = \int\int R\,C\,d\mathbf{x}\,dt = \int\int C^*\,Q\,d\mathbf{x}\,dt \tag{31}$$

where the source- and receptor-oriented approaches are represented by the first and second integral terms, respectively. A final goal of dispersion modeling is to calculate a certain characteristic of air pollution at a given receptor $\Phi[C]$, which can be defined, in general, as the integral of concentration, $C(\mathbf{x}, t)$ over the time and space modeling domain. The receptor function, R, determines the geometry (point, area or volume) and location of the receptor and the sampling time of concentration at the receptor. Therefore, this integral expresses averaging of the concentration field over the receptor and sampling time with the weight function R.

The traditional source-oriented approach consists of solving model equations forward in time for given emission sources of pollutant $Q(\mathbf{x}, t)$ to obtain a time- and space-distributed concentration field $C(\mathbf{x}, t)$. It allows us to calculate various air pollution characteristics, Φ, for any number of receptors located within the modeling domain. However, for any new emission scenario, the model solution must be in principle repeated in order to calculate Φ. This is not necessary for the LPD model if particles are tagged with their release sources.

In many practical applications when air pollution at the receptor is of primary interest, the alternate receptor-oriented modeling may be considered as a more effective approach. In this case, an influence function, $C^*(\mathbf{x}, t)$ is calculated backward in time for a given receptor. The influence function, C^*, is defined by the second integral term in equation (31). It depends on meteorology, deposition, and transformations of pollutant in the atmosphere but is independent of emission sources. The air pollution at the receptor, $\Phi[C]$, may now be calculated with the aid of the influence function directly from the emission field, $Q(\mathbf{x}, t)$. It should be pointed out that these calculations can be repeated for any emission field or emission scenario, Q, without additional solving of model equations. However, a new influence function must be determined for each receptor.

In the particular case when the emission field

$$Q = \sum_i e_i \, \delta(x - x_i) \, \delta(y - y_i) \, \delta(z - z_i) \tag{32}$$

consists of multiple point sources with coordinates x_i, y_i, z_i and constant emission rates e_i, the average concentration at the receptor may be rewritten in a simple form:

$$\Phi[C] = \sum_i e_i \int C^*(x_i, y_i, z_i(t)) \, dt \tag{33}$$

where the source vertical coordinate (effective stack height), z_i, may vary in time. The above expression indicates that the time integrated influence function may be used to characterize dispersion conditions for the receptor if the emission sources are constant during the period of simulation. If the model initial and boundary conditions are taken into account, the influence function allows us to express $\Phi[C]$ as a sum of contributions from:

- local sources within the modeling domain;
- distant sources outside the modeling domain in terms of pollution flux across the model boundaries; and
- initial pollution in the modeling domain.

For a sufficiently long period of simulation the contribution from the initial concentration field is negligible.

The influence function is calculated from backward trajectories of particles in the LPD model where particles are released from the receptor during an assumed sampling time. In the case of grid dispersion models governed by partial differential equations, the influence function is obtained as a solution of the adjoint equations

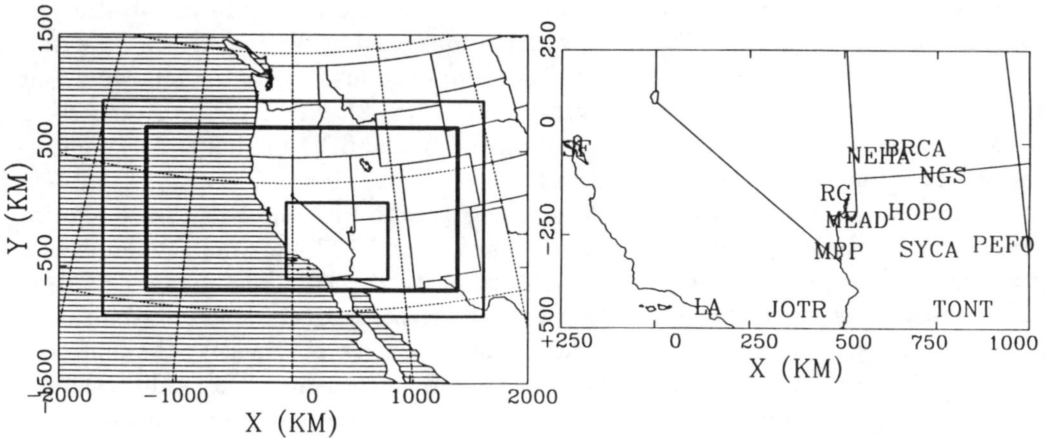

Figure 9: Modeling domain for the MOHAVE project: meteorological grids #1 and #2 - thin line rectangles, dispersion domain - thick line rectangle (left) and selected receptors and emission sources (right)

with the receptor function, R, as a source term. Applicability of the receptor-oriented option is limited to linear dispersion models. It is necessary to assume that all chemical reactions of pollutants are linear and that pollutants do not affect the atmospheric dynamics. The integral of concentration in equation (31) must also be linear, but it can be defined in an arbitrary form depending on application, particularly, it can involve concentrations of several pollutants.

8 Examples of applications

8.1 Project MOHAVE

The goal of the Measurement Of Haze And Visual Effect (MOHAVE) project is to assess the impact of the Mohave Power Project (MPP) and other potential sources of air pollution to specific Class I areas located in the desert southwest United States including the Grand Canyon National Park. The Colorado State University team is performing the daily meteorological and dispersion simulations for a year long study using a non-hydrostatic mesoscale meteorological model, the CSU RAMS coupled with the LPD model (Uliasz et al., 1993). Both the source- and receptor-oriented approaches are used. The modeling domain (Figure 9) covers the southwestern United States with its extremely complex terrain.

The design for daily dispersion simulations with an emphasis on influence function calculations using examples from the winter and summer intensive periods of the

MOHAVE project is discussed below. This very computationally intensive study is performed on two IBM RISC-6000/550 workstations, each equipped with 4 Gb external hard disks, dedicated to the project. The computations involve the following four steps: 1) meteorological simulations using the RAMS; 2) extracting and processing meteorological fields; 3) dispersion simulations using the LPD model; and 4) concentration and influence function calculations.

The daily meteorological simulations carried out with the RAMS use two interactive grids with horizontal grid increments of 50 and 12.5 km, and the number of gridpoints of $66 \times 38 \times 33$ and $74 \times 54 \times 33$, respectively. Each of the modeled days during the year is initialized with the National Meteorological Center initialization data and standard National Weather Service surface data obtained at the National Center for Atmospheric Research. These RAMS simulations do not use 4-dimensional data assimilation but update the model fields through the lateral boundaries with the conditions at the boundaries being linearly interpolated between the 12 hour observation times. For each day of the study period, 36 hours of simulation are performed with the first 12 hours being used to spin-up the atmosphere. The next 24 hour period is then used to represent the diurnal cycle of a given day.

The simplified version of the LPD model based on a fully random walking scheme and with neglected horizontal diffusion (LPD1c) was selected for this study. The computer efficiency of this particle model allows us to design a variety of dispersion simulations using daily output from the RAMS. All daily dispersion simulations are limited to the dispersion of a passive (conservative) tracer. Concentration or influence function fields are calculated from particle distributions using the uniform kernel (22) with $\alpha = 0.5\%$. The number of particles used in the study was determined empirically to be sufficient to calculate concentration using averaging boxes with sizes $50 \times 50 \times 0.2$ km. To obtain a better resolution of concentration fields, the release of a higher number of particles may be required.

Source-oriented simulations. Three series of forward in time particle simulations are currently being performed:

- Local emission sources. These simulations include three power plants: Mohave Power Project (MPP), Navajo Generating Station (NGS) and Reid Gardner (RG) power plant. Particles are released continuously from the effective stack height calculated for each power plant.

- Distant emission sources. These simulations are performed for the purpose of visualization of long range transport rather than to calculate actual concentration fields. Particles are released with a rate of 80 particles per hour continuously from $20 \times 20 \times 0.2$ km volumes located at Los Angeles, San Francisco, Phoenix and Salt Lake City. Additionally, Las Vegas is included as the closest urban area to the region of interest. A fictitious source is added in northwest Nevada in order to study the possible effect of the location of new emission sources in a so-called clean air corridor extending northwest from the Grand Canyon area. Selected particle animations from the forward in time simulations will be recorded on video tape.

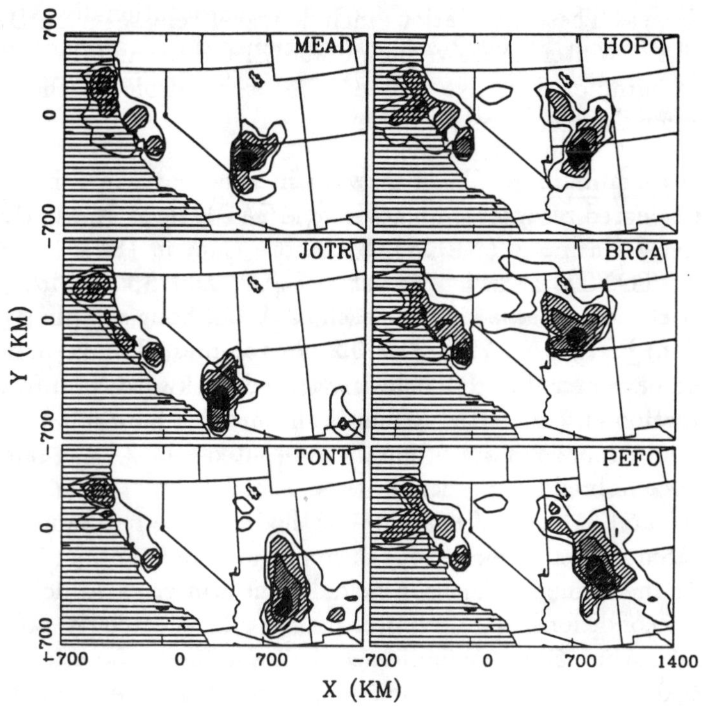

Figure 10: Time integrated influence function fields for the layer 0-500 m calculated for six selected receptors for the period January 16-31, 1992 (contours of $\log(C^*) = -12, -11.5, -11, \ldots s\,m^{-3}$).

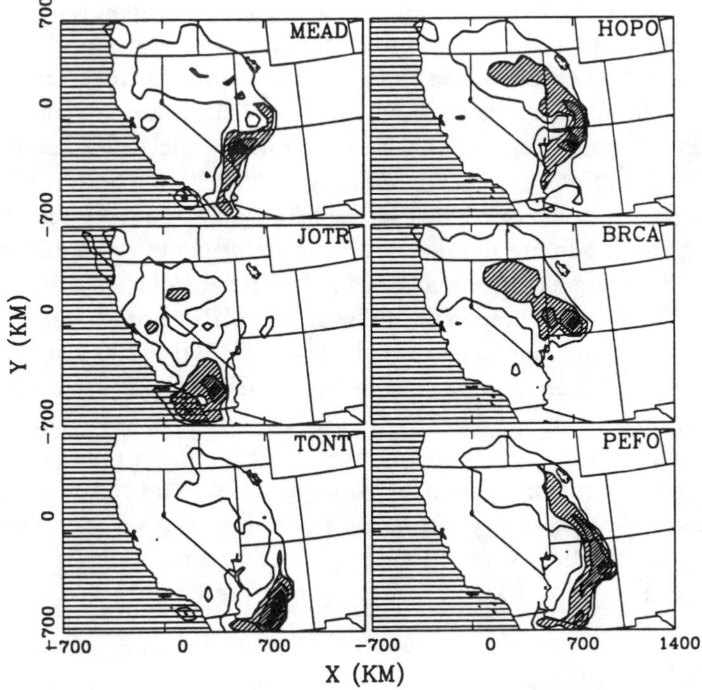

Figure 11: Same as Figure 10 but for the period July 16-31, 1992.

- Tracer releases. These simulations include tracer releases from Dangling Rope, UT during the winter intensive period and El Centro and Tehachapi Pass, CA during the summer intensive period. The simulation for the MPP is being performed for the whole year.

Receptor-oriented simulations. The backward in time particle simulations are performed for nine selected receptors: Meadview (MEAD), Hopi Point (HOPO), Joshua Tree (JOTR), New Harmony (NEHA), Sycamore Canyon (SYCA), Bryce Canyon (BRCA), Tonto (TONT), Petrified Forest (PEFO), and Spirit Mountain (SPMO) (Figure 9). Particles are released during each of the 12 hour sampling periods (6am-6pm and 6pm-6am PST) from $20 \times 20 \times 0.2$ km volumes. Twelve hundred particles are released from each receptor. Particles are traced backward from receptors for five days. The simulation is terminated earlier if all particles have left the modeling domain. The influence functions are originally computed for 12 hour sampling periods and then can be combined for any longer time period.

Figures 10 and 11 present examples of the influence functions for six selected receptors calculated for two week periods during winter and summer 1992. A compact circular shape of the influence function fields in the winter cases does not necessary indicate stagnant conditions. The receptor can be still affected by pollution releases at elevations higher than 500 m as indicated by the influence functions calculated for the layer 500-1000 m. The effective stack height for NGS varies from 300 to 700 m.

The influence functions integrated over the time of simulation do not provide information about age of particles arriving to the receptor from different sources. This information as well as meteorological conditions encountered by particles between the source and receptor are important for the modeling of chemical transformations and removal processes of pollutants in the atmosphere. Therefore, particles are used to trace several meteorological variables along their trajectories: mean relative humidity, maximum relative humidity, mean temperature, and time spent within clouds. Additionally, the travel time during daylight is determined. These characteristics together with the influence functions are calculated against travel time between the source and receptor (i.e. aging time of particles) for selected point sources and a grid of area sources covering the modeling domain. Figures 12 presents an example of this type of results from the dispersion simulations for the Hopi Point receptor in the Grand Canyon National Park and three emission sources: MPP, Los Angeles (LA) and San Francisco (SF). The urban sources are represented by 100×100 km area sources.

The program of daily simulations in the MOHAVE project is strongly limited by available computer resources. In the case of dispersion simulations, the disk space is the strongest constraint. The daily meteorological simulations are performed approximately in a real time on IBM RISC workstations. The computer time required by the particle simulation is very variable since it depends on meteorological conditions. Typically, the particle simulation for nine receptors, which is run for five days backward in time, needs from 1 to 2 hours of computer time.

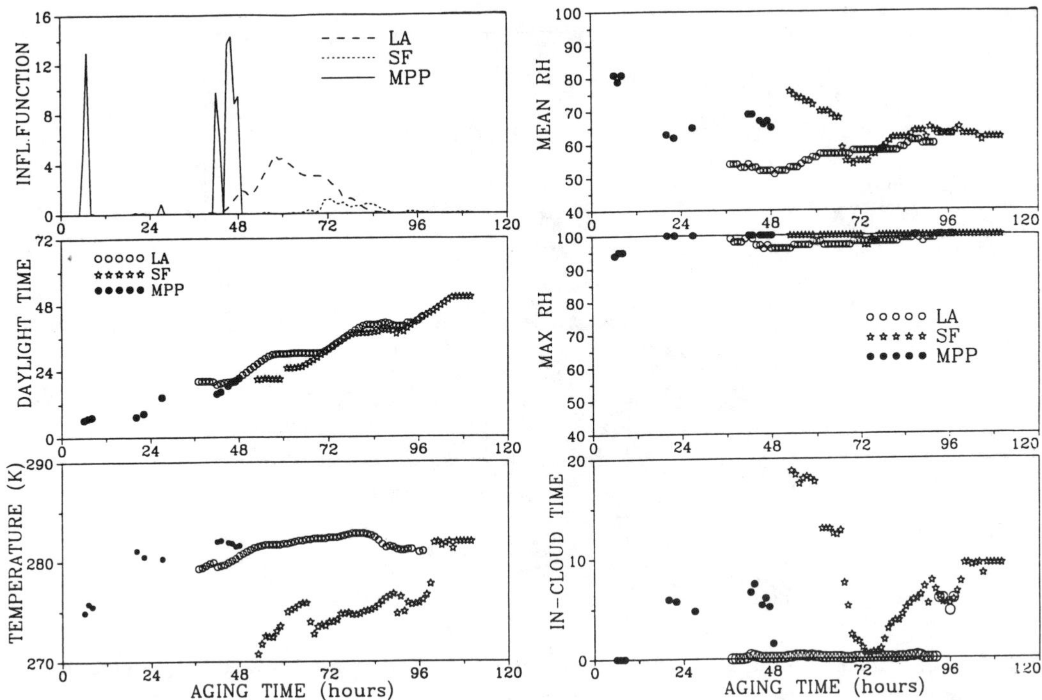

Figure 12: Distributions of influence function, daylight time, mean temperature, mean relative humidity, maximum relative humidity and in-cloud travel time with aging time of particles arriving at the Hopi Point receptor within sampling period 0600-1800 PST, February 10, 1992 and released from the MPP, Los Angeles and San Francisco sources.

8.2 Black Triangle simulations

The modeling methodology developed for the MOHAVE project is being applied to the region of eastern Europe where Poland, Germany and the Czech Republic meet (Figure 13). This region called "Black Triangle" is an area with an extensive mining of hard coal and lignite, mostly by opencast methods coupled with numerous coal-fired power plants and heavy industry. There are also some national parks and large forest areas in this region. However, they have been already seriously damaged by air pollution.

The hydrostatic meteorological mesoscale model MESO together with the random walk version of the LPD model is used for preliminary simulations for the Black Triangle region. The modeling domain $300 \times 300 \times 6$ km with $31 \times 31 \times 25$ gridpoints is centered in the Sudety Mountains. One soil type (loam) is assumed in the whole modeling domain but four land-use categories are distinguished (water, urban, forest and agriculture areas). A series of 36-hours 3-D meteorological simulations for May and June 1993 was performed assuming that synoptic fields vary in time but are horizontally homogeneous within the modeling domain. Additionally, 1-D simulations are

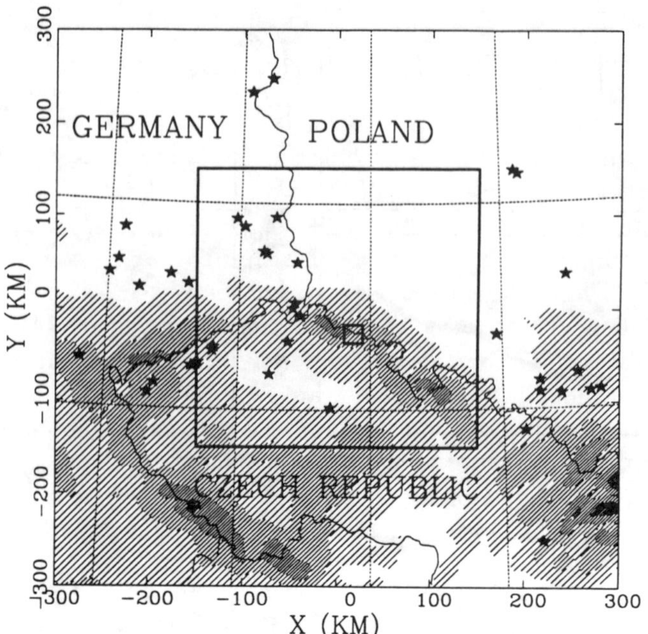

Figure 13: Black Triangle region: stars - major emission sources of SO_2, rectangles - modeling domain and receptor in the Karkonosze Mountain National Park, terrain elevation contours - $250, 500, 750, ...$ m.

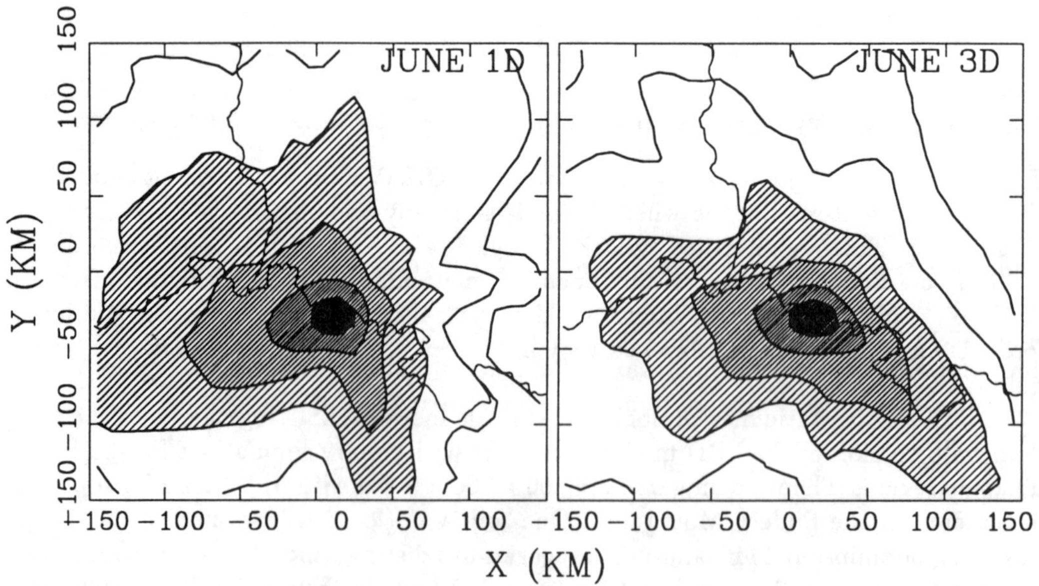

Figure 14: Time integrated influence function for June 1993 in the layer 0-500 m calculated for the Karkonosze Mountain National Park using 1-dimensional (left) and 3-dimensional (right) meteorological simulations.

run for a flat and homogeneous terrain in order to study the sensitivity of pollution transport patterns in respect to the representation of terrain and surface processes parameterizations in the meteorological model. One 3-D meteorological simulation needs 2 hours of computer time on an IBM RISC-6000/550 workstation.

The LPD model is used in a receptor-oriented mode for a $20 \times 20 \times 0.2$ km receptor at the Karkonosze Mountain National Park with the highest peak in the region (Śnieżka, 1602 m). The influence functions are calculated for 12 hour sampling periods and then combined for longer periods in a similar manner to what was done in the project MOHAVE. The influence functions were computed for the whole of June 1993 using series of 1-D and 3-D meteorological simulations (Figure 14). This comparison clearly demonstrates the importance of mesoscale circulations in this mountainous region for dispersion modeling although terrain features are rather poorly resolved with the horizontal grid spacing of 10 km used in the current meteorological simulations.

The traditional source-oriented dispersion simulations would be difficult to perform at this time because of problems with the availability of reliable emission data for this region. In further research, the modeling domain will be extended in W-E direction to cover Lower and Upper Silesia in Poland, the region around Ostrava and northern Bohemia in Czech Republic, and the surrounding of Cottbus and Leipzig in east Germany. This larger domain may be called a "Sulphur Triangle" due to high emissions of SO_2. An attempt will also be made to investigate pollution dispersion patterns as represented by the influence functions with the aid of a synoptic classification (Yu and Pielke, 1986).

9 Conclusions

This chapter demonstrates the value of using state-of-the-art meteorological and Lagrangian particle dispersion models on high performance workstations in order to assess air pollution impacts over mesoscale and regional areas. The possible simplifications of the Lagrangian particle model were examined and simple physical parameterizations were implemented in order to develop an efficient tool for mesoscale applications. The source- and receptor-oriented approaches provide a different insight into regional air pollution transport. The influence functions can be very useful in application to emission control problems, especially, in defining so-called clear air corridors for the Grand Canyon National Park and other receptors in the southwestern US and assessing the emission reduction scenarios in the eastern Europe.

Acknowledgments. The author wishes to thank Roger Stocker for reviewing this paper and providing valuable comments. Funding for this research was partially provided by the National Park Service through Interagency agreement #0475-4-8003 with the National Oceanic and Atmospheric Administration through agreement #CM0200 DOC-NOAA to the Cooperative Institute for Research in the Atmosphere (Project #5-31796). The physical parameterizations for the LPD model were developed under various contracts in ASTeR, Inc.

References

Andrén, A., 1990: Evaluation of turbulence closure scheme suitable for air-pollution applications. *J. Appl. Meteor.*, **29**, 224–239.

Anfossi, D., E. Ferrero, G. Brusasca, A. Marzorati, and G. Tinarelli, 1993: A simple way of computing buoyant plume rise in Lagrangian stochastic dispersion models. *Atmos. Environ.*, **27A**, 1443–1451.

Boughton, B. A., J. M. Delaurentis, and W. E. Dunn, 1987: A stochastic model of particle dispersion in the atmosphere. *Boundary-Layer Meteor.*, **40**, 147–163.

Cogan, J. L., 1985: Monte Carlo simulation of buoyant dispersion. *Atmos. Environ.*, **19**, 867–878.

Csanady, G. T., 1963: Turbulent diffusion of heavy particles in the atmosphere. *J. Atmos. Sci.*, **20**, 201–208.

Gaffen, D. J., C. Benocci, and D. Olivari, 1987: Numerical modeling of buoyancy dominated dispersal using a Lagrangian approach. *Atmos. Environ.*, **21**, 1285–1293.

Grossmann, P. A., 1989: Kernel density estimation applied to a Lagrangian particle dispersion model. Report 10/89, Chisholm Institute of Technology, Australia.

Grubb, D. P. and R. R. Borchers, 1991: Worstations emerge as major scientific resource. *Computers in Physics,* 571–573.

Hanna, S. R., 1982: Applications in air pollution modeling. In *Atmospheric Turbulence and Air Pollution Modeling*, Nieuwstadt, F. T. M. and H. Van Dop, Editors, D. Reidel Publ., Dordrecht, 275-310.

Hashem, A. and C. S. Parkin, 1991: A simplified heavy particle random-walk model for the prediction of drift from agricultural sprays. *Atmos. Environ.*, **25A**, 1609–1614.

Helfand, H. M. and J. C. Labraga, 1988: Design of a nonsingular level 2.5 second-order closure model for the prediction of atmospheric turbulence. *J. Atmos. Sci.*, **45**, 113–132.

Hunt, J. C. R. and P. Nalpanis, 1985: Saltating and suspended particles over flat and sloping surfaces. In *Proceeding of the International Workshop on the Physics of Blown Sand*, Barndorff-Nielsen, O. E., Editor, Aarhuis, Denmark, 9-36.

Hurley, P. and W. Physick, 1993: Lagrangian particle modelling of buoyant point sources: plume rise and entrapment under convective conditions. *Atmos. Environ.*, **72A**, 1579–1584.

Lamb, R. G., H. Hogo, and L. E. Reid, 1979: A Lagrangian approach to modeling air pollutant dispersion: Development and testing in the vicinity of roadway. Technical report, EPA Research Report EPA-600/4-79-023.

Legg, B. J. and M. R. Raupach, 1982: Markov-chain simulation of particle dispersion in inhomogeneous flows: the mean drift velocity induced by a gradient in Eulerian velocity variance. *Boundary-Layer Meteor.*, **24**, 3–13.

Lorimer, G. S., 1986: The kernel method for air quality modeling - I. Mathematical foundation. *Atmos. Environ.*, **20**, 1447–1452.

Luhar, A. K. and R. E. Britter, 1992: Random-walk modeling of buoyant-plume dispersion in the convective boundary layer. *Atmos. Environ.*, **26A**, 1283–1298.

Lyons, W. A., R. A. Pielke, W. R. Cotton, M. Uliasz, C. J. Tremback, R. L. Walko, and J. L. Eastman, 1993: The applications of new technologies to modeling mesoscale dispersion in coastal zones and complex terrain. In *Air Pollution*, P. Zannetti, C.A. Brebbia, J.E.G. and G. A. Milian, Editors, Computational Mechanics Publications, Southampton, 33-85.

Mellor, G. L. and T. Yamada, 1982: Development of a turbulence closure model for geophysical fluid problems. *Rev. Geophys. Space Phys.*, **20**, 851–875.

Monin, A. S., 1959: On the boundary condition on the earth surface for diffusing pollution. *Adv. Geophys.*, **6**, 435–436.

Netterville, D. D. J., 1990: Plume rise, entrainment and dispersion in turbulent winds. *Atmos. Environ.*, **24A**, 1061–1081.

Obukhov, A. M., 1959: Description of turbulence in terms of Lagrangian variables. *Adv. Geophys.*, **6**, 113–116.

Pielke, R. A., W. R. Cotton, R. L. Walko, C. J. Tremback, M. E. Nicholls, M. D. Moran, D. A. Wesley, T. J. Lee, and J. H. Copeland, 1992: A comprehensive meteorological modeling system - RAMS. *Meteor. Atmos. Phys.*, **49**, 69–91.

Pielke, R. A., W. A. Lyons, R. T. McNider, M. D. Moran, D. A. Moon, R. A. Stocker, R. L. Walko, and M. Uliasz, 1991: Regional and mesoscale meteorological modeling as applied to air quality studies. In *Air Pollution Modeling and Its Application VIII*, van Dop, H. and D. G. Steyn, Editors, Plenum Press, New York, 259-290.

Sawford, B. L. and F. M. Guest, 1991: Lagrangian statistical simulation of the turbulent motion of heavy particles. *Boundary-Layer Meteor.*, **54**, 147–166.

Shimanuki, A. and Y. Nomura, 1991: Numerical simulation of instantaneous images of the smoke released from a chimney. *J. Meteor. Soc. Japan*, **60**, 187–195.

Smith, F. B., 1968: Conditioned particle motion in a homogeneous turbulent field. *Atmos. Environ.*, **2**, 491–508.

Uliasz, M., 1990a: Development of the mesoscale dispersion modeling system using personal computers. Part I: Models and computer implementation. *Zeitschrift für Meteorologie*, **40**, 104–114.

Uliasz, M., 1990b: Development of the mesoscale dispersion modeling system using personal computers. Part II: Numerical simulations. *Zeitschrift für Meteorologie*, **40**, 285–298.

Uliasz, M., 1993: The atmospheric mesoscale dispersion modeling system. *J. Appl. Meteor*, **32**, 139–149.

Uliasz, M. and R. Pielke, 1990: Receptor-oriented Lagrangian-Eulerian model of mesoscale air pollution dispersion. In *Computer Techniques in Environmental Studies III*, Zannetti, P., Editor, Computational Mechanics Publications, Southampton and Springer-Verlag, Berlin, 57-68.

Uliasz, M. and R. Pielke, 1991a: Lagrangian-Eulerian dispersion modeling system for real-time mesoscale applications. In *Third Topical Meeting on Emergency Preparedness and Response*, Chicago, Illinois, April 16-19, 1991, 95-98.

Uliasz, M. and R. A. Pielke, 1991b: Application of the receptor oriented approach in mesoscale dispersion modeling. In *Air Pollution Modeling and Its Application VIII*, van Dop, H. and D. G. Steyn, Editors, Plenum Press, New York, 399-408.

Uliasz, M. and R. A. Pielke, 1993: Implementation of Lagrangian particle dispersion model for mesoscale and regional air quality studies. In *Air Pollution*, P. Zannetti, C.A. Brebbia, J. E. G. and G. A. Milian, Editors, Computational Mechanics Publications, Southampton, 157-164.

Uliasz, M., R. A. Stocker, and R. A. Pielke, 1993: Numerical modeling of atmospheric dispersion during the MOHAVE field study. In *Air Pollution*, P. Zannetti, C.A. Brebbia, J. E. G. and G. A. Milian, Editors, Computational Mechanics Publications, Southampton, 208-216.

Van Dop, H., 1992: Buoyant plume rise in a Lagrangian framework. *Atmos. Environ.*, **26A**, 1335–1346.

Walcek, C. J., R. A. Brost, J. S. Chang, and M. L. Wesely, 1986: SO_2, sulfate and HNO_3 deposition velocities using regional land use and meteorological data. *Atmos. Environ.*, **20**, 949–964.

Walklate, P. J., 1986: A Markov-chain particle dispersion model based on air flow data: extension to large water droplets. *Boundary-Layer Meteor.*, **37**, 313–318.

Walklate, P. J., 1987: A random-walk model for dispersion of heavy particles in turbulent air flow. *Boundary-Layer Meteor.*, **39**, 175–190.

Wang, L.-P. and D. E. Stock, 1993: Dispersion of heavy particles by turbulent motion. *J. Atmos. Sci.*, **50**, 1897–1913.

Yamada, T. and S. Bunker, 1988: Development of a nested grid, second moment turbulence closure model and application to the 1982 ASCOT Brush Creek data simulation. *J. Appl. Meteor.*, **27**, 567–578.

Yu, C.-H. and R. A. Pielke, 1986: Mesoscale air quality under stagnant synoptic cold season conditions in the lake powell area. *Atmos. Environ.*, **20**, 1751–1762.

Yudine, M. I., 1959: Physical consideration on heavy-particle diffusion. *Adv. Geophys.*, **6**, 185–191.

Zannetti, P., 1986: Monte-Carlo simulation of auto- and cross-correlated turbulent velocity fluctuations (MC-LAGPAR II model). *Environmental Software*, **1**, 26–30.

Zannetti, P., 1992: Particle modeling and its application for simulating air pollution phenomena. In *Environmental Modelling*, Melli, P. and P. Zannetti, Editors, Computational Mechanics Publications and Elsevier Applied Science, 211-241.

Zannetti, P. and N. Al-Madani, 1984: Simulation of transformation, buoyancy and removal processes by Lagrangian particle methods. In *Proc. of the 14th International Technical Meeting on Air Pollution Modeling and Its Application*, de Wispelaere, C., Editor, Plenum Press, New York, 733-744.

Chapter 5

Survey of long range transport models

G. Graziani

Commission of the European Community, Joint Research Centre, Environment Institute, I-21020 Ispra (VA), Italy

Abstract

This chapter presents an overview of state-of-art for long range dispersion models of pollutants in the atmosphere. Separate consideration is given to episodic and long term models, since each of these classes requires a different approach. The most important mechanisms of pollutant removal from atmosphere, such as dry and wet deposition and chemical reaction, are also described. Finally, results from a recent intercomparison study of such models applied to the Chernobyl episode are summarized.

Keywords: atmospheric dispersion, long range transport, deposition, nuclear and chemical releases, concentration data assimilation.

1. INTRODUCTION

In the last few years, the Chernobyl accident and the oilwell fires in the Gulf have indicated the fact that emissions into the air from human activities are transported over long distances. However, this problem was already apparent in the early sixties, when the first evidence of lake acidification due to air pollution transported over the North Sea was reported by Scandinavian scientists. The chemical composition of their lakes was found to be affected by emissions of sulphur and nitrogen compounds from the large industrial centres in Central Europe and British Isles, which were deposited by precipitation, thereby altering the biological equilibria among species and affecting the economical life of entire regions.

This fact indicated that airborne pollutants were a matter of concern, not only because of their concentration level, but also because of their deposition onto the earth's surface. The complexity of atmospheric chemical processes that must be taken into account, when considering travel times associated with continental distances, enhanced the difficulty of long range transport (LRT) simulations.

On 26 April 1986, the most serious accident ever at a nuclear power plant happened in Chernobyl (51°17'N, 30°15'E), when 4% of its entire reactor core inventory of radioactive material (more than 10^{18}Bq) was released. The emission continued until May 6. During the weeks after the initial explosion, radioactive material was observed almost everywhere in the Northern Hemisphere due to LRT in the air. Substantial contamination, caused mainly by wet deposition of the radioactive plume, continued for about two weeks after the accident at sites thousands of kilometers from Chernobyl, in particular in Scandinavia, the Alps, and mainland Greece.

These examples point to the necessity of having long range air pollution dispersion models capable of covering spatial scales from a few hundred kilometers up to global distances, with corresponding times ranging from hours to years. Such atmospheric transport and diffusion models are in general models in which the air motion itself is not calculated, i.e. they are kinematics models. Information on the motion is however derived from meteorological measurements or from meteorological model outputs.

Long range atmospheric models fall into two broad categories:

- those calculating concentration values over a short time periods (hours or days),
- those giving only average concentration values over long time periods (monthly to annual).

Models of the first type are often used to deal with accidental releases (Section 2), and a discussion of their use of data in real-time is an object of this section. Results from these models can be used to furnish useful information for longer times by averaging the short-term results. However, since the basic approaches are different, long-term models are treated here separately (Section 3).

As most air pollutants are emitted at or near the surface of the earth, long range atmospheric transport and related chemical formulations must deal in detail with a thin layer of about 2 km in depth, called the atmospheric boundary layer (ABL). A description of the evolution of ABL depth due to the diurnal solar energy cycle is then also an object of this chapter; in fact long range atmospheric transport models usually include an appropriate description of the ABL physical properties, such as its three-dimensional windfield (with appropriate time and space scale resolution) and its turbulence.

Description and interpretation of the main removal phenomena, such as chemical transformation, and dry and wet deposition of pollutants on the earth surface, is also essential in LRT modelling studies. A description of the physics of clouds and precipitation is thus also of great importance in such models. This chapter then also presents a summary of the main problems concerning the modelling of removal processes (Section 4).

Finally, Section 5 presents the main outcomes of a recent study aimed at comparing LRT model results with radioactivity measurements after the Chernobyl accident.

2. SHORT-TERM DISPERSION MODELS

Two main categories of LRT models can be distinguished: Eulerian and Lagrangian. Eulerian models describe the dispersion of pollution in a fixed frame of reference (fixed with respect to a point on the earth surface). In Lagrangian models, the evolution of a polluted air parcel is described relative to a mobile reference system associated to the parcel from its initial position as it moves along its trajectory.

Of course both descriptions are equivalent, i.e. the wind velocity in the Eulerian frame of reference u(x,t) is related to the Lagrangian velocity dx/dt by:

$$\frac{dx(t)}{dt} = u(x,t) ,$$

where, for simplicity, only one dimension has been considered.

The choice of model category depends on various aspects of the desired application, e.g. numerical facilities, as discussed below. Eulerian and Lagrangian transport models formulation starts with a definition of mass conservation, in which a balance is made between what is created, is destroyed, enters and leaves each air volume.

The dry atmosphere can be considered as a mixture of ideal gases. The dynamics of atmospheric flow is given by the Navier-Stokes equations, which together with the continuity and energy equations complete the set of atmospheric equations. An important simplification made in LRT models is the Boussinesq assumption, under which fluctuations in density must be taken into account only in combination with the acceleration of gravity. This implies that the atmosphere can be considered incompressible, and thus the continuity equation reduces to:

$$\frac{\partial u}{\partial x} + \frac{\partial v}{\partial y} + \frac{\partial w}{\partial z} = 0 , \qquad (1)$$

where (u,v,w) are wind velocity components in the fixed frame of reference (x,y,z).

Denoting χ as the contaminant concentration (in units of mass per unit volume of fluid), the equation of mass conservation is:

$$\frac{\partial \chi}{\partial t} + \frac{\partial u\chi}{\partial x} + \frac{\partial v\chi}{\partial y} + \frac{\partial w\chi}{\partial z} = S ,$$

where S denotes non-surface sources and sinks.

Setting all instantaneoues values equal to average plus fluctuation terms, e.g. $\chi = \overline{\chi} + \chi'$, $u = \overline{U} + u'$, and using Reynolds averaging (in which the average fluctuation equals zero) the conservation mass equation yields:

$$\frac{\partial \overline{\chi}}{\partial t} + \overline{U}\,\frac{\partial \overline{\chi}}{\partial x} + \overline{V}\,\frac{\partial \overline{\chi}}{\partial y} + \overline{W}\,\frac{\partial \overline{\chi}}{\partial z} = -\frac{\partial}{\partial x}\,\overline{u'\chi'} - \frac{\partial}{\partial y}\,\overline{v'\chi'} - \frac{\partial}{\partial z}\,\overline{w'\chi'} + S , \qquad (2)$$

where the eddy correlation terms $\overline{u'\chi'}$, etc., have been moved to the right and the continuity equation (1) has been used.

To solve this equation, assumptions have to be made for the eddy correlation terms (also called contaminant turbulent fluxes). A usual approach is to assume that they are proportional to concentration gradients (i.e. K-theory), as follows:

$$-\overline{u'\chi'} = K_x \frac{\partial \bar{\chi}}{\partial x}, \quad -\overline{v'\chi'} = K_y \frac{\partial \bar{\chi}}{\partial y} \quad \text{and} \quad -\overline{w'\chi'} = K_z \frac{\partial \bar{\chi}}{\partial z} .$$

This is called first order closure or gradient transport theory.
Use of this assumption in (2) yields:

$$\frac{\partial \bar{\chi}}{\partial t} + \bar{U} \frac{\partial \bar{\chi}}{\partial x} + \bar{V} \frac{\partial \bar{\chi}}{\partial y} + \overline{W} \frac{\partial \bar{\chi}}{\partial z} =$$

$$= \frac{\partial}{\partial x} \left(K_x \frac{\partial \bar{\chi}}{\partial x} \right) + \frac{\partial}{\partial y} \left(K_y \frac{\partial \bar{\chi}}{\partial y} \right) + \frac{\partial}{\partial z} \left(K_z \frac{\partial \bar{\chi}}{\partial z} \right) + S . \qquad (3)$$

The coefficients K_x, K_y and K_z parametrize the strength of the turbulent exchanges and are functions of the coordinate location in heterogeneous flows. First order K-theory closure is almost always used to describe the transport and dispersion equation for the long range diffusion. Only recently higher order closure schemes are employed.

2.1. Eulerian Models

Equation (3) is the basis for a number of applications in LRT atmospheric dispersion calculations. Contrary to what happens over short distances, in LRT applications the non-homogeneity and non-stationarity of the wind and turbulence introduce such complexities, that only numerical approximations can provide solutions to this equation (Van Dop and De Haan, 1983).

Long range models are also often used in complex terrains, in which the vertical mean velocities cannot be neglected. In these cases (see for example Carmichael and Peters, 1984), the topography can be included by transforming the vertical coordinate in a dimensionless terrain following coordinate, such as:

$$\sigma = \frac{z - h(x,y)}{H(x,y,t) - h(x,y)} ,$$

where z is height above some reference level, h(x,y) is surface height above the same level and H(x,y,t) is the top of the modelling region. Under this transformation the atmospheric dispersion equation becomes:

$$\frac{1}{\Delta H} \left\{ \frac{\partial(\Delta H \bar{\chi})}{\partial t} + \frac{\partial(\bar{U} \Delta H \bar{\chi})}{\partial x} + \frac{\partial(\bar{V} \Delta H \bar{\chi})}{\partial y} + \frac{\partial(\bar{W} \bar{\chi})}{\partial \sigma} \right\}$$

$$= \frac{1}{\Delta H} \left\{ \frac{\partial}{\partial x} \left(K_H \Delta H \frac{\partial \bar{\chi}}{\partial x} \right) + \frac{\partial}{\partial y} \left(K_H \Delta H \frac{\partial \bar{\chi}}{\partial y} \right) + \frac{\partial}{\partial \sigma} \left(\frac{K_z}{\Delta H} \frac{\partial \bar{\chi}}{\partial \sigma} \right) \right\} + S ,$$

where

$$\Delta H = H(x,y,t) - h(x,y) \ .$$

In this equation, K_H is the homogeneous horizontal component of the K-tensor and K_z is its vertical component. No cross components are taken into account and the terms containing space derivatives of h and ΔH have been assumed to be zero. This last assumption indicates that rapid variations of topography are not considered, a good approximation in LRT models with their large horizontal mesh sizes.

The horizontal eddy-flux term in the σ coordinate system is often neglected, since experience indicates that horizontal turbulent diffusion is much less important than indirect lateral mixing caused by wind shear and vertical diffusion (Kimura and Yoshikawa, 1988).

Other modellers have used reduced pressure coordinates ($\sigma = p/p_s$) as the vertical coordinate (where p_s is the pressure at surface). In such models the topography enters only indirectly, since its effects are included in the variations of p. In this cases the input data are derived from global or regional circulation models, which are written in spherical coordinates. It is therefore necessary to solve the LRT equations in spherical coordinates, which requires the solution of an equation of the form:

$$\frac{\partial \bar{\chi}}{\partial t} = - \frac{\bar{U}}{R \cos \Phi} \frac{\partial \bar{\chi}}{\partial \lambda} - \frac{V}{H} \frac{\partial \bar{\chi}}{\partial \Phi} - \sigma \frac{\partial \bar{\chi}}{\partial \sigma} + \frac{1}{R^2 \cos^2 \Phi} \frac{\partial}{\partial \lambda} \left(K_\lambda \frac{\partial \bar{\chi}}{\partial \lambda} \right) + $$

$$+ \frac{1}{R^2 \cos^2 \Phi} \frac{\partial}{\partial \Phi} \left(K_\Phi \cos \Phi \frac{\partial \bar{\chi}}{\partial \Phi} \right) + \Gamma^2 \frac{\partial \bar{\chi}}{\partial \sigma} + S \ ,$$

where R is earth's radius, Φ is latitude and λ longitude. Also:

$$\Gamma = g\sigma / R_A T \ ,$$

where g is acceleration due to gravity, T is the absolute temperature and R_A is the ideal gas constant.

The transport-diffusion equation is completed with an initial field, $\bar{\chi}(x,y,z,t=0) = \chi_o$ and with boundary conditions which define in- and outflow rates through the model boundaries. The boundary conditions frequently specify a zero flux on the top of the domain and constant derivatives at its lateral edges. On the bottom of the domain, the flux is set equal to the algebric sum of area sources and dry deposition.

Wet deposition is included in term S, which describes all internal sources and sinks. In general, wet deposition is treated in Eulerian models as the product of a coefficient (that parameterizes air to water transfer) times of air pollutant concentration and precipitation intensity at each vertical level. However, since precipitation is only measured at ground level, it is usually assumed a constant precipitation rate with height and thus a uniform scavenging coefficient (see also Section 4.2).

A chemical transformation or radioactive decay term is also included in source term S. The form of the matrix describing the transformation between chemical substances should require the concurrent consideration of the advective, diffusive, and transformation processes for each group of elements. However, this cannot generally be achieved with present computational facilities, and hence transformation calculations are treated separately. The situation is much simpler with radionuclides, as the analogous transformation matrix is sparse. Thus a linear radioactive decay term is usually included in S.

Representation of point sources in a Eulerian grid is a numerical modelling problem, as the plume or puff size of the emitted substance remains as sub-grid for a significant time. This is particularly true in LRT models, where differences between the source scale and that resolved by the model are several orders of magnitude. Many solutions exist to this problem. One is based on the multigrid method discussed by Pudykiewicz et al. (1984), in which releases are simulated in a series of nested grids with decreasing resolution. This method also has the advantage of using detailed information about deposition patterns in the vicinity of the source during short time periods after an accident.

In a different approach, Bompay in his contribution to the ATMES study (Klug et al., 1991) described the early growth of a pollutant source of intensity Q in a sub-grid source mesh using the following Gaussian distribution:

$$\frac{d\bar{\chi}}{dt} = \frac{Q(t)}{2\pi\,\sigma^2{}_n\,H_c}\exp\left(-\frac{d^2}{2\,\sigma^2{}_n}\right),$$

where d is distance from the source, H_c the vertical spread of the cloud and $\sigma^2{}_n$ the mesh surface area from which the source has been emitted. This phase of the simulation continues until the plume dimension grows to be the same order as the numerical grid size.

The number of vertical levels in Eulerian models varies from case to case; it is usually taken between 11 and 30, dependending on available computer resources. The spacing between these levels depends on the nature of the source. Sources located close to the ground surface will require high vertical resolution within the atmospheric boundary layer (ABL), whereas high stratospheric sources require a more uniform positioning of model levels. The first layer, however, is usually always located in the surface boundary layer (SBL). A correct description of these lower layers is also important for the correct definition of effective turbulent roughness length scales (André and Blondin, 1986) and therefore of SBL turbulence parameters.

Under the assumption that horizontal advection exceeds diffusion and that, in the vertical direction, diffusion is the most important phenomenon, a frequently used numerical technique to solve the transport equation consists of splitting the advective and diffusive processes according to:

$$\frac{\partial\bar{\chi}}{\partial t} + \bar{U}\,\frac{\partial\bar{\chi}}{\partial x} + \bar{V}\,\frac{\partial\bar{\chi}}{\partial y} = 0. \tag{4a}$$

and

$$\frac{\partial \bar{\chi}}{\partial t} = \frac{\partial}{\partial z}(K_z \frac{\partial \bar{\chi}}{\partial z}) + S \ . \tag{4b}$$

When $\bar{\chi}$ represents the concentration of a chemically reactive species, a third equation must be added to describe its production and depletion. The advantage of this method is that one applies numerical techniques appropriate to the mathematical nature of each equation (Table 1).

Table 1
Summary of algorithms tested by Chock and Dunker (1983).

Algorithm	Time-Step Procedure*	Splitting Method*
Flux-corrected transport	Explicit Euler	Yes
Multidimensional flux-corrected transport (fourth-order finite difference)	Explicit leapfrog-trapezoidal	No
Multidimensional flux-corrected transport (fourth-order finite difference)	Modified Euler predictor-corrector	No
Orthogonal collocation on finite elements	Modified Euler predictor-corrector	No
Orthogonal collocation on finite elements	Implicit Crank-Nicolson	Yes
Orthogonal collocation on finite elements	Implicit backward Euler	Yes
Second moment (Egan-Mahoney)	Explicit	No
Pseudospectral (Fourier expansion: periodic boundaries)	Explicit leapfrog	No
Chapeau function	Implicit Crank-Nicolson	Yes
Chapeau function with a dissipative term	Implicit Crank-Nicolson	Yes
Chapeau function with a dissipative term	Modified Euler predictor-corrector	Yes

* See Equations (4a) and (4b).

As the solutions of the transport equation can only be obtained by numerical methods, the practical limitations of computer resources frequently determine the numerical method followed and the mesh sizes used. For the partial differential equation (3), the basic problem is to keep numerical errors smaller than the other uncertainties involved in the LRT problem.

In particular, with Eulerian grids, one faces the problem of undesired computational diffusion associated with numerical integration of the advection equation. The larger the mesh size, the greater the numerical dispersion. The magnitude of the artificial numerical diffusion ε for a first order approximation is given for the one-dimensional transport equation as:

$$\varepsilon = \left[\frac{\bar{U}\,\Delta x}{2} \left(1 - \frac{\bar{U}\,\Delta t}{\Delta x} \right) \right] \frac{\partial^2 \bar{\chi}}{\partial x^2}.$$

This is a diffusion-type term, with a diffusion coefficient given by the expression in square brackets. Especially in long range models, where large values of Δx are used, the term can reach values at least of the same order as the physical diffusion.

Several methods for reducing computational diffusion exist, one of which is based on the hypothesis that the mass distribution within each grid element is characterized by both the center of mass (first moment) and the radius of inertia around the center of mass (second moment). These parameters are then used to recalculate the advection and diffusion terms (Egan and Mahoney, 1971).

Another approach uses the Crank-Nicolson-Galerkin finite element technique to approximate the differential equation (Fairweather, 1978) which has the advantage of respecting the material distribution structure. Time integration by such a scheme also has the advantage of leading to tridiagonal linear systems that may be solved by rapid factorization methods.

A third solution to the problem is the Smolarkiewicz (1983) scheme. Numerical diffusion occurs with a magnitude characterised by a diffusion coefficient K_n dependent on the Courant number ($\lambda = U\ \Delta t/\Delta x$); an artificial "anti-diffusion" velocity $U_A = -(K_n/\bar{\chi})\partial\bar{\chi}/\partial x$ is introduced to balance the numerical diffusion.

Another possibility for reducing the computational dispersion is provided by spectral or pseudospectral methods, in which the spatial dependence of concentration is expanded in a series of orthogonal functions. Consider for simplicity, the one-dimensional case for which concentration $\chi(x,t)$ may be written as the following finite Fourier series:

$$\bar{\chi}(x,t) = \Sigma_n A_n(t) e^{iK_n x}. \qquad (5)$$

Substitution of this series into the partial differential equation, and use of the orthogonality property of the Fourier components, give a finite set of ordinary differential time-dependent equations that can be solved for the coefficients $A_n(t)$. Concentration $\overline{\chi}(x)$ can then be evaluated at any time t from the Fourier expansion.

Fourier expansions can also be used to obtain accurate numerical estimates of the spatial derivatives of $\overline{\chi}$ in the differential equation (3). Such spatial derivatives are obtained directly as:

$$\frac{\partial \overline{\chi}}{\partial x} = \sum_{n} iK_n A_n e^{iK_n x} \quad .$$

The differential equation (3) can thereafter be integrated using only simple first order time-differencing schemes. A limitation of the method is that it requires periodic boundary conditions, which means that spurious inflow must be reduced to negligible values.

Time discretization of (3) can also be performed by successively applying the various space operators to advance a time step Δt and then successively applying these operators in reverse order to advance a further timestep (Pudykiewicz, 1988). This particular method of operator splitting, called flip-flop, has the advantage of giving only second-order time truncation errors. As noted by McRae et al. (1982), the sequence of operator splitting is somewhat arbitrary and can be chosen on the basis of coordinate direction and/or on the relative importance of physical processes described by these operators.

Some of the above mentioned numerical schemes were tested by Chock and Dunker (1983) (Table 1).

To summarize, the main problems related to Eulerian models are the initial source description and the spurious numerical dispersion. However, their advantages include great flexibility in the incorporation of complex meteorological processes, such as wind velocity variation with height, vertical exchange processes, and treatment of deposition and source emissions. Additionally, Eulerian models can treat non-linear chemical processes between species, and can use arbitrarily small grid sizes, which reduces their inherent numerical errors.

2.2. Lagrangian Models

The mass balance equation (2) for pollutants within the ABL can be considerably simplified if a total emission is considered to be the sum of contributions from individual emitted air parcels, and if an integration along their individual trajectories is thus performed to obtain concentration profiles within the moving parcels. In this case, concentration is given by:

$$D\overline{\chi}/Dt = S \quad ,$$

where the total derivative operator (D/Dt) calculated along the trajectories is equivalent to $(\partial/\partial t + U.\text{grad})$.

In Lagrangian formulations of long range transport, air parcels are assumed to remain as physical entities. A volume is associated with each parcel, within which concentration fluctuations due to turbulent exchange processes can be neglected. In the calculation of each trajectory, every air parcel is displaced independently from all the others, advected by the wind velocity at its particular time and position. As each particle moves along a trajectory dependent on mean wind plus its fluctuations, its position at any time is determined by the various displacements experienced in previous times.

A simple class of long range Lagrangian models are those using diffusing Gaussian puffs travelling along trajectories computed as it will be described below. For these models, concentration at a surface point (x, y) at time t is given by:

$$\bar{\chi} = \frac{2QE_1E_2}{(2\pi)^{3/2}\,\sigma_h^2\,\sigma_z} \exp\left\{-\frac{(x-x_p)^2}{2\sigma_h^2} - \frac{(y-y_p)^2}{2\sigma_h^2} - \frac{(-z_p)^2}{2\sigma_z^2}\right\}, \tag{6}$$

where E_1 and E_2 are terms representing chemical depletion (or decay) and depletion due to soil deposition, respectively; σ_h and σ_z the diffusive standard deviations; and (x_p, y_p, z_p) the puff centre-of-mass coordinates at time t. The resulting concentration at time t at the observation point is calculated by adding the contributions of all puffs at the point.

Parcel trajectories can be calculated using the following Eulerian-Lagrangian relationship:

$$d\mathbf{X}(t) = \mathbf{U}\,dt\ , \tag{7a}$$

where vector **X** indicates the three-dimensional position of a parcel at time t+dt and **U** is the Eulerian velocity field calculated at some point intermediate in space and time between **X**(t) and **X**(t+dt).

Eulerian velocity data are usually derived from the pressure field (e.g. ApSimon et al., 1985), or obtained directly from regional or limited area meteorological models, e.g. European Centre for Medium range Weather Forecasts (ECMWF) fine mesh model for Europe or the RAMS model (Tripoli and Cotton, 1982). Such models provide regional forecasts; they calculate a number of meteorological fields (such as wind, temperature, pressure, precipitation, and cloud cover) with spatial and temporal resolutions suitable for input into LRT models, provided that correct spatial and temporal interpolations are performed.

Usually, the pollutant cloud is advected with a velocity U obtained by linear interpolation of wind values at the closest mesh points, but in some models quadratic interpolation is also used (Desiato, 1992). In any case, the value of U in (7a) depends also on X(t + dt), which requires the use of very small timesteps.

To avoid this difficulty, the algorithm of Reap (1972) can be used; this approach iterates the interpolation procedure to calculate the new parcel positions. In the following, these calculation steps are shown below for an explicit 3-D scheme.

At each time-step, the first approximation of the new parcel position **X'** at time t+dt is obtained using the wind velocity $\mathbf{U_0}$ in $\mathbf{X_0}$:

$$\mathbf{X}' = \mathbf{X_0} + \mathbf{U_0}\,dt\ .$$

Then **U'**, the new velocity at **X'**, is interpolated, and the parcel position re-evaluated from:

$$\mathbf{X''} = \mathbf{X_0} + 0.5\,(\mathbf{U_0} + \mathbf{U'})\,dt\,.$$

When convergency is achieved in the parcel position, the iterative procedure is stopped (Bonelli et al., 1992).

Detailed definition of wind velocity as a function of height is crucial, and thus a large number of vertical levels is required (e.g. a shear of 180° between two levels does not mean that a parcel is not moving). This, however, requires large meteorological data sets as input, and thus long data transmission times, a limitation in emergency situations.

To reduce the need for large input data files in the evaluation of trajectories, some models use simple 2-D horizontal wind vector fields from regional circulation models. In this case, vertical interpolation and extrapolation schemes frequently use a power law wind profile for speed and direction (e.g. Verver and Scheele, 1988). The required exponent p can be obtained from $p = \ln(V(z_1)/V(z_2))/\ln(z_1/z_2)$, where z_1 and z_2 are the nearest two heights for which wind vectors are available and V is the horizontal wind intensity. Wind speed at the puff centre of mass $V(z_m)$ can then be calculated from $V(z_m) = (z_m/z_1)^p.V(z_1)$, while the corresponding direction can also be obtained by vertical interpolation using the weighting factor of $(z_m/z_1)^p$.

For a pollutant cloud of considerable dimensions, different parts experience different wind values and thus follow different trajectories (synoptic dispersion). It is therefore important to couple 2-D trajectory calculations with diffusion models capable to describe parcel dimension growth with time to vertically and horizonally split the parcel into different parts that will move in different directions. Particular care is thus required in such models to the description of the diurnal variation of mixing height and of the vertical variation of the flow field (Gifford, 1984).

For the long ranges considered, horizontal dispersion of air parcels increases for two reasons: wind shear and atmospheric turbulence. In the horizontal direction, eddy size is considered to be unbounded in LRT, in contrast to the vertical direction where eddies are limited by the finite mixed layer depth. Therefore, turbulence at these large scales has a non-zero contribution to horizontal cloud dispersion. A second mechanism for horizontal cloud dispersion is due to the already discussed fact that different vertical portions of the cloud experience different horizontal trajectories.

Both of these effects implicitly appear in the horizontal dispersion cloud measurements obtained over travel times of up to four days (a uniquely long observational period) by Carras and Williams (1988). The relationship for lateral dispersion (Figure 1) as function of time that can be derived from this data set has the same form of the equation frequently used for shorter travel times, i.e.

$$\sigma_h = a\,t^b\,.$$

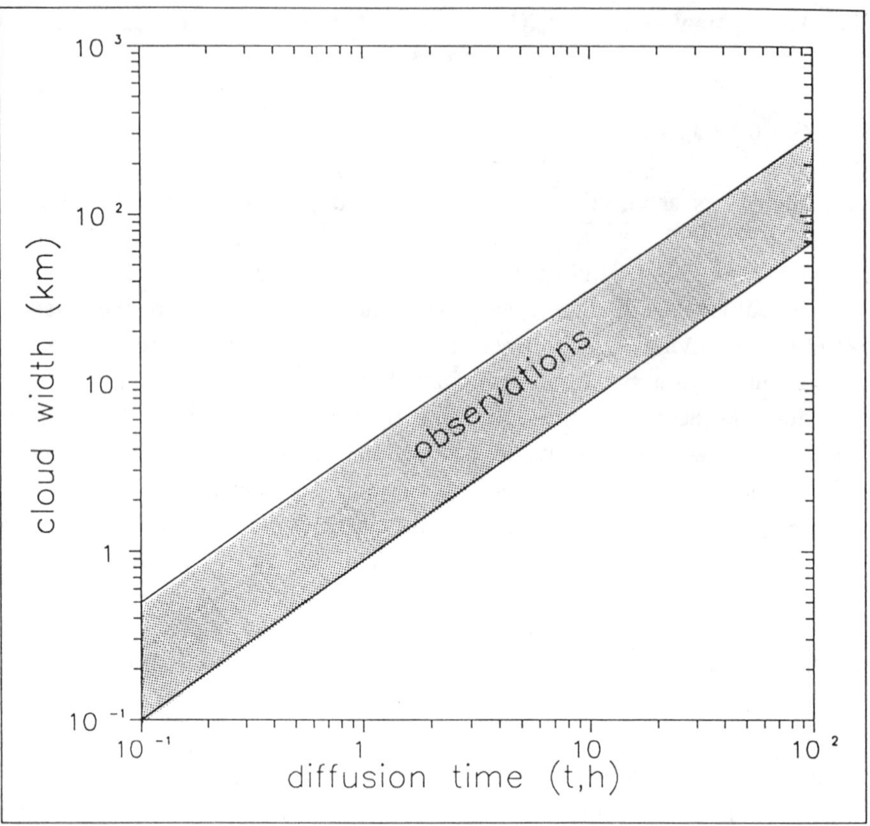

Figure 1. Maximum and minimum cloud width σ_y as a function of diffusion time, t, from various data (Carras and Williams, 1988).

The a and b values (i.e. 0.5 and 1.0, respectively) are appropriate to long distances, in that they are comprehensive of the effects of horizontal turbulence, as well as of vertical wind shear, in combination with the diurnal cycle of mixed layer depth.

Due to the ABL, vertical diffusion produced by turbulence is treated differently from that in the horizontal. For short and medium travel distances, vertical cloud growth can be estimated using the same empirical laws established by Pasquill (1974), which are dependent on stability classes. When time after emission is larger than 6-10 hours, however, mass emitted within the ABL can be considered homogeneously distributed in the vertical through the extent of the mixed layer, except close to the surface (due to dry deposition).

Above the mixed layer, vertical turbulence can be assumed to have very low values and the vertical diffusion coefficient can be given as:

$$\sigma_z^2 = 2 K_z t ,$$

with K_z having a value less than 0.5 m^2/s (Maryon and Heasman, 1988).

In more advanced Lagrangian models, a multi-particle Monte Carlo approach is used to simulate sub-grid scale vertical turbulence and shear effects. In the ADPIC model of Lange (1978), dispersion is treated by adding the following term to the average wind vector,

$$U_D = -\frac{K}{\overline{\chi}}\nabla\overline{\chi} ,$$

which is proportional to the local concentration gradient.

At each step, concentration values are used to calculate this term, called the diffusivity velocity, only for cells with non-zero concentration. New particle coordinates are then obtained by advecting the particle by U_D. Finally a new concentration distribution $\overline{\chi}$ is re-calculated from the new particle position.

Another way of describing diffusivity in multi-particle Monte Carlo models is by the addition of a random motion term to the equation of motion and then by repeating trajectory calculations many times to create ensemble averages. Thus the position of each particle is assumed to be displaced according to a first order Markov chain:

$$d\mathbf{X}(t) = \mathbf{U}\,dt + \mathbf{X}''r , \tag{7b}$$

where **X"** is a vector dependent on location and meteorological conditions and r an independent random variable. **X"** has the physical meaning of a displacement due to diffusion; in terms of K-theory, it can be identified with $(2\,dt\,K)^{1/2}$, with K as an average diffusivity.

Comparison of this equation with similar ones used for short and medium range dispersion, shows that the correlation term is not present here due to the large temporal and spatial scales involved. The random variable r in the equation is generally assumed normally distributed; however, it can be chosen from any a-priori distribution (even rectangular), since over many timesteps the Central Limit Theorem will play its role.

The fluctuation term **Xr"** must only be applied within the ABL, where turbulence is strong. Some multi-particle Lagrangian models, however, also use this term (properly reduced in magnitude) for short distances above the inversion at the top of the boundary layer, where stratified flows are established. This is done, to introduce some subgrid scale effects and to improve the statistical reliability of particle ensembles.

With the use of (7b) in full three-dimensional LRT models, due to the presence of a temperature inversion at the top of the ABL, after a few steps particles invade the entire boundary layer. It is, therefore, more economical to treat (7b) as a 2-D equation, and at each timestep to arbitrarily re-assign the vertical position of each particle within the ABL. This re-assignment allows particles to sample the wind at all levels, as it would naturally occur in convective situations, when they experience vertical wind shear.

The main drawback of particle model approach is the large number of particles that must be used. This may be circumvented, however, by attributing a Kernel density to each particle, which requires some assumptions on the mass distribution within each Kernel.

Some parameters necessary to describe ABL physics can be derived from output data of circulation models. Those which depend on the nature of the earth surface can be evaluated separately, but can then be used for many calculations.

This is the case for the roughness length z_0, which can usually be found in tables as a function of soil. Its values vary from about $5\ 10^{-4}$ m for water to about 2 m for densely populated areas. In models, z_0 must be representative of the influences of soil characteristics on the circulation within the entire ABL and for areas on the order of 10^4 km^2, sometimes including the effects of hills or high mountains. Different methods exist on how to obtain effective roughness length values (Kondo and Yamazawa, 1986; André and Blondin, 1986; Taylor, 1987). In some cases, however, it is sufficient to only distinguish between flat, hilly, mountainous and water areas.

To conclude, the main limitation of Lagrangian models is that, in order to maintain their computational advantages, linear or step-linear chemistry must be used, which limits either result accuracy or the length of timestep used to that stated. The above also indicates the importance for Lagrangian models of a correct simulation of the evolution of boundary layer depth in various meteorological situations. In the following, some indication will be given on how to tackle this problem.

The troposphere is usually divided into three layers:

- the surface boundary layer (SBL), where the effect of the friction is predominant.
- the mixed-layer, from the top of surface layer to the top of ABL;
- the free troposphere, where turbulence is low and wind is mainly geostrophic.

In SBL turbulence is determined by the friction velocity u_* (which, when squared and multiplied by average density, is equal to the turbulent shear stress caused by the horizontal wind) and Obukhov length L (which indicates the relative importance of mechanical to thermally induced turbulence).

The Obukhov length is defined by:

$$L = -\, u_*^3 / k\, B_s \ ,$$

where k is the Von Karman constant (0.4) and B_s is the buoyancy flux that expresses the buoyancy induced by cooling or heating of the ground surface:

$$B_s = \frac{g}{T_s} \frac{H_s}{c_p\, \rho_s} + 0.608\, g \frac{E_s}{\rho_s} \ ,$$

ρ_s being air density, T_s air temperature near the surface, and H_s and E_s are the vertical turbulent fluxes of sensible heat and water vapour, respectively.

Usually the second term in the expression is neglected. In LRT models, however, where large fractions of the domain considered are covered by water it is important to take it into account.

ABL depth varies considerably with the diurnal cycle, its maximum development occurs in the period between sunrise and sunset. Depending on time of day, a fraction of any emission will be within the ABL and the rest will be sited above it, where the characteristics of the transport and dispersion are different and decoupled from surface effects. Many modellers (e.g. Verver and Scheele, 1988; Desiato,1992) therefore distinguish between the boundary layer and what is above it (sometimes called reference, buffer or reservoir layer) where turbulence is much reduced.

The mass in both layers is affected by chemical reactions (or radioactive decay) and wet deposition, with depletion due to dry-deposition only important in the lowest part of the boundary layer. The only way in which material can cross the top of the boundary layer is through a variation of the mixed layer depth h.

In this case, the material flux across the boundary will equal the product of concentration χ at the boundary times the depth time derivative:

$$F = \overline{\chi}(h)(dh/dt) \quad ,$$

where $\overline{\chi}$ is taken just above the boundary when the derivative is positive and viceversa.

In some models, ABL depth is determined using temperature profiles data from regional meteorological models. One option, for example, is to determine the evolution of h by temporal changes of parameters such as potential temperature, wind velocity components and specific humidity (Maryon and Heasman, 1988). In other models, however, semi-empirical approaches are used.

During the night, a surface based radiation inversion usually develops. Then vertical motions are considerably damped and momentum exchange between different layers is low ($L > 0$). The mixed layer depth h to be used in this case in LRT can be given by (Zilitinkevich, 1972):

$$h = c\, u_*^2\, [f\, B_s]^{-1/2} \quad ,$$

where f is the Coriolis parameter ($= 2\,\Omega \sin \Phi$), Ω the earth's angular velocity, and the dimensionless constant c is of order unity.

During the day, when convective motions exist, different approaches for mixed layer depth h can be used to determine h. For example, in the APOLLO model (Desiato, 1992) estimation of h in convective conditions requires the 1200 UTS vertical temperature profile available in real time from the Global Transmission System (GTS). From this profile, a single linear interpolation scheme is carried out for the rest of the day (Figure 2).

More detailed models based on physical considerations have however been developed. In the following an example is presented for the development of ABL depth as a function of several basic turbulence parameters.

In the model of Batchvarova and Gryning (1991), some restrictions are however imposed on the system. For example, all contributions to the balance due to condensation and

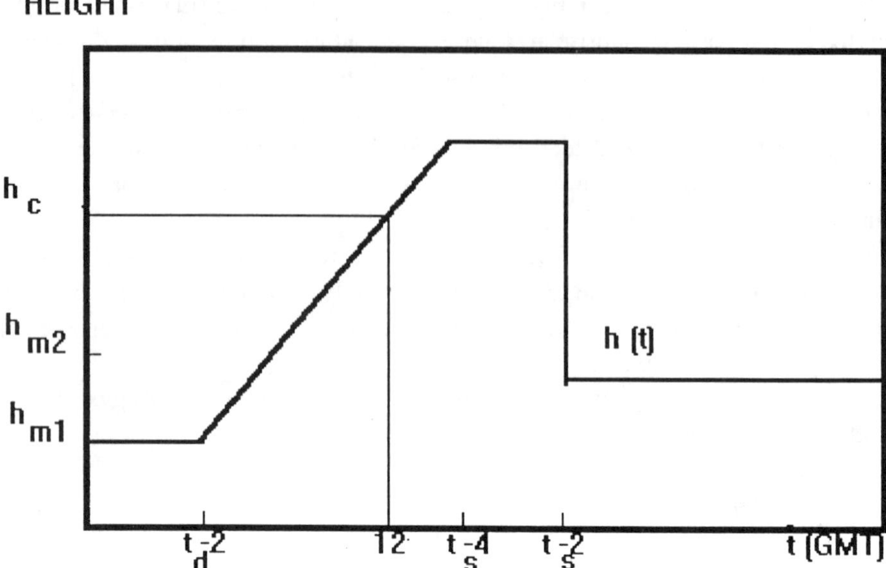

Figure 2. Diurnal variation of mixing depth as modeled in APOLLO, where t_d and t_s at dawn and sunset respectively; h_{m1} and h_{m2} are the early morning and late afternoon mechanical mixing depths respectively; h_c is the convective mixing height derived from vertical temperature soundings at 12:00 GMT.

evaporation are neglected. Also the fraction of the atmosphere considered to be within the ABL is assumed characterized by horizontal homogeneity, while the atmosphere outside the ABL is considered stable and stratified. Further, it is assumed that the mixed layer is limited on the top by an infinitesimally thin layer, above which the potential temperature gradient (assumed zero within ABL) has a positive value due to a thermal inversion (Figure 3). This infinitesimally thin layer identifies an Entrance Zone (EZ).

In these hypotheses, the turbulent energy balance must take into account the contribution of energy from below, which generates ABL depth development, and the consequent entrainment of cold stable air from above. Under these assumptions, the turbulent vertical flux at the top of ABL can be represented as proportional to the time derivative of its depth, i.e.:

$$-(w'\Theta')_h = \Delta dh/dt \ , \tag{8}$$

where the factor of proportionality Δ is not constant, but is function of time:

$$d\Delta/dt = \gamma(dh/dt) - (d\Theta/dt)_{ml} \ . \tag{9}$$

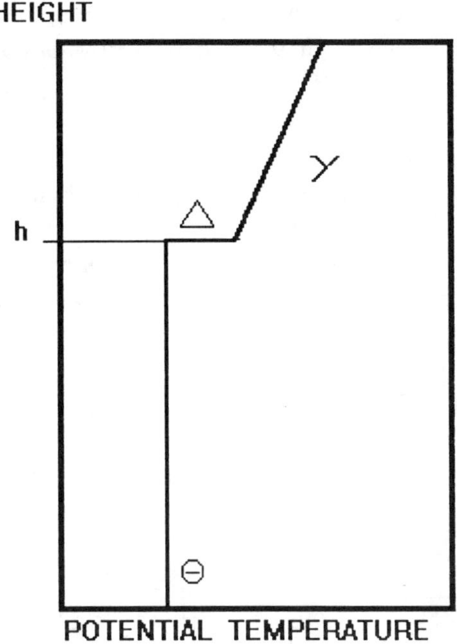

Figure 3. Idealized vertical profile of the potential temperature used in the model of Batchvarova and Gryning (1991).

According to (9) this factor increases with dh/dt (entrainment rate) and it is limited as the mixed layer is heated. In the equation γ is the potential temperature gradient above the mixed layer and Θ is the potential temperature. In this scheme, the heating rate of mixed layer is proportional to the difference of the kinematic heat fluxes at ground and at its top, i.e.:

$$\left(\frac{d\Theta}{dt}\right)_{ml} = \frac{\overline{(w'\Theta')}_s}{h} - \frac{\overline{(w'\Theta')}_h}{h} \quad . \tag{10}$$

To solve for h(t), it is necessary to give an expression for the balance of vertical heat flux at the top of the mixed layer. The balance equation derived (Gryning and Batchvarova, 1990a) is:

$$-\frac{gh}{T}\overline{(w'\Theta')}_h + Cu_*^2 \frac{dh}{dt} = A\frac{gh}{T}\overline{(w'\Theta')}_s + Bu_* \quad , \tag{11}$$

which equates the reduction of potential and kinetic energies due to entrainment of cold and stable air through the EZ at the inversion level [left side of (11)] with the production of thermal and mechanical kinetic energy within the ABL (two terms on right hand side).

The non-dimensional constants A, B and C have been set equal, respectively, to 0.2, 2.5 and 8. It is interesting to note that in the case of small values of depth h, one gets:

$$dh/dt = (B/C)\, u_* \ ,$$

which indicates the proportionality between the growth of small mixed layers and friction velocity.

Substituting (8) and (10) into (9), a differential equation is obtained that gives the dependence on h of the proportionality factor Δ:

$$\left(h \frac{d\Delta}{dh} + \Delta - \gamma h \right) \frac{dh}{dt} = -(\overline{w'\Theta'})_s \ . \qquad (12)$$

From (8) and (11), using the definition of L, another differential equation for the dependence on h of the variable Δ can be obtained:

$$\frac{d\Delta}{dh} + \Delta \left(\frac{1}{h} + \frac{1}{Ah - BkL} \right) = \gamma \ . \qquad (13)$$

The solution of this equation can be found considering its limit in nearly neutral conditions (when L tends to infinity from unstable conditions $L < 0$):

$$\Delta = \frac{Ah - BkL}{(1 + 2A)h - 2\,BkL} \gamma h \ .$$

Substituting this expression into (8) and (10) and taking into account the balance equation (11), one finally deduces the differential equation for h, that describes boundary layer depth growth in convective conditions:

$$\left\{ \frac{h^2}{(1 + 2A)\, h - 2\, BkL} + \frac{C u_*^2 T}{\gamma g\, [(1 + A)\, h - BkL]} \right\} \frac{dh}{dt} = \frac{(\overline{w'\Theta'})_s}{\gamma} \ ,$$

where all physical quantities can be obtained from output data from circulation models using a meteorological preprocessor ([e.g., based on the work of Van Ulden and Holtslag (1985) or Holtslag and De Bruijn, (1988)]. Use of this equation in a LRT Lagrangian model has improved simulation results (Galmarini et al., 1992).

2.3. Concentration-Data Assimilation

During emergency situations following either nuclear or chemical emissions, it is very important to immediately identify the source term parameters (such as location, intensity, duration, height) and at the same time to estimate the long range evolution of the cloud to assess in advance useful counter-measures. To do this, use should be made of all available information consisting not only of meteorological forecasts but also of concentration data.

Pollution observations in real-time are gathered from observational networks. In case of LRT, it is particularly important to make full use of the dense networks installed by many countries for airborne radioactivity measurements, that are immediately available after an accident. Recent use of this kind of information has been made on the continental scale, using experience gained by the various weather services on the assimilation of meteorological data after Chernobyl.

Among the different techniques employed to couple LRT models results and the observations of pollutant levels, it is worth mentioning the Adjoint Technique that has been applied to a simplified single-layer version of a 3-D regional dispersion model (Persson and Robertson, 1990). In the Adjoint Technique, a penalty functional is introduced, that is a measure of the model misfit. This is expressed by the sum of square differences between model results and observations. The optimization problem is to find the minimum of the functional in terms of source parameters, such as position and intensity, that can be unknown in case of an accident. The technique works as follows: a regional model starts with an initial guess of source parameters. At each iteration, the dispersion model compares the results obtained with the observations, re-adjusts the source term and performs another simulation from the beginning.

The basic idea underlying this approach is that a long range dispersion model can reasonably well describe pollutant concentration evolution in the air, once the source term has been properly introduced. Its limitations are linked to the long computer times required and to the simplifications necessary on the model structure due to the amount of data required in real cases. Particularly difficult to assess, is the initial vertical spread of the source, which would require a multi-layer simulation.

More recently, another technique for pollutant data assimilation has been worked out (Raes et al., 1991) based on the idea that a monitoring network is able to completely characterize the cloud evolution and that the major source of uncertainty in LRT model outputs lies in the use of a forecasted windfield as input data. Moreover, in emergency situations, it is more important to forecast the future cloud displacement, than to correctly describe the accidental emission that has already occurred.

Following this idea, the procedure starts to predict a concentration field at the times $t_0 + \Delta t$, $t_0 + 2\Delta t$, etc., where t_0 is the estimated time of the accident. However, when at time $t_0 + j\Delta t$, the observation field becomes available, a function describing this field is fitted to these values and a new area source is used to update the predictions at times $t_0 + (j+1)\Delta t$, etc. To limit the number of parameters to be found, a double Gaussian expression was selected for the interpolation, with a cut-off at a low concentration level (comparable with the detection limit). The procedure has been tested using data from the Chernobyl accident and it has been shown to give substantial improvement in the prediction of cloud location and air concentration levels.

3. LONG-TERM DISPERSION MODELS

When the effect of continuous emissions of chemical substances into the atmosphere is considered, long term average values of air pollutant concentrations and deposition are required. Two ways are here considered to obtain these seasonal or annual mean values. One possibility is to use relatively simple LRT models with seasonally averaged input data and mean values of the appropriate parameters. An alternative method is based on the application of LRT models in hourly steps, computing from the hourly output data the required averages. Both methods present advantages and drawbacks.

For the first method, selection of average values is a very delicate point, as they should be representative of all weather situations in the period considered. Long time period models must then be validated and parameters optimized by comparing model results with experimental data. Their values are re-adjusted minimizing some statistical functions, (such as Minimum Least Square Error). These models are therefore built upon an empirical basis and have the limitation that parameters are determined for typical environmental conditions and hence, they cannot be applied to different situations. Nonetheless they are simple to understand and to use. Furthermore they can be of a Eulerian type and can include explicit non-linear chemistry modelisation schemes. The second approach requires large amounts of input data and computer time, which in general limits their use to simple trajectory Lagrangian models where the chemical scheme must be linearized at each time-step (e.g. Eliassen and Saltbones, 1983).

In the past, long term LRT models used a simple linear chemistry. One example of models of the first type is the two-dimensional Eulerian box model of Klug (1982); there are however no reasons that inhibit use of more complex non-linear reactions, such as those of the Nitrogen containing pollutants. Its main assumption is an instantaneous mixing in the horizontal and vertical direction within each box of the domain. Pollutants are advected by the mean wind in each box, which covers an area on the order of 100 km x 100 km and has a constant height (mixed layer depth) of 1000 m. Therefore, the concentration are assumed to be constant in each box both for the primary and secondary pollutant, $\overline{\chi}_1$ and $\overline{\chi}_2$ respectively. Their time-variations are caused by a uniform area source term S, an advection term, and a decay term:

$$\frac{\partial \overline{\chi}_1}{\partial t} = (1 - \alpha) \frac{S}{h} - v.\nabla \overline{\chi}_1 - \left(\frac{V_{d2}}{h} + \kappa_c + k_{r2}\right) \overline{\chi}_1$$

$$\frac{\partial \overline{\chi}_2}{\partial t} = \alpha(M_2/M_1) \left(\frac{S}{h} + k_c \overline{\chi}_1\right) - v.\nabla \overline{\chi}_2 - \left(\frac{V_{d4}}{h} + k_{r4}\right) \overline{\chi}_2 ,$$

with α as the fraction of secondary pollutant in the emission. The first and second pollutant have molecular masses M_1 and M_2 and are depositing with V_{d2} and V_{d4} dry deposition velocities; k_c is chemical transformation rate and k_{r2} and k_{r4} are wet deposition rates. Different wind values are taken into account according to their distributions within eight sectors of direction. Precipitation probabilities in each sector are then used, while wind velocity v, transformation rate and dry deposition velocities are constant within each sector.

Other models of the first type using linear chemistry are the LRT Wet and Dry Model developed independently by Smith (1981) and Venkatram et al. (1982). In these models, the basic assumptions are a separation of wet and dry periods in which the behaviour of the pollutant is rather different. A single wind and precipitation distribution rose is applied throughout the grid area. In these cases the above mentioned simple assumption of pollutants advection by a constant wind speed is made and trajectories are considered as straight lines. A constant mixing depth (= 1000 m) is also assumed; a continuous and constant dry deposition is introduced, and the oxidation of primary pollutant occurs at different rates for the wet and dry periods.

The probability of any pollutant parcel moving from a dry period to a wet period or vice versa is given by different timescales (S_W and S_D respectively). Using the suffix W for wet and D for dry period, four equations for the fractions of primary and secondary pollutant in both wet and dry periods respectively can be written:

$$\frac{d}{dt}\bar{\chi}_{1D} = S_w\bar{\chi}_{1w} - S_D\bar{\chi}_{1D} - \frac{V_{d2}}{h}\bar{\chi}_{1D} - \alpha_D\bar{\chi}_{1D}$$

$$\frac{d}{dt}\bar{\chi}_{1w} = S_D\bar{\chi}_{1D} - S_w\bar{\chi}_{1w} - \alpha_w\bar{\chi}_{1w} - A\bar{\chi}_{1w}$$

$$\frac{d}{dt}\bar{\chi}_{2D} = S_w\bar{\chi}_{2w} - S_D\bar{\chi}_{2D} + \alpha_D\bar{\chi}_{1D} - \frac{V_{d4}}{h}\bar{\chi}_{2D}$$

$$\frac{d}{dt}\bar{\chi}_{2w} = S_D\bar{\chi}_{2D} - S_w\bar{\chi}_{2w} + \alpha_w\bar{\chi}_{1w} - A\bar{\chi}_{2w} ,$$

where α_w is the chemical transformation rate in wet periods (and α_D in dry periods) and the rate of removal A is linked to precipitation rate.

The system can be easily solved analytically to yield:

$$\bar{\chi}_{1D}(t) = c\,e^{r_1 t} + d e^{r_2 t}$$

$$\bar{\chi}_{1w}(t) = a\,e^{r_1 t} + b e^{r_2 t}$$

$$\bar{\chi}_{2D}(t) = F\,e^{s_1 t} + G\,e^{s_2 t} + K\,e^{r_1 t} + L\,e^{r_2 t}$$

$$\bar{\chi}_{2w}(t) = f\,e^{s_1 t} + g\,e^{s_2 t} + K\,e^{r_1 t} + l\,e^{r_2 t} ,$$

where a, b, c, d, f, g, k, l, F, G, K, L, r_1, r_2, s_1 and s_2 are constants dependent on input parameters. It must be noted that as the equations are written in a Lagrangian reference frame, the advection and diffusion terms should be calculated separately, using trajectories with dispersion around them. The chemistry is considered linear within each time-step and vertical dispersion is usually eliminated assuming well mixed ABL concentrations. As trajectories are straight lines from source to receptor points, horizontal dispersion can be evaluated using the expression $\varphi/(2\pi\, r\, V)$ where φ is the relative frequency of the particular source-receptor direction, r the source-receptor distance and V the average windspeed.

A more complex chemical scheme is present in the LRT Lagrangian model TREND, in which a statistical long range version of a Gaussian plume model is used (van Egmond and Kesselboom, 1986 or van Jaarsveld, 1988). Long term averages are calculated by consideration of the occurrence of 72 classes (12 wind directions times 6 stability classes), while meteorological data are averaged over a 10 year period (1977-1986). Mixed layer depth is evaluated as a function of stability class and statistically increased with distance from the source. In the model, terms due to wet deposition are calculated assuming a Poisson distribution of precipitation events, where the mean duration and intensity are a function of the wind direction class. The chemical scheme does not only take into account the SO_2^- SO_4 transformation, but it includes the NH_3 to Ammonium aerosol transformation and the oxidized Nitrogen compounds, modelled as NO_x and NO_3.

Models considering many trajectories and then averaging the results are in general of a Lagrangian type, i.e. each air parcel is followed from its emission until it arrives at the final (or receptor) point. Such models usually include linearised chemical schemes of the more important pollutants like NO_x, NH_3 and their oxidation processes. Mass balance equations are integrated along each trajectory, taking into account the emission inputs, chemical reactions and dry and wet deposition rates.

The equations to be solved along each trajectory represent the Lagrangian solution of the transport and diffusion equation (3), writing explicitly, the source term as:

$$\frac{D}{Dt}(\overline{\chi}_i) = (p_i - L_i)\,\overline{\chi}_i + E_i - \frac{v_d^{\,i}}{h}\,\overline{\chi}_i - \frac{\Lambda_i\, p_r}{h}\,\overline{\chi}_i \;,$$

$\overline{\chi}_i$ is the concentrations of species i; h is the mixed layer depth; p_i and L_i are the chemical production and destruction rates; E_i is the source emission; $v_d^{\,i}$ is the dry deposition velocity; Λ_i is the scavenging ratio; p_r is the integrated precipitation rate in the time interval considered. Long term behaviour at a given point is obtained by weighting the results of many trajectories. Input meteorological data are required by the model at predetermined time intervals.

The main differences between models of this type lie in the quantification of depletion parameters and in the chemical transformation equations for each reaction. Three of these models are briefly summarized here. The EMEP (European Monitoring and Evaluation

Programme) Lagrangian model (Eliassen et al., 1988) assumes instantaneous mixing of the pollutant in meshes of 150 km x 150 km with a vertical mixed layer depth obtained from radiosonde data. Trajectories of 96 hours are calculated using 925 hPa winds at six hour time intervals at the centre points of the grid meshes and at the EMEP observation stations. Precipitation data at some intervals are used to estimate wet deposition. Recently (Simpson, 1992) the model has been used with success to calculate ozone concentrations over the month of July 1985 in northwest Europe.

In the Harwell model (Derwent and Nodop, 1986), vertical dispersion is also based on the assumption of a constant mixed layer depth of 800 m. Horizontal dispersion is calculated by assuming long-term average straight line trajectory segments of 96 hours arriving at receptor points. Resulting trajectories are weighted with wind direction frequency at each receptor. Calculation of wet deposition is derived from historical average precipitation rates. The model considers NO_x chemistry, in which concentration of Ozone and OH radicals are assigned as input data.

The LRT long term IE model (Asman et al., 1988) is similar, with an improved spatial resolution (75 km x 75 km) and with a treatment of wet deposition based on actual values of precipitation. In the model, long term average concentration and deposition values are obtained by using actual 96 h trajectories.

The brief review of this paragraph indicates the assumptions that are usually made to calculate long term average transport of chemical pollutants over long distances. A large number of such models exists. Comparison among results from some of them was carried out in the past (Klug and Lüpkes, 1985) and more recently (Derwent et al.,1989). Their main limitations are the difficulty to correctly describe transport and dispersion processes with simple schemes, their necessity to introduce complex chemical transformations and to cope with uncertainties in emission source inventories.

4. REMOVAL PROCESSES

Removal processes of pollutants from the atmosphere are determined either by deposition processes or by chemical transformations and radioactive decay. As airborne pollutants can be in the form of gas or aerosols, different phenomena must be considered. As an example, it is worth mentioning that the removal properties of aerosol particles depend strongly on particle size, and that the size distribution of aerosols is a function of humidity.

The processes by which aerosol and gases may be transferred to the earth surface can be split into two main categories:

- Dry deposition, by which gases or particles are deposited directly into the soil.
- Wet deposition in which pollutants are incorporated into clouds, rain, snow, and then transferred to the ground by precipitation.

In the following a brief description of these processes and of the basis for chemical reaction modelization are given, indicating their particular application in LRT models.

4.1. Dry Deposition

Dry deposition in LRT models does not include the gravitational settling of particles which occurs mainly at short distances from emission sources and can then be neglected. In the case of LTR models, dry deposition processes are particularly important over night, when the mixed layer depth is low. Dry deposition processes are different for gaseous and for aerosol pollutants.

Transfer of gases from atmosphere to the ground can be separated in three stages:

- first they are transported from the surface layer to a few millimetres thick viscous sub-layer above the surface
- subsequently they diffuse through the viscous sublayer
- finally they are captured by the surface.

If one writes the general transport equation of gaseous pollutant (3) including the terms of molecular diffusivity which is usually neglected, one gets:

$$\frac{\partial\bar{\chi}}{\partial t} + \bar{U}\,\frac{\partial\bar{\chi}}{\partial x} + \bar{V}\,\frac{\partial\bar{\chi}}{\partial y} + \bar{W}\,\frac{\partial\bar{\chi}}{\partial z} = -\frac{\partial}{\partial x}(\overline{u'\chi'}) - \frac{\partial}{\partial y}(\overline{v'\chi'}) - \frac{\partial}{\partial z}(\overline{w'\chi'}) +$$

$$+ K_M\left(\frac{\partial^2}{\partial x^2}\bar{\chi} + \frac{\partial^2\bar{\chi}}{\partial y^2} + \frac{\partial^2\bar{\chi}}{\partial z^2}\right) + \bar{S}\ ,$$

where K_M is the molecular diffusion coefficient. With the further assumptions of horizontal homogeneity, steady-state and absence of sources and chemical sinks, this equation reduces to:

$$0 = -\frac{\partial}{\partial z}(\overline{w'\chi'}) + K_M\,\frac{\partial^2\bar{\chi}}{\partial z^2}\ .$$

Integrating this expression from height 0 to z in the surface layer and considering that the turbulent transport exceeds molecular transport and that surface turbulence vanishes, one gets:

$$\overline{w'\chi'} = -K_M\left(\frac{\partial\bar{\chi}}{\partial z}\right)_0 = F\ ,$$

which indicates that the vertical turbulent flux is proportional to the vertical concentration gradient. Integration of this equation between 0 and z yields to:

$$F = \frac{\bar{\chi}(z) - \bar{\chi}(0)}{\int_0^z \frac{d_z}{K_M}}\ .$$

This introduces the concept of deposition velocity V_d: the vertical difference in concentration values is equal to the product of the flux F times a surface resistance r_d. When assuming $\overline{\chi}(0)=0$ at the surface one obtains:

$$F = V_d\, \overline{\chi}(z) \quad .$$

By analogy with Ohm's law, one can assume that $1/V_d$ is the total surface resistance consisting of a series of separate resistances, the aerodynamic resistance, r_a, laminar resistance, r_b, and canopy resistance r_c, so that:

$$r_s \equiv 1/V_d = r_a + r_b + r_c$$

or

$$V_d = \frac{1}{r_a + r_b + r_c} \quad .$$

Aerodynamic resistance is determined by turbulent transport properties that are time dependent. In the surface layer, turbulence can be expressed in terms of Obukhov length L, friction velocity u_* and surface roughness z_0:

$$r_a = \frac{\left[\ln\left(\frac{z}{z_0}\right) - \psi_h\left(\frac{z}{L}\right) + \psi_h\left(\frac{z_0}{L}\right)\right]}{k u_*} \quad .$$

The parameter r_a is therefore strongly dependent on the diurnal cycle and on the effective roughness z_0 used in long transport models. The integrated stability corrective function ψ can be taken from many references, as for example Van Ulden and Holtslag (1985) for unstable situations. Laminar resistance can be expressed as function of Reynolds and Schmidt numbers. However, in most cases, a simple parameterization can be used in long range transport models:

$$r_b = 2.6/(\alpha\, u_*) \quad ,$$

where α is thermal diffusion constant of air (Sheih et al., 1979).

Canopy resistance r_c depends on the state and nature of the surface vegetation and on the chemical properties of the contaminant. One may obtain r_c-values for some gases by scaling them to the values of a certain pollutant according to their relative reactivity and solubility. It is important to consider that this resistance depends strongly on surface moisture and therefore on the time and period in the year.

Another possibility for determination of deposition velocity to be used in LRT models is to rely simply on deposition velocity measurements. Dry deposition measurement techniques are mainly based on flux gradient and eddy correlation methods (see for example Lindberg et al., 1990).

In the eddy correlation method, the turbulent fluctuating component of the vertical wind velocity w' and of the airborne containment concentration χ' are measured simultaneously above the canopy; the turbulent deposition flux is calculated as the time average of the product $w'\chi'$. In the vertical gradient method, airborne contaminant concentration is measured at two or more heights above the canopy and the flux is inferred as $-K(d\overline{\chi}/\overline{d}z)$ where K is contaminant eddy diffusivity.

Tables of dry deposition velocity values for a number of pollutants and radioactive isotopes are normally given in atmospheric physics publications. Some average values for the deposition velocity of some chemical species are reported in Table 2 to indicate the effect of the ground and of the time of the day (from Endlich et al., 1984).

Table 2
Deposition velocities (cm/sec.).

		SO_2	$SO_4^=$	NO_2	PAN	NO_3^-	HNO_3	NO
Night	*Sea*	0.55	0.32	0.07	0.07	0.66	0.07	0.07
	Land	0.07	0.07	0.07	0.07	0.66	0.07	0.07
Day	*Sea*	0.55	0.32	0.2	0.25	0.6	1	0.2
	Land	0.35	0.62	0.2	0.25	0.6	1	0.2

The deposition velocity of particles is based on the fact that they are subject to Brownian motions and their diffusivity decreases with increasing particle diameter (for aerosol particles having a diameter of 1 mm or lower). Touching earth's surface, they tend to stick to it, particularly when they are of small size, since molecular forces can be larger than particle momentum. As the necessity of separating laminar and soil resistances no longer exists, they can be combined into a single term and thus one can re-write deposition velocity resistance as:

$$V_d = \frac{1}{r_a + r_s} .$$

The value of r_s can be assumed constant and equal to 100 s/m (Sheih et al., 1979).

In the long range models, due to the variation of atmospheric resistance as indicated before, the concentration close to the surface is lower than above. In that region close to the surface, both for Eulerian and Lagrangian models, a constant flux layer can be assumed. Then the concentrations at a certain height (e.g. 10 m, $\overline{\chi}_{10}$) can be estimated by the linear relation:

$$\bar{\chi}_{10} = \bar{\chi}_{1L} \cdot v_{g_{1L}} / v_{g_{10}} \ ,$$

where the suffix $_{1L}$ indicates the average value in the first layer.

4.2. Wet Deposition

Each type of precipitation presents different simulation problems. Convective showers, for example, have the effect of both depositing pollutants at ground level and venting air from the inside the ABL to the free troposphere. For convective storms, complex 3-D models as well as one-dimensional models exist, that are capable of representing the evolution of the vertical column (ApSimon et al., 1991). In this case, water molecules are assumed to move with rising air. They are therefore cooled by expansion and subsequently captured by rain and ice crystals. Chemical aerosols are either captured by cloud droplets or are moving with the air and subsequentely scavenged to the soil.

Models describing these phenomena are able to maintain budgets for air, vapour, cloud droplets, ice crystals, and aerosols of different size. In the following, the impact of aerosol particles with precipitation droplets falling from above, and absorption of gaseous pollutant in raindrops below clouds are sketched. In both cases, the fall velocity of raindrops can be assumed a function of drop radius. The probability distribution of raindrops with a certain radius R, f(R) can be fitted by a gamma distribution (Yangang, 1993). Rainfall rate J of raindrops with velocity V_r is:

$$J = - \frac{4\pi N}{3} \int_0^\infty R^3 \, V_r(R) \, f(R) \, dR \ ,$$

where N is the number of raindrops per unit volume.

For aerosol scavenging, the impact of a raindrop (of size R and velocity V_r) falling through an aerosol polluted layer must be considered. The mass transfer rate must be proportional to droplet volume per unit of time ($\pi R^2 V_r$), to the aerosol concentration in the air $\bar{\chi}_a$, and to the collection efficiency, E(R,a), defined as the probability (covering all the microphysical processes by which an aerosol particle enters a raindrop) that an aerosol particle of radius a is caught by a raindrop of radius R. Therefore, the transfer rate to one raindrop of the aerosol mass is given by:

$$\pi R^2 \, V_r(R) \, E(R,a) \, \bar{\chi}_a \ .$$

Multiplication by the number of raindrops per unit volume and summation over all R values, yields to a mass transfer rate per unit volume:

$$\partial \bar{\chi}_a / \partial t = -\pi N \int_0^\infty R^2 \, V_r(R) \, E(R,a) \, f(R) \, dR \, \bar{\chi}_a \ .$$

Defining the wash-out coefficient for aerosols as:

$$\Lambda = -\pi N \int_{0}^{\infty} R^2 V_r(R)\, E(R,a)\, f(R)\, dR \ ,$$

the mass transfer rate equation can be written as:

$$\partial\bar{\chi}_a/\partial t = - \Lambda\bar{\chi}_a \ ,$$

which represents the decrease in air concentration during a precipitation event; this can be easily integrated. In practical applications for LRT simulations, an expression is often used for the scavenging coefficient as a function of rain intensity J:

$$\Lambda = \Lambda_0 \, J^n \ .$$

Different values for Λ_0 and n are used in the various models (e.g. n varies from 0.6 to 1.0 for various substances, see Klug et al., 1991).

The main difference from particles, when dealing with gas scavenging, is that with gases there is a reversible transfer; in fact gases may both be absorbed by raindrops and desorbed from them. The two phenomena can be described by a single scavenging flux, defined as:

$$F = v_E(\bar{\chi} - H\bar{\chi}_w) \ ,$$

where v_E is an exchange velocity between the gas and the aqueous phases (H is the Henry's law gas constant) determined by the equilibrium concentrations in air and in raindrops ($\bar{\chi}_w$).

Using an analogy with the dry deposition approach, the exchange resistance (inverse of the exchange velocity), can be considered as split into an atmospheric resistance, and an aqueous resistance. However, when gas solubility in water is large, the second existance can be neglected and concentration variation in time can be obtained integrating the raindrop size distribution times the droplet surface:

$$\frac{\partial\bar{\chi}}{\partial t} = - 4\pi N \int_{0}^{\infty} R^2\, v_E(R)\, \bar{\chi}\, f(R)\, dR \ .$$

Again one gets:

$$\frac{\partial\bar{\chi}}{\partial t} = -\Lambda\,\bar{\chi} \ ,$$

with the scavenging coefficient:

$$\Lambda = -4\pi N \int_0^{\infty} R^2 \, v_E(R) \, f(R) \, dR \ ,$$

where v_E is the exchange velocity coefficient.

The addition of chemical transformation in the rain droplets further complicates wet deposition process, because it feeds back into the transfer rates. The description of these effects are, however, outside the scope of this chapter.

In LRT models, it is sometimes necessary to consider that scavenging processes can occur within clouds. In fact, at some distance from the source (usually > 50 km) the pollutant flux may enter into clouds. Beyond such distances, the fraction of pollutant undergoing in-cloud processes increases. In this case, the formulation described above does not hold, as the spectrum of cloud droplet size differs strongly from the final raindrop spectrum. Other differences are due to the fact that continuous condensation and evaporization processes occur, thereby increasing the complexity of the simulation. Furthermore, as the relative velocity differences between airborne pollution and cloud droplets are smaller than in rain scavenging, the importance of impaction is reduced. For a more extensive treatment of this subject, the reader is referred to Reinking (1987) and Lee (1992). In practical cases, in-cloud scavenging (or rain-out) can be accounted for by assuming higher wash-out coefficients in the higher vertical layers.

4.3. Chemical Transformation

Transformation processes, which subtract pollutants from the mass balance equations, are basically thermochemical, photochemical and aerosol reactions. These reactions are all strongly dependent on temperature, and some of them on solar radiation. Therefore, their correct simulation is particularly crucial in LRT models where the diurnal cycle must necessarily be taken into account.

Starting from the conservation equation, and assuming that the concentration gradients of all species i are zero at all points in space, a set of i coupled differential equations can be written indicating the net source term of Eq. (3), which govern chemical changes in a completely isotropic air mass:

$$\frac{d\overline{\chi}_i}{dt} = \sum_{l,m} k_{lm} \cdot \overline{\chi}_l \cdot \overline{\chi}_m - \sum_j k_{ij} \cdot \overline{\chi}_i \cdot \overline{\chi}_j + Ph_i \ ,$$

where the terms on the right are the chemical formation, destruction and photochemical formation or loss.

Numerical methods are available for integrating fairly large sets of coupled differential equations that involve a great range of reaction rates values. When the initial conditions and the other physical parameters (temperature, pressure, light intensity) are specified, the concentrations $\overline{\chi}_i$ of reactants and products can be obtained as functions of time. This assumes that the chemical reaction mechanisms are correctly modeled, and that the rate constants k_{lm} and k_{ij} as well as the photochemical rates are well-known.

However, in real cases atmospheric chemical models are neither complete nor rigourously correct. This is particularly true because the number of reactions than can be simulated is limited and the reaction rate constants are approximated.

In chemical models, the atmosphere is acting as a chemical reactor, where the vertical dimensions to be considered in long range transport of pollutants are those of the ABL (1-2 km). In this chemical vessel, the pressure decreases in the vertical direction (approximately -11%/km), while the mean vertical temperature gradient is approximately -1°C/100 m. Partial water pressure also decreases rapidly from its ground level value of around 20 hPa to less than 1 hPa. The atmospheric vessel is periodically heated from the bottom and illuminated by the sun, depending on time of day and season and on weather conditions. The solar spectrum is cut off below about 300nm by the absorbing ozone layer in the stratosphere. These are, in essence, the physical boundary conditions under which chemical transformations occur in the troposphere.

The space and time scales of dispersion and transformation of some of the chemical species in the atmosphere are given in Table 3 (from Hov, 1983).

Table 3
Characterization of dispersion and transformation of some chemicals in terms of spatial and temporal scales.

Species	**Scales in**	
	Space	**Time**
O_3	1000 km	10 d
PAN	1000 km	5 d
NO_2	some 100 km	2 d
NO_3	1000 km	5 d
CO	10000 km	30 d
HCHO	100 km	10 d
CH_4	global	10 y
NMHC	10-1000 km	1-100 h

The values show the minimum modelization domain and time interval required in order not to produce results dependent on boundary and initial conditions.

Reactions in the atmosphere frequently consist of a chain of many elementary reactions, started often by free radicals, each described by a non-linear first order differential equation. However, in order to reduce the complexity of the simulation, they are frequently lumped together in a single reaction scheme. Even so, chemical models can include up to 40 to 60 different species involved in some 100 reactions.

The mechanism leading to thermochemical reactions between atmospheric components is based on collision density which is proportional to the number density of the reactants and to the probability that the collision is reactive. This product gives the rate constant of the reaction.

Photochemical reactions are those for which the energy of photons in the near ultra-violet spectrum is sufficient to break certain molecular bonds, provided the photons can be absorbed. This occurs when the absorption spectrum overlaps with the solar spectrum. The photochemical dissociation reaction:

$$AB + h\nu \rightarrow A + B$$

has a rate given by the equation:

$$\frac{dA}{dt} = j\,[AB] \ ,$$

where the photolysis frequency of the reaction j (sec^{-1}), is a function of the solar angle and is given by integrating the spectrum light intensity times the absorption cross-sections of species AB and the quantum yield of the dissociation reaction.

LTR models usually account for (Dunker et al., 1984):

- the sulphur cycle.
- the NO cycle, known to react with ozone.
- the ammonium cycle, starting from NH_3 emission and including ammonium aerosol formation.
- the hydrocarbon cycle, as the methane cycle.

Nitrogen monoxide (NO) emitted during the combustion processes is rapidly oxidizes by O_3 and RO_2 radicals:

$$NO + O_3 \rightarrow NO_2 + O_2$$

$$NO + H_2O \rightarrow NO_2 + OH$$

$$NO + RO_2 \rightarrow NO_2 + RO \ .$$

NO_2 is then removed either by photodissociation:

$$NO_2 + h\nu \rightarrow NO + O \ ,$$

where subsequently atomic oxygen is used for ozone formation, or by other oxidation reactions leading to formation of nitric acid and nitrates.

The main reaction leading to the formation of nitric acid is the NO_2 oxidation in the homogeneous phase by OH radicals. The nitric acid is then removed from the atmosphere by

NO_3 with homogeneous phase reactions (with formation of NH_4NO_3) or with a gas-particle reaction.

Ozone is generated only from the reaction of atomic oxygen with molecular oxygen that can be produced by photodissociation of NO_2, as indicated above. The main reaction leading to ozone removal is its reaction with NO. However, if NO oxidation would depend only on this reaction, steady-state condition could be rapidly established. In real cases, the presence of methane and other hydrocarbons, and of CO, produces radicals that reduce the real rate of ozone removal. Therefore, ozone concentration tends to increase.

The reactions leading to ozone formation and destruction are highly non-linear, therefore the control on ozone concentration is difficult to achieve. Figure 4 indicates qualitatively the complex dependence of ozone on NO_x and reactive organic gas emissions: reducing the NO_x emissions only, ozone concentration can decrease (point A), or increase (point C), depending on their level and on the corresponding level of reactive organic gas emissions.

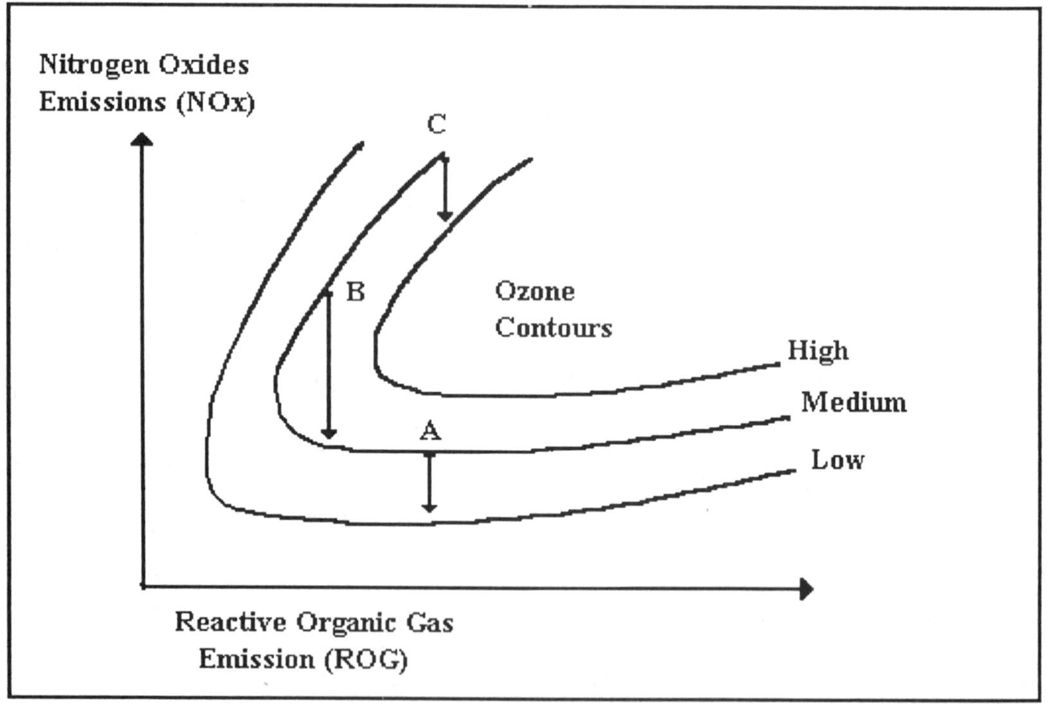

Figure 4. Dependence of ozone concentration on emission levels.

Sulphur dioxide present in emissions can react with OH radicals and with H_2O, as follows:

$$SO_2 + OH \rightarrow HOSO_2$$

$$HOSO_2 + O_2 \rightarrow HO_2 + SO_3$$

$$SO_3 + H_2O \rightarrow H_2SO_4 \ .$$

Sulphur acid can then be partially or totally removed by ammonia.

5. CHERNOBYL CASE STUDY

Following the Chernobyl accident and the recommendations of various groups of experts concerning the capability of numerical models in simulating long range atmospheric transport and deposition of radionuclides, the Atmospheric Transport Model Evaluation Study (ATMES) was performed under the sponsorship of the International Atomic Energy Agency (IAEA), the World Meteorological Organization (WMO), and the Commission of the European Communities (CEC).

Its primary aim was to compare concentration and deposition estimates with corresponding values measured in Europe after the accident. The study area included the majority of Western European countries, while the time period considered was the two weeks after the beginning of the release (26 April 1986).

To avoid extra sources of variation between model results that could have made the comparison even more difficult, it was decided that participants in the study had to use the same meteorological data-sets made available from the European Centre for Medium Range Weather Forecasts (ECMWF). They consisted of two data sets: one of analyzed data covering the period from April 25, 12:00 UTC to May 10, 1986, with 6 hour timestep and about 1 degree of spatial resolution in latitude and longitude; the other of 120 hour forecasts with the same characteristics.

The radiological data set was prepared at Joint Research Centre (JRC) Ispra and it was not released until the end of the exercise. Nonetheless a large part of the information contained had already appeared in the open literature. It consists of time histories of I-131 and Cs-137 air concentrations for approximately 100 localities; Cs-137 daily depositions for a limited number of localities (27) and gridded average values of cumulated deposition for the same isotope.

The 21 participants, from either national Meteorological Offices or Nuclear Safety Establishments, come from various countries worldwide: Japan, Russia, USA, Canada, Israel, South Africa, and the majority of European Countries. Some of the models were Eulerian, and others Lagrangian with various degrees of refinement.

The information supplied by participants indicated the different degree of model complexity. Further, some of them were planned to be used in nuclear emergencies, while others were still in development stage.

Model evaluation was performed using a number of basic statistical parameters, such as the measures of central tendency (mean, median, mid-range), measures of spread (standard deviation, bias, cumulated distribution), correlations (Pearson's and rank correlation coefficients) and more advanced statistical criteria, such as normalized mean square error and figure of merit in time and space.

Results of the comparisons are described in detail in the original report (Klug et al., 1991). Here only the main findings are summarized.

The reproduction of I-131 and Cs-137 air concentration time histories in the various localities was reproduced with a certain accuracy only by a limited number of LRT models. In fact only between five and seven models (out of 21) had more than 50% of their output concentration results within a factor five from the observations (Figure 5).

The situation is worst for daily deposition values, where difficulties in the simulation of wet deposition added to the errors in the input data used for precipitation. Cs-137 cumulated deposition results show in general better agreement than daily values, as when dealing with time integrated values, errors usually compensate.

Of those models that gave results for deposition (not all in fact presented a scheme for pollutant depletion), only five presented distributions of their results similar to that of the measurements. Figure 6 shows the comparison between deposition estimates of a particular model and of the observations as contour plots for a given contamination level.

Among the reasons for these discrepancies, it is important to mention the fact that model removal parameters used by the participants were varying over a relatively large range. This means that an effort must be made in data unification even considering that the deposition schemes used in long range models can be considered satisfactory, at least for dry deposition.

The uncertainty in the vertical wind velocity in the ABL was indicated as one of the major sources of difference. The windfield supplied was criticised as a probable source of errors in computed trajectories, due to the large size of the spatial meshes employed and the relatively long time interval (6 hours).

Other sources of errors, which could have a considerable influence on long range transport model performances could be the implementation and development of the initial stage of the source term, subgrid phenomena (processes such clouds and fronts), and numerical errors. The necessity of improving the mixed layer description, the simulation of the material entrainment through it and detrainment out of it was recognized to be important in particular in the presence of fronts. No evidence was found, within the limits of the study, that the model complexity improved their performances. No systematic difference between Lagrangian and Eulerian model results was found.

Finally, since atmospheric transport models are kinematic, their results depend on input meteorological fields. Thus, it is not surprising that their behaviour deteriorates considerably when using 120 hour forecast data. This indicates that there is then an urgent need to improve meteorological forecasts, in particular in view of using such models if another nuclear emergency should occur.

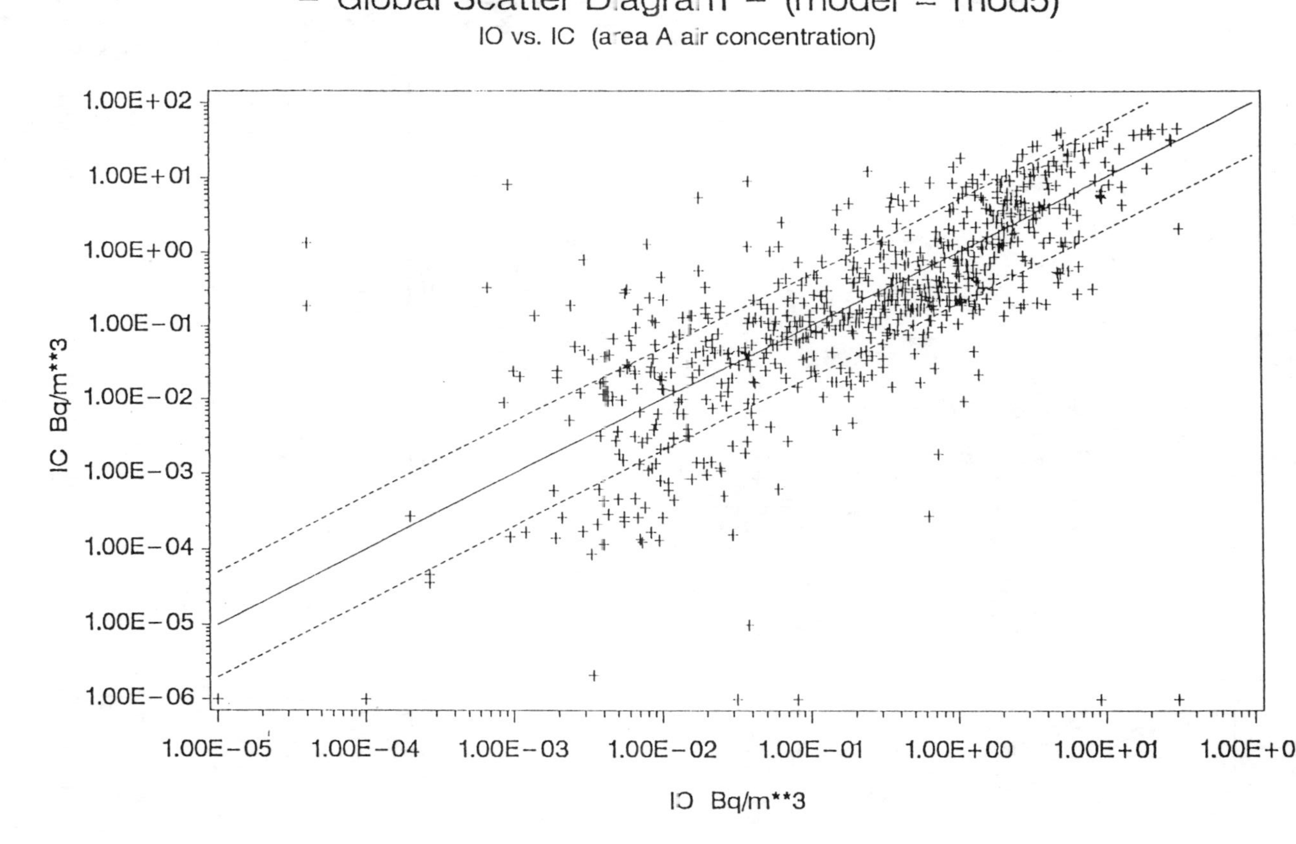

FA2 = 34 % FA5 = 69 % FOEX = 46 %

Figure 5. Global scatter diagram. Observed IO vs. calculated (IC) I-131 air concentration

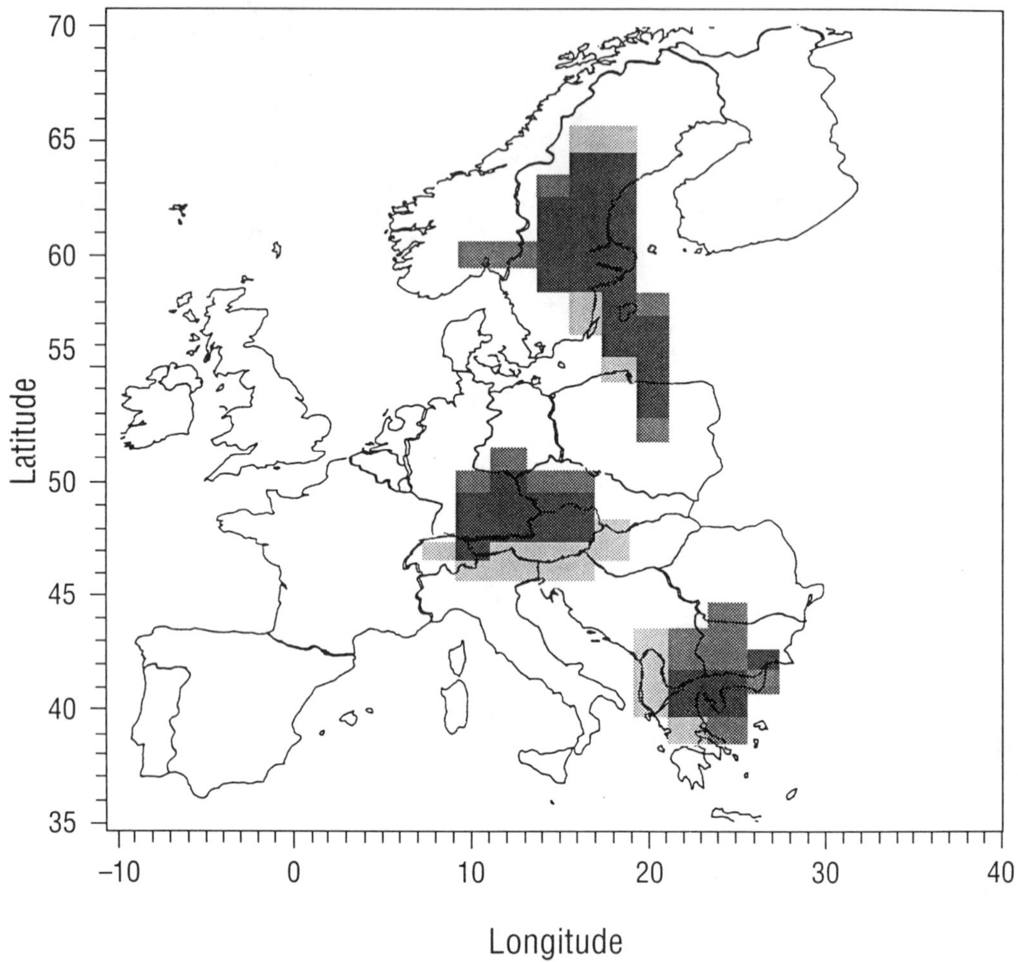

Coverage = 52%
Significant level = 5 KBq/m**2
Light gray = meas. Medium gray = pred. Dark gray = overlap

Figure 6. Space Figure of Merit for Cs-137 cumulative deposition.

ACKNOWLEDGEMENTS

Thanks are due to Prof. R.D. Bornstein (San Jose State University, USA), who thoroughly reviewed the paper, discussed the arguments in detail and made a number of useful suggestions to improve its quality.

REFERENCES

André J.C. and Blondin C., 1986. On the effective roughness length for use in numerical 3-D models. Boundary Layer Met., 35, 231-245.

ApSimon H.M., Goddard A.J.H. and Wrigley J., 1985. Lagrangian atmospheric dispersion of radioisotopes-I. The MESOS model. Atmo. Env. 19, 99-111.

ApSimon H.M., Barker B., Kayin S., Wilson J.J.N., 1991. Characterizing cloud processes and wet depositions in long range transport models. Paper presented at the 19th ITM on Air Pollution Modelling and its Applications, Ierapetra, 29 September - 4 October.

Asman W.A.H., Drukken B. and Janssen A.J., 1988. Modelled historical concentrations and depositions of ammonia and ammonium in Europe. Atmo. Env. 22, 725-735.

Batchvarova E. and Gryning S.E., 1991. Applied model for the growth of the daytime mixed layer. Boundary Layer Met. 56, 261-274.

Bonelli P., Calori G. and Finzi G., 1992. A fast long-range transport model for operational use in sulphirode simulation. Application to the Chernobyl accident. Atmo. Env. 26A, No. 14, 2523-2535.

Carmichael G.R. and Peters L.K., 1984. An Eulerian transport/transformation/removal Model for SO_2 sulphate. Atmo. Env. Vol. 18 No. 5, pp. 937-951.

Carras J.N. and Williams D.J., 1988. Measurements of relative sigma, up to 1800 km. Atmo. Env., 22, 1061-1069.

Chock D.P. and Dunker A.M., 1983. A comparison of numerical methods for solving the advection equation. Atmo. Env., 17, 11-24.

Derwent R.G. and Nodop K., 1986. Long range transport deposition of acidic nitrogen species in North West Europe. Nature, 356-358.

Derwent R.G., Hov Ø., Asman W.A.H., Van Jaarsveld J.A. and De Leeuw F.A.A.M., 1989. An intercomparison of long term atmosphere transport models; the budgets of acidifying species for the Netherlands. Atmo. Env., 23, No. 9, 1893-1909.

Desiato F., 1992. A long-range dispersion model evaluation study with Chernobyl data. Atmo. Env. 26A, No. 15, 2805-2820.

Dunker A.M., Kumar S. and Berzins P.H., 1984. A comparison of chemical mechanisms used in atmospheric models. Atmo. Environ. 18, 311-321.

Egan B.A. and Mahoney J.R., 1971. Numerical modelling of advection and diffusion of urban area source pollutants. J. Appl. Meteor., 11, 312-322.

Eliassen A. and Saltbones J., 1983. Modelling of long range transport of sulphur over Europe: a two years run and some model experiments. Atmo. Env. 17, 1457-1473.

Eliassen A., Hov Ø., Iversen T., Saltbones J. and Simpson D., 1988. Estimates of airborn transboundary transport of sulphur and nitrogen over Europe in 1985. EMEP/MSCW Note 1/88, The Norwegian Meteorological Institute, Oslo, Norway.

Endlich R., Nitz K., Brodzinsky R. and Bhumralkar C., 1984. A long range air pollution for Eastern North America I-sulphur oxides. Atmo. Env. 18,(11),2345-2360.

Fairweather G., 1978. Finite element Galerkin methods for differential equations. Lecture Notes in Pure and Applied Mathematics, Vol. 34, Marcel Dekker, New York.

Galmarini S., Graziani G. and Tassone C., 1992. The atmospheric long range transport model LORAN and its application to Chernobyl release. Environmental Software 7, 143-154.

Gifford F.A., 1984. The random force theory: application to meso- and large-scale atmospheric diffusion. Boundary Layer Met., 30, 159-175.

Gryning S.E. and Batchvarova E., 1990a. Simple model of the daytime boundary layer height. Ninth Symposium of Turbulence and Diffusion. American Meteorological Society, April 30 - May 3, 1990. Roskilde, Denmark, 379-382.

Gryning S.E. and Batchvarova E., 1990b. Analytical model for the growth of the coastal internal boundary layer during onshore flow. Quart. J.R. Met. Soc., 116, 187-203.

Holtslag A.A.M. and De Bruijn H.A.R., 1988. Applied modelling of the nighttime surface energy balance over land. J. Appl. Met. 27, 689-704.

Hov Ø., 1983. Numerical solution of a simplified form of the diffusion equation for chemically reactive atmospheric species. Atmo. Env. 17, 551-562.

Kimura F. and Yoshikawa T., 1988. Numerical simulation of global scale dispersion of radioactive pollutants from the accident at the Chernobyl nuclear power plant. J. of the Met. Soc. of Japan, Vol. 66, No. 3, 489-495.

Klug W., 1982. Physical transport or the problem how to model air pollution. T. Schneider and L. Grant (Edit.): Air pollution by Nitrogen Oxides; Elsevier Scientific Publishing Company, Amsterdam.

Klug W. and Lüpkes, 1985. A comparison between long term interregional air pollution models. Institut für Meteorologie, Technische Hochschule, Darmstadt, Germany.

Klug W., Graziani G., Grippa G. and Tassone C., 1991. ATMES, Atmospheric Long Range Transport Model Evaluation Study. Elsevier Scientific Publishing Company, London.

Kondo J. and Yamazawa H., 1986. Aerodynamic roughness over an inhomogeneous ground surface, Boundary Layer Met., 35, 331-348.

Lange R., 1978. ADPIC - A three-dimensional particle-in-cell model for the dispersal of atmospheric pollutants and its comparison to regional tracer studies. J. Appl. Meteorol., 17, 320-329.

Lee I.Y., 1992. Comparison of cloud microphysics parametrization for simulation of mesoscale clouds and precipitation. Atmo. Env. 26A, 15, 2699-2712.

Lindberg S.E., Page A.L. and Norton S.A., 1990. Acid precipitation (Advances in environmental science). Springer-Verlag, N.Y.

Maryon R.H. and Heasman C.C, 1988. The accuracy of plume trajectories forecast using the U.K. Meteorological Office operational forecasting models and their sensitivity to calculation schemes. Atmo. Env., Vol. 22, pp. 259-272.

McRae G.J., Goodin W.R. and Seinfield J.H., 1982. Numerical solution of the atmospheric diffusion equation for chemically reacting flows. J. Comp. Phys., 45, 1-42.

Pasquill F., 1974. Atmospheric diffusion. 2nd edition. Chichester, England: John Wiley and Sons.

Persson Ch. and Robertson L., 1990. An operational Eulerian dispersion model appliable to different scales. Proc., 18th Technical Meeting of NATO-CCMS on Air Pollution and its Applications, May 14-17, Vancouver, Canada.

Pielke A.R., 1984. Mesoscale meteorological modelling. Academic Press, London.

Pudykiewicz J. and Stanforth A., 1984. Some properties and comparative performance of the semi-Lagrangian method of Robert in the solution of the advection-diffusion equation. Atmosphere-Ocea, 22, 283-308.

Pudykiewicz J., 1988. Numerical simulation of the transport of radioactive cloud from the Chernobyl nuclear accident. Tellus, 40B, 241-259.

Raes F., Tassone C., Grippa G., Zarimpas N. and Graziani G., 1991. Updating long range transport model predictions using real-time monitoring data. Atmo. Env. 25A, 2807-2814.

Reap R.M., 1972. An operational three-dimensional trajectory model. J. App. Meteorology, Vol. 11, 1193-1201.

Reinking R.F., 1987. Perspectives for research in wet chemistry and cloud modification. Boundary Layer Met. 41, 381-465.

Ritchie H., 1987. Semi-Lagrangian advection on a Gaussian grid. Mon. Wea. Rev., 115, 608-619.

Sheih C.M., Wesely M.L. and Hicks B.B., 1979. Estimated dry deposition velocities of sulphur over Eastern U.S. Atmo. Env. 13, 1361-1368.

Simpson D., 1992. Long-period modelling of photochemical oxidants in Europe. Model calculations for July 1985. Atmo. Env., 26A, (9),1609-1634.

Smith F.B., 1981. The significance of wet and dry synoptic regions on long-range transport of pollution and its deposition. Atmo. Env., Vol. 15.

Smolarkiewicz P.K., 1983. A simple positive definite advection scheme with small implicit diffusion. Mon. Wea. Rev., 111, 479-486.

Taylor P.A., 1987. Comments and further analysis on effective roughness lengths for use in numerical 3-D models. Boundary Layer Meteo 39, 403-418.

Tripoli G.J. and Cotton W.R., 1982. The Colorado State University three-dimensional cloud/mesoscale model. Journal de Recherches Atmosphériques, No. 3, Vol. 16, Juillet-Septembre 1982.

Van Dop H. and De Haan B.J., 1983. Mesoscale air pollution dispersion modelling. Atmo. Env. 17, 1449-1456.

Van Egmond N.D. and Kesselboom H., 1986. Mesoscale air pollution dispersion models - I: Eulerian grid-model. Atmo. Env., 17, 257-265.

Van Jaarsveld J.A., 1988. Modelling of the NO_y budgets over the Netherlands. Paper presented at ECE-WHO workshop on modelling transformation processes and transport of air pollution, Potsdam, 21-25 March, 1988.

Van Ulden A.P. and Holtslag A.A.M., 1985. Estimation of atmospheric boundary layer parameters for diffusion applications. J. Clim. Appl. Met. 24, 1196-1207.

Venkatram A., Ley B.E. and Wong S.Y., 1982. A statistical model to estimate long-term concentrations of pollutants associated with long-range transport. Atmo. Env. 16, 249-257.

Verver G.H.L. and Scheele M.P., 1988. Influence of non-uniform mixing heights on dispersion simulation following the Chernobyl accident. NATO/CCMS 17th ITM on Air Pollution Modelling and Applications, Cambridge (UK).

Yangang L., 1993. Statistical theory of the Marshall-Palmer distribution of raindrop. Atmo. Env., 27A, 1, 16-19.

Zilitinkevich S.S., 1972. On the determination of the Height of the Ekman Boundary Layer. Boundary Layer Met. 3, 141-145.

Chapter 6

A new method of rainfall-runoff modelling and its applications in catchment hydrology

I.G. Littlewood,[a] A.J. Jakeman[b]

[a] *Institute of Hydrology, Wallingford, Oxon OX10 8BB, UK*

[b] *Centre for Resource and Environmental Studies, Institute of Advanced Studies, Australian National University, Canberra, A.C.T. 0200, Australia*

Abstract

A new rainfall—runoff modelling method is described and several of its applications are demonstrated and discussed. Unit Hydrographs (UHs) can be identified for total streamflow from rainfall, streamflow and temperature time series, thereby circumventing the need for subjective prior hydrograph separation into baseflow and direct flow as in more traditional methods of runoff (including flood) event UH analysis. In many cases the modelled streamflow can be separated into dominant quick and slow flow responses corresponding to conceptual storages acting in parallel. The Unit Hydrograph model, no longer restricted in practice to mainly events, can now be considered for a much wider range of applications. This still includes flood event analysis but embraces also: land-use change and climate change scenario impact studies for gauged catchments; characterisation of low-flow regimes by a Slow Flow Index (SFI); and streamflow hydrochemical investigations. The potential for information-transfer to ungauged catchments (regionalisation), using statistical relationships linking physical catchment descriptors (PCDs) and dynamic response characteristics (DRCs) derived from UH parameters, is also discussed.

Key words

Unit hydrograph, flood event analysis, Slow Flow Index, hydrograph separation, regionalisation, land-use change, climate change.

1 Introduction

1.1 Rainfall—runoff models

Rainfall—runoff models play a central role in catchment hydrology, assisting in a wide range of investigations: assessment of the hydrological impacts of land-use and possible climate changes; real-time flood forecasting and 'design flood' estimation; assessment of the reliability of natural water resources; and river water quality investigations. A frequent focus of interest in streamflow hydrochemical studies is the relationship between observed streamflow—concentration dynamic behaviour and a hypothesised continuously variable degree of mixing of water from two or more sources with different chemical characteristics (e.g. rain water and catchment soil water drainage).

The literature describes a wide range of models and modelling approaches, reflecting personal preferences and a 'horses for courses' strategy adopted by practising hydrologists. Time series of precipitation, streamflow and some surrogate for evapotranspiration (e.g. temperature or pan evaporation) are generally used to calibrate the model (i.e. estimate its parameters). The most simply-structured models operate at catchment-scale, converting an input of areal rainfall to an output of streamflow. These are spatially 'lumped' parameter models, whereas those which include representation of processes at sub-catchment scale (e.g. for slopes) are known as 'distributed' parameter models. Distributed models require additional data inputs (e.g. describing topography and soil properties), mainly to define elemental areas or volumes on which to distribute the parameters. They generally employ idealised equations of mathematical physics, assuming that the elemental areas or volumes to which they apply are relatively homogeneous. Although there does not appear to be a single rainfall—runoff model or modelling approach which can answer all questions which arise, some types of model are used more than others. Lumped rainfall—runoff models are particularly attractive because, compared with more distributed models, they are simple to understand, require minimal input data and perform at least as well in predicting streamflow. In principle, therefore, a lumped rainfall—runoff model is the obvious choice for extracting, with least effort, hydrological information from rainfall and streamflow data.

1.2 Unit Hydrograph theory

The theory of the Unit Hydrograph (UH), the type of lumped model of interest in this chapter, is described extensively in the literature (e.g. Wheater *et al.* 1993, and references cited therein) and requires, therefore, only a brief introduction here. For a historical perspective of the development of UH theory, including development of the systems identification approach employed in this chapter, the reader can consult the existing literature (e.g. Jakeman *et al.* 1990, Wheater *et al.* 1993 and references cited therein). The continuous-time convolution of rainfall excess $u(s)$ with an instantaneous unit hydrograph $h(t)$, to give continuous flow $y(t)$, is given by eqn (1).

$$y(t) = \int_0^t h(t-s)\, u(s)\, ds \qquad (1)$$

In the light of recent developments, the following general points concerning the derivation and application of UHs are important in a review of the role of the UH model and its applications.

1.3 Unit Hydrograph practice

First, note that a UH is the streamflow response to an isolated unit input (over a specified sampling interval) of rainfall excess, the residual when evapotranspiration losses have been deducted from rainfall. Ignoring any imports and exports of groundwater, or bulk transfer for water supply, across the topographical boundary of a catchment, rainfall excess is that part of rainfall which contributes (eventually) to streamflow at the catchment outlet. Unit Hydrograph derivation, therefore, involves the prior application of a rainfall 'loss' model to calculate rainfall excess. Although UH *theory* does not limit itself to modelling any particular component of streamflow, or to individual runoff events, applications of it have been largely thus restricted. One reason for this has been the difficulty of identifying the ordinates (or parameters) of a UH for *total* streamflow. In general practice, UHs have been identified only for a 'direct flow' component of streamflow from short periods of record describing an individual runoff event (or a few events over a fairly short period). A UH which characterises the direct flow regime of a catchment has had to be estimated by averaging the UHs derived from each of several individual runoff events for which baseflow has been deducted.

In established methods for identifying UHs, rainfall excess and direct flow are derived from rainfall and streamflow respectively by intuitively reasonable but fairly arbitrary separation techniques. For example, baseflow is often estimated by invoking a parameterised geometric interpretation of the streamflow hydrograph. It is then subtracted from streamflow, leaving direct flow. An alternative way of prescribing baseflow is to pass the streamflow data through a low-pass numerical filter. Volumes of direct flow and rainfall excess are forced to be equal over the model calibration period. When dealing with individual runoff events, however, this attempt to comply with the principle of mass conservation might be only partially successful because the volume of water stored in the catchment may not be the same at the start and end of the calibration period. A related practical problem with event-based UHs for direct flow is that it is often necessary to truncate the streamflow recession curve (e.g. before the next hydrograph rise). Careful selection of suitable runoff events is required, therefore, to minimise the effects of these problems.

In traditional methods, parameters of the rainfall 'loss' model (for estimating rainfall excess), and for the baseflow separation model, are varied within the constraint of equal volumes over the calibration period until a best UH for direct flow is found. Clearly, the integrity of a UH for direct flow depends on the integrity of the rainfall excess *and* of the baseflow, where both these components are arguably rather poorly defined. On the other hand, a UH for *total* streamflow would depend only on the integrity of the rainfall excess. A methodology (described later) does now exist for systematically identifying UHs for total streamflow, from time series rather than runoff event data, opening-up a much wider range of applications for UH models than before.

There can be other problems in identifying UHs from rainfall excess and streamflow data. UHs derived by the matrix inversion technique, for example, often have to be smoothed and constrained in some way to ensure a hydrologically sensible result. A key problem with this approach is over-parameterisation of the UH (i.e. the first n ordinates, where n might be 10 or more) with respect to the information in the rainfall excess and direct flow data. In contrast, the UH model for total streamflow described in this chapter usually has four (or less) well-identified parameters; these parameters define the magnitude and shape of a 'mixed exponential decay' UH.

The main objective of this chapter is to discuss the expanded role of UH models now that they can be identified for total streamflow. This is preceded by an introduction to the UH modelling methodology first presented by Jakeman *et al.*, (1990) which overcomes many of the problems described above. It identifies UHs for total streamflow from only rainfall, streamflow and temperature data (the latter as a surrogate for evapotranspiration). In this methodology, the UH for total streamflow is parameterised in a flexible but efficient way, and a powerful technique of parameter estimation is employed. By optionally including a suitable rainfall loss model as an integral part of the overall procedure, it enables rainfall—streamflow modelling on a variety of catchment sizes and types, using time series data at intervals ranging from less than hourly to monthly, depending on the nature of the hydrological regime. A strong feature of the methodology is its general ability, for permanently flowing streams, to identify a UH for total streamflow which itself is the sum of separate UHs for dominant quick and slow flow components. This leads to a new method of hydrograph separation with a wider range of applications than other methods.

Section 2 of this chapter describes the new modelling methodology in detail, followed by Section 3 which discusses several of the possible applications of UHs for total streamflow and separate UHs for dominant quick and slow components of streamflow. The chapter is drawn to a close in Section 4.

2 Identification of unit Hydrographs And Component flows from Rainfall, Evaporation and Streamflow (IHACRES)

2.1 Model structure

The overall model structure employed in the IHACRES computer program (Jakeman *et al.* 1991a) comprises a non-linear loss module which generates rainfall excess from rainfall and temperature, followed by a linear rainfall excess—streamflow module. They will be described in reverse order here. Full details are given elsewhere by Jakeman *et al.* (1990, 1991b).

2.1.1 The rainfall excess—streamflow (linear) UH module

The discrete-time version of eqn (1) for the convolution of a time series of rainfall excess up to and including time step k $(\ldots, u_{k-3}, u_{k-2}, u_{k-1}, u_k)$ with a UH $(h_0, h_1, h_2, \ldots, h_i, \ldots)$ to estimate streamflow y_k at time step k is given by

$$y_k = h_0 u_k + h_1 u_{k-1} + h_2 u_{k-2} + \ldots + h_i u_{k-i} + \ldots + \zeta_k \tag{2}$$

which can be written as

$$y_k = (h_0 + h_1 z^{-1} + h_2 z^{-2} + \dots + h_i z^{-i} + \dots)\, u_k + \zeta_k \tag{3}$$

The operator z^{-i} in eqn (3) is the backward shift operator such that $z^{-i}\, x_k = x_{k-i}$, where x_k is any variable at time step k. The stochastic error compensation term ζ_k represents the sum of all uncertainties arising from sampling, measurement and model errors, the latter including unrepresented model inputs.

If the terms within brackets in eqn (3) are represented by the linear operator $H(z^{-1})$ then

$$y_k = H(z^{-1}) u_k + \zeta_k \tag{4}$$

Thus $H(z^{-1})u_k$ is the noise-free streamflow x_k so that

$$y_k = x_k + \zeta_k \tag{5}$$

The coefficients of the polynomial $H(z^{-1})$ are the ordinates of the UH for total streamflow. Clearly, the number of ordinates for such a UH can be large, in which case it will be difficult to estimate them all well from the information in rainfall excess and streamflow data. Fortunately, in practice, $H(z^{-1})$ can be approximated closely by a rational transfer function of the form

$$H(z^{-1}) = \frac{B_m(z^{-1})}{A_n(z^{-1})} \tag{6}$$

where the polynomials $B_m(z^{-1})$ and $A_n(z^{-1})$ are defined as

$$B_m(z^{-1}) = b_0 + b_1 z^{-1} + b_2 z^{-2} + \dots + b_m z^{-m} \tag{7}$$

$$A_n(z^{-1}) = 1 + a_1 z^{-1} + a_2 z^{-2} + \dots + a_n z^{-n} \tag{8}$$

with both n and m small. As will be seen later, a further advantage of the rational transfer function is that for the case when $n=2$ and $m=1$, i.e. a (2,1) transfer function, it can yield separate UHs for quick and slow response components of streamflow.

The expanded form of $x_k = H(z^{-1})u_k$ is, therefore,

$$x_k = -a_1 x_{k-1} - a_2 x_{k-2} - \dots - a_n x_{k-n} + b_0 u_k + b_1 u_{k-1} + \dots + b_m u_{k-m} \tag{9}$$

A pure time delay δ can be incorporated by replacing all subscripts k in the right hand side of eqn (9) with $k-\delta$.

Consider further, now, the case when $n = 2$ and $m = 1$, i.e. the (2,1) transfer function given by

$$x_k = \left[\frac{b_0 + b_1 z^{-1}}{1 + a_1 z^{-1} + a_2 z^{-2}} \right] u_k \tag{10}$$

This second-order model can be written as the sum of two first-order (1,0) transfer functions, i.e.

$$x_k = \left[\frac{\beta^{(1)}}{1 + \alpha^{(1)} z^{-1}} \right] u_k + \left[\frac{\beta^{(2)}}{1 + \alpha^{(2)} z^{-1}} \right] u_k \tag{11}$$

Eqn (11) corresponds to a parallel configuration of storages in the linear part of the model, as shown in Figure 1. If b_0, b_1, a_1 and a_2 in eqn (10) are known, then $\beta^{(1)}$, $\beta^{(2)}$, $\alpha^{(1)}$ and $\alpha^{(2)}$ in eqn (11) can be calculated.

First-order (1,0) transfer functions of the form $\beta/(1 + \alpha z^{-1})$ have the following properties. For an isolated unit input over a single time interval (i.e. zero input in all other time intervals) the output from a first-order transfer function is of magnitude β for that time interval, followed by an exponential decay with an approximate time constant τ given by

$$\tau = \frac{-\Delta}{\ell n(-\alpha)} \qquad (\Delta \text{ is the data time interval}) \tag{12}$$

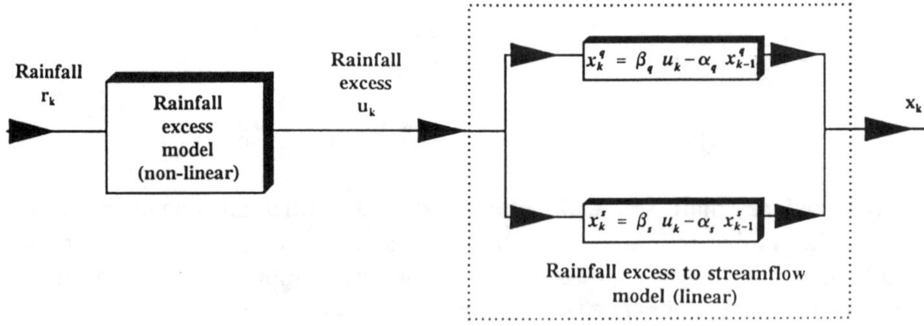

Figure 1: Model structure—two parallel storages in the linear module.

For an isolated unit input to a transfer function $\beta/(1 + \alpha z^{-1})$, τ in eqn (12) is the time over which the output decays from its maximum value β to a value of $e^{-1}\beta$ (i.e. to 37% of β). For a unit step input (i.e. zero input for all time intervals up to time step k followed by unit input for all time intervals thereafter), the output increases from zero at time step k asymptotically to its steady-state gain (*ssg*) given by

$$ssg = \frac{\beta}{1 + \alpha} \tag{13}$$

It follows from eqn (12) that the component first-order transfer function in eqn (11) with the value of $\alpha^{(.)}$ nearer to -1 ($-1 < \alpha^{(.)} < 0$) represents slow flow, while the other first-order transfer function represents quick flow. Separate UHs for the quick and slow flow components of streamflow are the output responses from the quick and slow first-order transfer functions respectively when subjected to an isolated unit input over a single time interval. Subsequently in this chapter, superscripts q and s will be used. We have, therefore,

$$x_k = x_k^{(q)} + x_k^{(s)} \tag{14}$$

$$x_k = \left[\frac{\beta^{(q)}}{1 + \alpha^{(q)} z^{-1}}\right] u_k + \left[\frac{\beta^{(s)}}{1 + \alpha^{(s)} z^{-1}}\right] u_k \tag{15}$$

For further background, and for other hydrological applications of the systems approach using transfer functions, the reader can refer to work by Young (1984, 1992), Young and Lees (1993), Dietrich *et al.* (1989), Jakeman *et al.* (1989), Whitehead (1979) and Whitehead *et al.* (1986).

2.1.2 The rainfall—rainfall excess (non-linear) loss module

In the spirit of parsimony in the total number of parameters, the non-linear rainfall—rainfall excess module in IHACRES has just two parameters. (For the case of a (2,1) transfer function in the linear part of IHACRES, therefore, there is a total of six parameters to be identified from observations of rainfall, streamflow and temperature.) The underlying conceptualisation for the rainfall loss model is that rain falling on a wet catchment produces more streamflow than if the catchment were relatively dry. Rainfall r_k is converted to rainfall excess u_k using a running catchment wetness index s_k, itself derived from past rainfall (and temperature) data. Several variations of this basic strategy have been tried. A simple case, first used by Whitehead *et al.* (1979), is that u_k is the product of s_k and r_k^*, where r_k^* is rainfall adjusted for evapotranspiration loss e_k using temperature t_k. The sequence of operations is given by

$$r_k^* = \left[1 - \frac{t_k}{T}\right] r_k \tag{16}$$

$$\text{where } e_k = r_k \left[\frac{t_k}{T}\right] \tag{17}$$

$$s_k = \left[\frac{\tau_w^{-1}}{1 - (1 - \tau_w^{-1}) z^{-1}}\right] r_k^* , \qquad s_0 = 0 \tag{18}$$

$$u_k = K_1 s_k r_k^* \tag{19}$$

The parameter τ_w in eqn (18) determines the rate at which s_k decays exponentially in the absence of rainfall, while the parameter T in eqns (16) and (17) is a reference temperature greater than any of the observed temperatures t_k. The constant K_1 in eqn (19) is selected to give equal volumes of rainfall excess u_k and observed streamflow y_k over the model calibration period. To complement this volume-forcing, model calibration periods are selected to start and end with similar levels of low flow on the tails of hydrograph recessions, when catchment storages are low and can reasonably be assumed to be approximately the same. Under these assumptions the steady-state gains (ssg) of the quick and slow first-order transfer function components of eqn (15) sum to unity (taking measurement units into account). Provided that catchment storages at the start and end of model calibration periods are approximately the same, K_1 is not a model parameter in the same sense as the others.

Although the loss model described by eqns (16) to (19) has been demonstrated to work well for a number of catchments (Jakeman *et al.* 1990, 1991a,b; Littlewood and Jakeman 1991, 1993) it certainly does have conceptual weaknesses. For example, evaporation e_k is not related to catchment wetness s_k, i.e. no account is taken of the availability of water for evaporation during non-rainfall periods. A conceptually more acceptable two-parameter loss model, where e_k *is* dependent on catchment wetness s_k, is given by the sequence

$$\frac{s_k}{K_2} = r_k \left(1 - s_{k-1}\right)_+ + \frac{s_{k-1}}{K_2}\left[1 - \frac{1}{\tau_w(t_k)}\right], \qquad s_0 = 0 \tag{20}$$

$$\tau_w(t_k) = \tau_w \exp(20f - t_k f) \tag{21}$$

$$e_k = \frac{s_{k-1}}{K_2 \tau_w(t_k)} \tag{22}$$

$$u_k = r_k s_{k-1} \tag{23}$$

The '+' subscripted term in eqn (20) is included only if it has a positive value (i.e. $s_{k-1} < 1$). The parameter τ_w defines the time constant of decay in s_k due to evapotranspiration in the absence of rainfall and at a constant temperature, arbitrarily 20°C. The temperature modulation factor f in eqn (21) quantifies how $\tau_w(t_k)$ varies per unit change of temperature from 20°C. K_2 in eqns (20) and (22) is a volume-forcing constant similar to K_1 in eqn (19). Preliminary assessments of different loss models used in conjunction with the linear module described in Section 2.1.1 are presented by Littlewood and Post (1993) and Chen *et al.* (1993).

2.2 Parameter estimation for the linear module

A key to the success of IHACRES in separating hydrographs into quick and slow flow components is the method it employs to estimate the parameters in the linear module, e.g. a_1, a_2, b_0 and b_1 in eqn (10).

2.2.1 The Simple Refined Instrumental Variables (SRIV) method

A general feature of recent approaches for estimating transfer function parameters is the use of instrumental variable (IV) techniques. Jakeman *et al.* (1990) summarise the features of the simple refined instrumental variable (SRIV) technique, the algorithm employed in IHACRES, with respect to those of its major competitors. As with all IV techniques, SRIV yields consistent parameter estimates. Unlike the well-known least squares and basic IV (BIV) techniques, SRIV can identify quick *and* slow components of a UH for total streamflow, and it is largely this feature to which IHACRES owes its success. While SRIV is conceptually and computationally simpler than other IV techniques, it allows more general assumptions on the stochastic term ζ_k in eqn (4) than asymptotically efficient IV versions, and still has their major useful properties.

A 'blind' test of the SRIV technique was undertaken by the first author as follows. A time series assumed to be rainfall excess was supplied to a colleague who (a) prescribed a second-order transfer function (the sum of hydrologically reasonable quick and slow response first-order transfer functions), then (b) independently of the IHACRES computer program, generated a synthetic streamflow series by passing the rainfall excess series through the prescribed transfer function. The rainfall excess and synthetic streamflow series, both 'double precision' (18 significant figures), were returned for SRIV analysis by the IHACRES program. The parameters of the prescribed first-order transfer functions were identified to better than one part in a million. Interestingly, using the same data set, the BIV technique did not yield the correct parameters. Although BIV gave a fairly good model-fit, the second-order transfer function it identified was not physically realistic in terms of quick and slow response components configured either in series or parallel. Comprehensive Monte Carlo experimentation by the second author has shown that SRIV can identify both quick and slow flow components even when the zero-mean noise term ζ_k has high variance.

2.2.2 Model-fit criteria

For a prescribed configuration of storages in the linear module (usually two in parallel as described in 2.1.1) the two parameters in the *non*-linear loss model are optimised by repeatedly running the IHACRES program with different values of the non-linear parameters, searching for optimal values of statistics. These include a maximum (or near-maximum) coefficient of determination D concurrent with a minimum (or near-minimum) 'average relative parameter error' ARPE. D and ARPE are defined as

$$D = 1 - \frac{\sum_{k=1}^{N} (y_k - x_k)^2}{\sum_{k=1}^{N} (y_k - \bar{y})^2} \tag{24}$$

where N is the number of time steps in the model calibration period, and

$$ARPE = \frac{\sum_{i=1}^{n} \frac{\hat{\sigma}_i^2}{\hat{a}_i^2} + \sum_{i=1}^{m} \frac{\hat{\sigma}_{i+n+1}^2}{\hat{b}_i^2}}{m + n + 1} \tag{25}$$

Each $\hat{\sigma}_i$ in eqn (25) is the estimated variance of the ith element in the set $(a_1, a_2, \ldots, a_n, b_o, b_1, \ldots, b_m)$; these variances are by-products of the SRIV algorithm. The trade-off between D and %ARPE when calibrating a model for a small upland catchment in Wales, using hourly data, is shown in Table 1. Note that in this case there was no adjustment of r_k to r_k^*; only eqns (18) and (19) were used, so the loss module had just one parameter τ_w.

2.2.3 Flow components in series and/or parallel?

Different configurations of individual transfer function components $\beta/(1 + \alpha z^{-1})$, and their corresponding overall system transfer functions, are shown in Figure 2. When the two parameters in the non-linear rainfall excess module have been identified using a prescribed configuration of storages in the linear module (e.g. a (2,1) structure), other structures of linear module can be prescribed, leaving the non-linear module parameters unchanged. As shown in Table 2, a trade-off between D and ARPE (corresponding to the same catchment modelling exercise as in Table 1) is employed to establish the most appropriate structure of flow components in series and/or parallel. If this indicates that a different configuration of storages in series and/or parallel is required to the one chosen initially, it may be necessary to repeat the parameter estimation for the non-linear loss model (though the optimal parameters τ_w and T (or f) may not be too different).

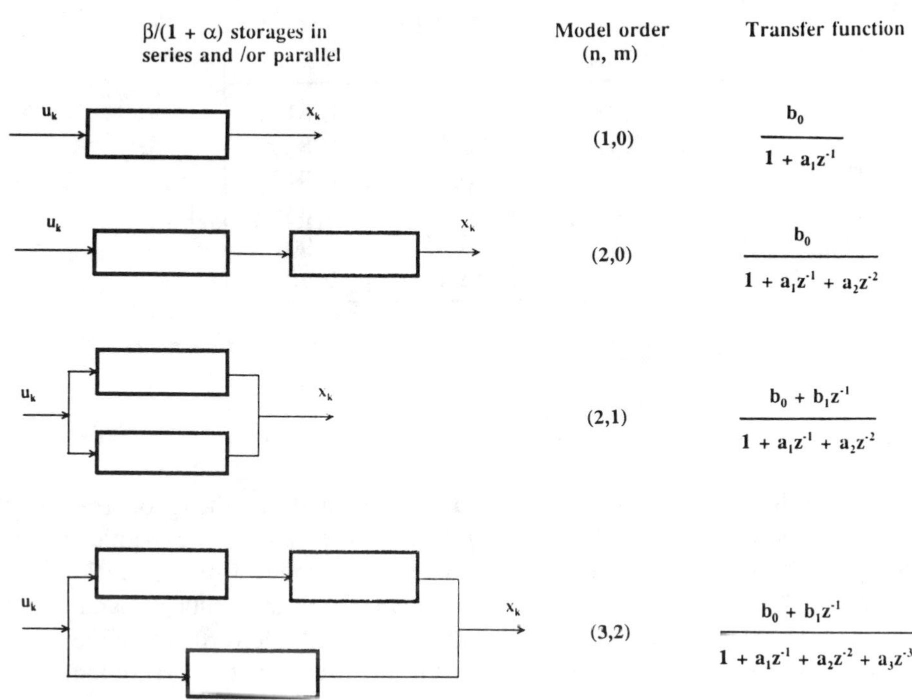

Figure 2: Configuration of storages and corresponding transfer functions.

Table 1: Trade-off between D and %ARPE while calibrating a non-linear module

τ_w (hours)	%ARPE	D	τ_w (hours)	%ARPE	D
5	0.0305	0.9047	86*	0.0214*	0.9455
10	0.0280	0.9164	88	0.0215	0.9461
15	0.0285	0.9110	90	0.0215	0.9467
40	0.0255	0.9099	102	0.0223	0.9484
50	0.0244	0.9178	104*	0.0225	0.9485*
60	0.0231	0.9277	106	0.0227	0.9485
70	0.0220	0.9365	108	0.0229	0.9484
80	0.0215	0.9429	110	0.0232	0.9483
82	0.0214	0.9438	120	0.0247	0.9470
84	0.0214	0.9447	130	0.0266	0.9448

The asterisks mark an optimal range of values for τ_w

Table 2: Trade-off between D and %ARPE to identify the most appropriate linear module structure

Parameterisation (n, m)	%ARPE	D
(1,0)	0.0319	0.8191
(2,0)	21.2976	0.8206
(2,1)*	0.0214	0.9455
(3,0)	0.3637	0.8449
(3,1)	0.8216	0.9014
(3,2)	38.1116	0.9078

The asterisk marks the optimal parameterisation

2.3. Dynamic response characteristics

For a (2,1) configuration of storages in the linear module, the properties of the component first-order transfer functions in eqn (15) lead to dynamic response characteristics (DRCs) for the dominant quick and slow components of streamflow: decay response times $\tau^{(q)}$ and $\tau^{(s)}$; relative average volumetric throughputs $\upsilon^{(q)}$ and $\upsilon^{(s)}$; and relative magnitudes of the separate UHs $\lambda^{(q)}$ and $\lambda^{(s)}$. Eqns (26) to (28) define these DRCs for the quick flow component (superscript s should be used for the corresponding slow flow DRCs).

$$\tau^{(q)} = \frac{-\Delta}{\ln(-\alpha^{(q)})} \tag{26}$$

$$\upsilon^{(q)} = \frac{\beta^{(q)}}{(1 + \alpha^{(q)})g} \tag{27}$$

$$\lambda^{(q)} = \frac{\beta^{(q)}}{p} \tag{28}$$

Because the volume of rainfall excess is set equal to the volume of streamflow over the model calibration period (see 2.1.2), g and p in eqns (27) and (28) respectively, are unity if the measurement units of the rainfall excess and streamflow data are the same (e.g. both in millimetres). When the measurement units are not the same g and p are given by eqns (29) and (30) respectively.

$$g = \frac{\beta^{(q)}}{1 + \alpha^{(q)}} + \frac{\beta^{(s)}}{1 + \alpha^{(s)}} \tag{29}$$

$$p = \beta^{(q)} + \beta^{(s)} \tag{30}$$

The non-linear rainfall loss module also has DRCs, e.g. τ_w, f, and K_2 in eqns (20) and (21). For this loss module, and the common two storages in parallel configuration conceptually transporting rainfall excess to the catchment outlet, we can write the DRCs as a vector $\underline{\tau} = (\tau_w, f, K_2, \tau^{(q)}, \tau^{(s)}, \upsilon^{(q)})$. (Note that once K_2 is set, $\upsilon^{(s)} = 1 - \upsilon^{(q)}$, and both $\lambda^{(q)}$ and $\lambda^{(s)}$ are known if $\upsilon^{(q)}$ is known.) The vector $\underline{\tau}$ therefore defines a *minimal* set of streamflow response properties of the catchment. Jakeman and Hornberger (1993) argue that this minimal set may also be *sufficient*, i.e. it will often be difficult to identify (with any reasonable statistical significance) more parameters than in this set, solely from rainfall, temperature and streamflow data. For the range of climatic conditions represented in modern records these properties may be *characteristics* in the sense that they are essentially invariant with respect to model calibration period (Jakeman *et al.*, 1992, 1993b).

Only in the case of error-free input and output data, and for a prototype system of known structure, will the DRCs of the prototype be estimated very precisely by the SRIV technique, as discussed in 2.2.1. In practice, the input and output data will contain errors, leading to statistical uncertainty in the DRCs. Furthermore, the prescribed model structure will always be a simplification of the prototype, both in its linear and non-linear modules, leading to an additional level of uncertainty in the DRCs. In this context it should be recalled that the methodology identifies *dominant* flow components, not processes.

The SRIV method yields information on the probability distributions for parameters b_0, b_1, a_1, and a_2, whence the $\hat{\sigma}$ in eqn (25). By repeatedly taking random samples from these distributions, calculating $\alpha^{(q)}$, $\alpha^{(s)}$, $\beta^{(q)}$ and $\beta^{(s)}$, and substituting in eqns (26) to (28) (with relevant superscripts), indicative statistical uncertainties associated with the DRCs can be derived (Jakeman *et al.*, 1993a). Alternatively, DRC uncertainties can be assessed by calibrating the model on different periods and observing the variations in DRC values (Jakeman *et al.*, 1993b).

3 Applications

Consider now some of the applications of a UH for total streamflow, and of separate UHs for quick and slow flow, where these can be identified from, and used in simulation mode with, time series rather than event data.

3.1 Flood and low-flow hydrology

As outlined in the Introduction, conventional UH analysis has been restricted largely to a rather loosely defined *direct flow* component and for runoff *events*. Traditionally, direct runoff UHs play a vital role in studies of floods, including regionalisation for design flood estimation at ungauged (flow) sites using statistical relationships between UH parameters and physical catchment descriptors (PCDs), such as basin area and stream channel slope, (e.g. NERC, 1975). The type of UH used in any overall procedure for design flood estimation is just one component of that scheme. Any change to the type of UH used in the procedure might, therefore, require adjustment or changes to other elements of the procedure. It should not be expected that a UH for total streamflow (or for quick flow) can simply be substituted for the UH for direct runoff in existing procedures. Nevertheless, IHACRES-derived UHs are arguably superior to direct runoff UHs because the conceptual definition of its flow components is less ambiguous. It would seem an area worthy of investigation to establish whether or not IHACRES UHs could lead to improvement in regional flood design estimation procedures. A UH for total streamflow derived from a single runoff (flood) event for a small headwater catchment is presented in 3.1.1 to illustrate the potential of the IHACRES approach for flood event analysis.

At the other end of the hydrological spectrum, a popular way of characterising low flow regimes is 'recession curve analysis'. This involves selection of suitable periods of record covering low flows from which a 'master recession curve' is derived. Another method in widespread use, which has the advantages that (a) it uses the whole streamflow record and (b) its derivation can be automated fairly easily, is the Base Flow Index (BFI). Here, baseflow is defined as the envelope of 'turning points' selected from relatively low-flow points on the hydrograph such that the result is intuitively reasonable. BFI is simply the proportion of streamflow volume over time which comprises baseflow thus defined. When derived for tens or hundreds of catchments from a national database of daily mean streamflows, BFI can be a very useful statistic for regional studies of low flows (e.g. Gustard *et al.*, 1992). Clearly, the BFI for a given catchment can depend on the nature of the hydrological record used for its derivation (e.g. length, wet/dry periods) but it is reckoned to be fairly stable, for UK catchments at least, provided the length of record is one year or more. In less humid catchments than those of the UK, one year of record will probably not be sufficient.

Unit Hydrographs have been employed mainly for flood events; they have not been used systematically for analysis of low flows. Identification of a UH for the slow flow component of streamflow is difficult and, until the application of the SRIV parameter identification technique in IHACRES, has not been a practical option. A new low-flow statistic, the Slow Flow Index (Littlewood and Jakeman, 1993) is a by-product of IHACRES. It is analagous to the BFI and is discussed further in 3.1.2.

3.1.1 A flood event UH for total streamflow

Figure 3 shows hourly rainfall, and observed and modelled streamflow, for a 3.9 km^2 headwater catchment in the Plynlimon experimental area (Wales) over the six days from 17 to 22 February, 1990. In this case, rainfall excess was derived solely from rainfall by eqns (18) and (19), i.e. there was no adjustment, using eqn (16), for seasonal variations in evaporation rate over this fairly short period (144 hours). The goodness of model-fit exhibited for the dominantly single event in Figure 3 ($D=0.98$) indicates strongly, however, that there may be considerable scope for UH analysis of floods by

IHACRES. Note how, in addition to the reasonable match for the timing and magnitude of the main flood peak in Figure 3, there are also good matches for the subsidiary events on the recession limb of the main event.

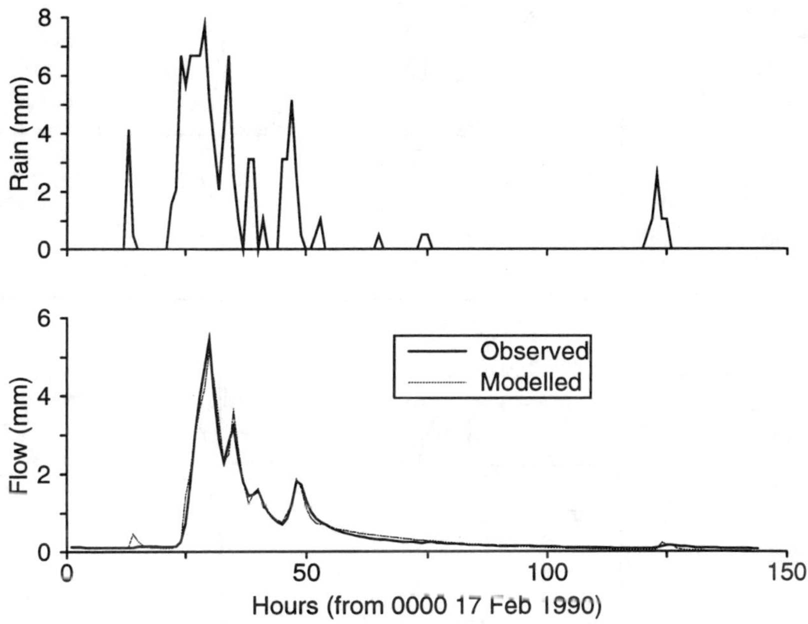

Figure 3: A runoff event for the Gwy 17-22 February, 1990.

3.1.2 Low flows and the Slow Flow Index (SFI)

A UH for total streamflow, and separate UHs for quick and slow flow components, can be derived as outlined in 2.1.1. A streamflow hydrograph can be estimated by convoluting a time series of rainfall excess with a UH for total streamflow, e.g. using the (second-order) transfer function given by eqn (10), or its recursive form given by eqn (9). Similarly, hydrographs for separate quick and slow flow components can be estimated using the first or second component (first-order) transfer function respectively in eqn (15). This modelling and hydrograph separation procedure is summarised in Figures 4 and 5 which show results for the Nant y Gronwen, a 0.7 km^2 headwater catchment in Wales (the first catchment to which IHACRES was applied).

Figure 6 shows modelled daily total streamflow and its slow flow component over about three years for the Teifi at Glan Teifi, an 894 km^2 catchment in Wales. The Slow Flow Index (SFI) can be compared with the BFI. However, rather than calculate the SFI from the ordinates of the total and slow flow hydrographs in Figure 6 (or from similar hydrographs for some other period of record), it can be derived simply as $\upsilon^{(s)}$ (i.e. eqn (27) with s superscripts). Since SFI $= \upsilon^{(s)}$ it is evident that it is a hydrological response characteristic of the catchment ($\upsilon^{(s)}$ is calculated from UH parameters), whereas BFI is simply a streamflow statistic.

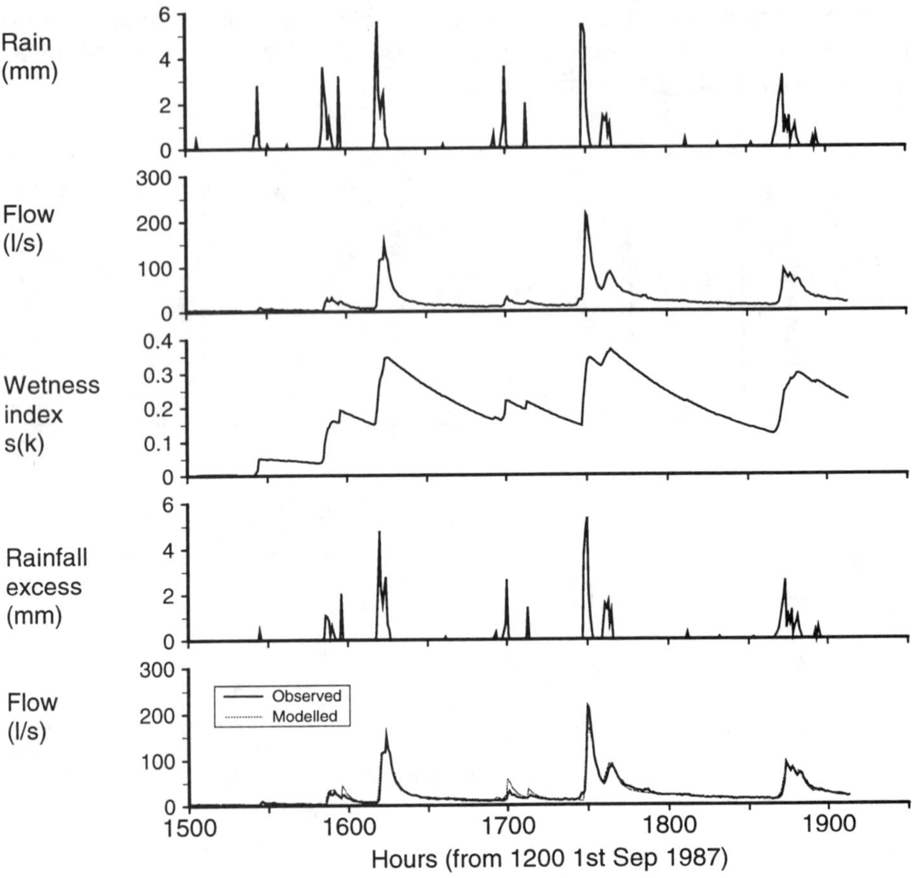

Figure 4: A sequence of events, Nant y Gronwen, September 1987; model calibration.

A preliminary comparison of SFI and BFI for the Teifi and one other UK catchment, the 601 km² Exe at Thorverton in south west England, was undertaken by Littlewood and Jakeman (1993). BFIs for the specified Teifi and Exe catchments are 0.53 and 0.51 respectively, where BFI represents ' ... the proportion of the river's runoff that derives from stored sources ' (Gustard *et al.*, 1992). Employing the loss module given by equations (16) to (19) Littlewood and Jakeman (1993) estimate the SFI for both catchments to be 0.33, and discuss the question 'Which statistic/characteristic (BFI or SFI) is more representative of the flow from deep storage?' They argue that the BFIs for these catchments are overestimates of the proportion of streamflow from deep storages because BFI is derived from streamflow data only. Streamflow in both catchments exhibits strong seasonality since, for each one, about 60% of the annual rainfall usually occurs during the winter months when evapotranspiration losses are

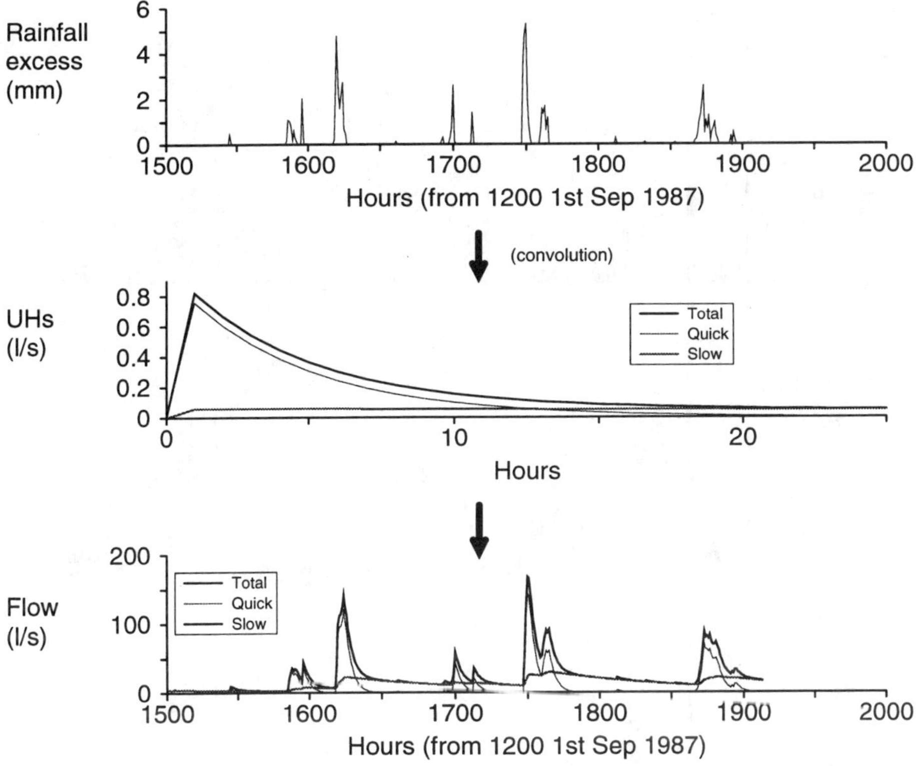

Figure 5: A sequence of events, Nant y Gronwen, September 1987; hydrograph separation.

relatively small. Winter rain falls on a relatively wet catchment, producing higher runoff yields, so, on average, winter flows are much higher than summer flows. The BFIs for the Teifi and the Exe are, therefore, influenced (towards a high value) by the persistence of hydrologically effective rainfall in winter, while SFI is not.

Provided that model-fits are good enough to simulate streamflow recessions well (as in Figure 6), the SFI appears to have much to offer as a catchment *characteristic*, whereas the BFI is simply a flow *statistic*. The comparison of the SFI and BFI outlined above is limited to just two catchments and is, therefore, very restricted. It is possible, for example, that the choice of rainfall excess (loss) module influences SFI. Clearly, a more thorough investigation is required to establish the relative merits of the two Indexes.

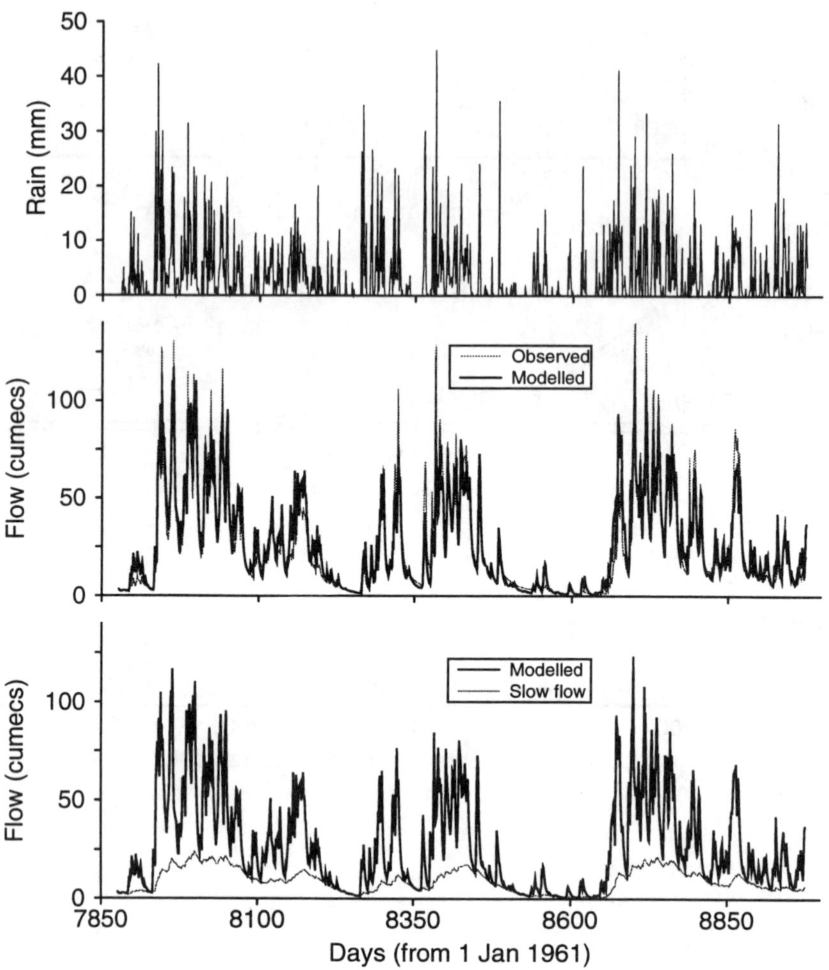

Figure 6: Teifi at Glan Teifi: Model-fit and hydrograph separation 26 July 1982 - 30 July 1985.

3.2 Characterisation of the catchment-scale response of gauged basins

There is increasing concern that natural environmental systems have, or might, become affected adversely by anthropogenic influences. The reliability of the engineered water supply resource in many developed parts of the world depends on the basic nature of river flow regimes; anthropogenic changes superimposed on natural variability may induce extra costs if the reliability and quality of water supplies are to be sustained. We need, therefore, to monitor and predict the effects on streamflow regimes of factors such as land-use changes, climatic variability and possible climate changes (principally rainfall and temperature).

The time series analysis methodology outlined in Section 2 can characterise dynamic hydrological behaviour in terms of a rainfall excess transformation and, for many catchments, simple parametric UHs for quick and slow flow components acting in parallel. The characterisation applies over time only if the physical features of the catchment which control the rate at which water passes through the catchment remain the same. In principle, any change in physical factors such as drainage density or dominant vegetation type (e.g. afforestation, deforestation) will cause a change in the streamflow regime and, therefore, a change in the UHs. It may, however, be difficult to detect the effect of spatially restricted physical changes within the catchment in terms of UHs derived from rainfall and streamflow data for the whole catchment. For catchments where there have not been significant changes in physical features, recorded climate forcing inputs to a model (rainfall and temperature time series) can be perturbed by modest amounts to investigate the sensitivity of streamflow to natural or possible anthropogenic systematic changes in those climate forcing variables.

The next section discusses an application of UHs for assessing the impact on mean streamflow of a scenario climate change shift in air temperature. Section 3.2.2 discusses application of the UH approach for assessing the impacts of land-use change. Only cases where streamflow data are available are dealt with here; transfer of information from (flow) gauged to ungauged catchments is discussed in 3.4.

3.2.1 Impacts of climate change scenarios on streamflow regimes

The sensitivity of mean streamflow in two catchments to increases in temperature was assessed by Jakeman *et al.* (1993b). One catchment was the Teifi at Glan Teifi referred to previously, for which daily rainfall and streamflow data were available from 1961 to 1989. Daily temperatures in any month were all set to the mean observed for that month. For the second catchment, the 767 km^2 French Broad River at Blantyre, North Carolina, daily rainfall, streamflow *and* temperature data, from 1953 to 1988, were available. Ten sub-periods, each of approximately three years duration, were selected for the French Broad River. Nine sub-periods, each of approximately two years duration, were selected for the Teifi. In each case the selected sub-periods covered the full period of available record reasonably well.

Each model derived from a particular sub-period for a catchment was then applied in simulation mode to all other sub-periods for that catchment, using observed rainfall but with all of the input temperatures increased by either 1°C, 2°C, 3°C or 4°C. (A rainfall excess model similar to that described by eqns (20) to (23) was used.) The difference between mean modelled streamflows using the unperturbed and perturbed temperature data was adopted as a measure of the change in runoff which would accompany a given rise in temperature. The spread of values obtained from applying each of the nine (or ten) models in this way gave an indication of the uncertainty in sensitivity of mean streamflow to an assumed temperature increase, *based on the information in the time series*. The results for a 1°C increase are shown in Figure 7. The indication is that a prescribed 1°C increase in temperature would cause a decrease in the mean flow of the Teifi of between about 2 and 6 per cent, and between about 4 and 7 per cent in the mean flow of the French Broad River. These levels of uncertainty are not surprising given the size of the catchments being modelled and problems in estimating areal rainfall. Partly as a result of successive reviews of the United Kingdom raingauge network, the number of raingauge records available to calculate daily areal rainfall for the Teifi catchment varied from about 12 in 1961, to 20 in 1973, to 12 in 1989. For the French Broad River only one raingauge record was used.

Similar analyses could be undertaken perturbing only the rainfall data, or both rainfall and temperature, according to agreed climate change scenarios.

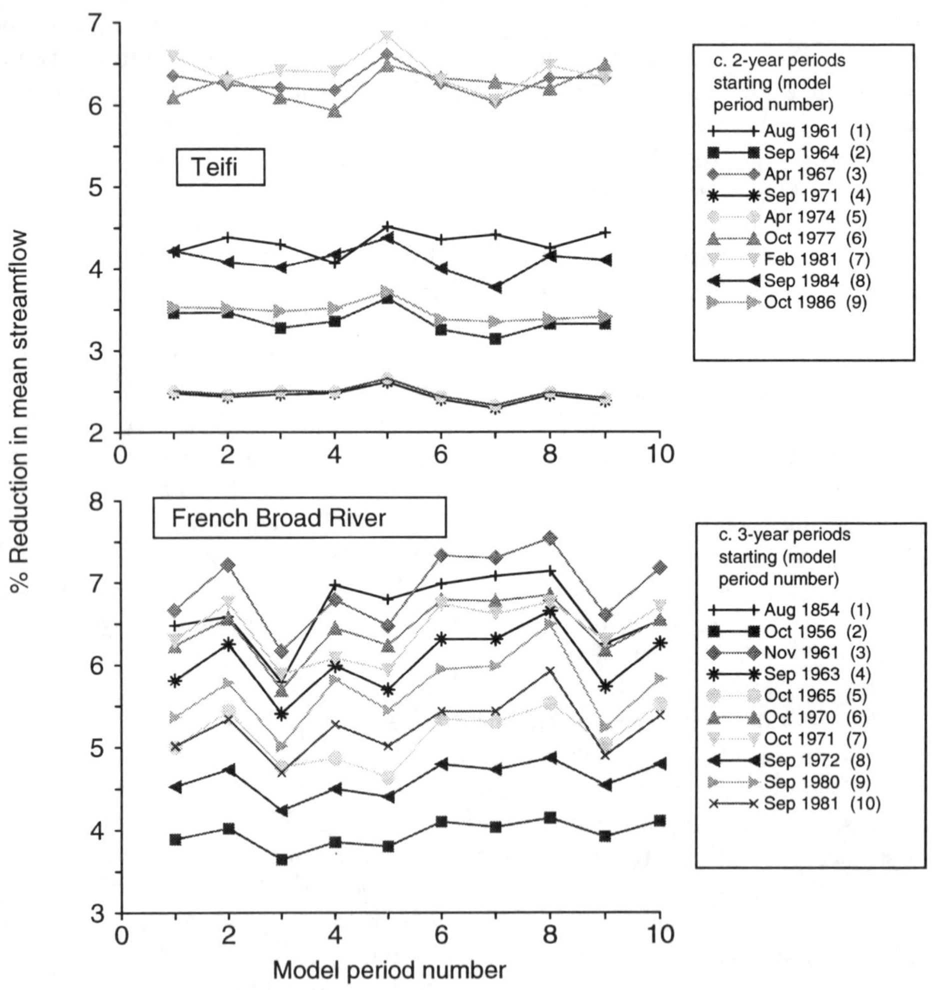

Figure 7: Range of per cent reduction in streamflow corresponding to a 1°C rise in temperature; Teifi at Glan Teifi and French Broad River.

3.2.2 Impacts of land-use change on streamflow regimes

The following is an outline of an attempt to detect changes in UHs for the quick and slow flow components of streamflow from a pair of adjacent catchments in Scotland during land-use changes associated with forestry (ditching, planting and clear-felling) (Jakeman *et al.*, 1993a). Starting in 1986, about 30 per cent of the moorland Monachyle catchment (7.7 km^2) was afforested; ditching in association with the planting affected only about 6 per cent of the catchment. Also starting in 1986, the partially afforested Kirkton catchment (39% of its 6.8 km^2) underwent clear-felling, so that the area with trees fell to about 16% of the catchment by summer 1990.

For each catchment, daily rainfall and streamflow data were employed to derive models for successive water-years (October to September) from 1984/1985 to 1988/1989; corresponding DRCs and their uncertainties were calculated as described in 2.3. Simple visual comparison of hydrographs for the two catchments (Figure 8) indicates that the Monachyle has a flashier response than the Kirkton, and this is quantified by the mean $\tau^{(q)}$ for the Monachyle over all years (1984/1985 to 1988/1989) of about 0.9 days compared with about 1.4 days for the Kirkton (Figure 9). In contrast, the mean values of $\tau^{(s)}$ for the two catchments, also shown in Figure 9, are not significantly different from each other at the 90% confidence level. The wide confidence intervals for $\tau^{(s)}$ are due to

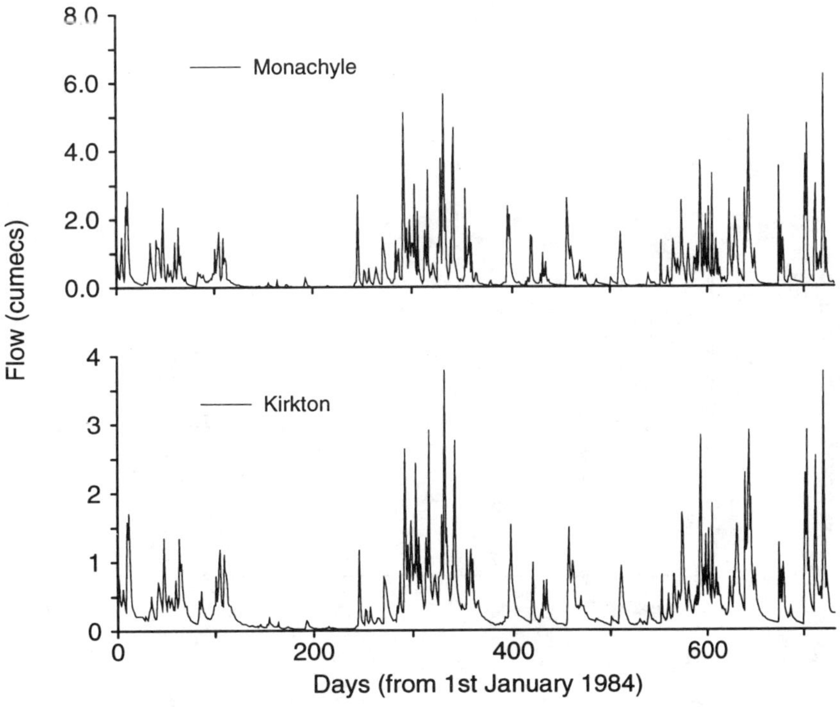

Figure 8: Comparison of hydrograph for the Monachyle and Kirkton, 1984-1985.

inherent difficulties in modelling the slow flow component for these catchments and the simple loss module used (no account was taken of different antecedent conditions or snowmelt). For the Monachyle, the technique was unable to identify separate quick and slow flow components from the data for water-year 1984/1985, probably because the spring and summer low flows were particularly small that year (Jakeman *et al.*, 1993a).

Figure 9 exhibits no trend in the individual mean water-year values for $\tau^{(q)}$ and $\tau^{(s)}$ for either catchment, as was the case for the other DRCs ($\upsilon^{(q)}$, $\upsilon^{(s)}$, $\lambda^{(q)}$ and $\lambda^{(s)}$). The effects of clear-felling in the Kirkton, and of ditching and planting in the Monachyle, were not detectable from the daily precipitation and streamflow data at the catchment outlets. The location and areal extent of a land-use change within a catchment are clearly important factors when trying to detect hydrological impacts at catchment-scale.

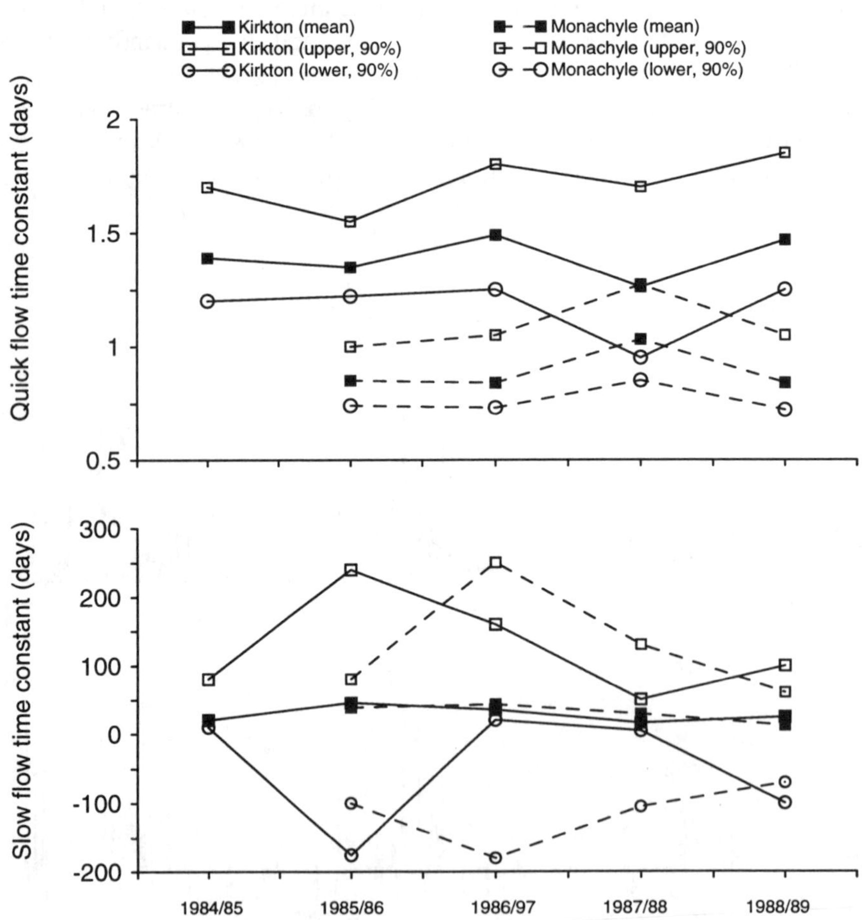

Figure 9: Quick and slow flow time constants ($\tau^{(q)}$ and $\tau^{(s)}$) and their uncertainties, for water-years 1984/85—1988/89; Kirkton and Monachyle.

3.3 Hydrograph separation for water quality studies

The concentration of many physical and chemical constituents of streamflow (determinands) varies with streamflow in a more-or-less systematic way, indicating the mixing of waters from different sources (e.g. event rain water and pre-event catchment water). In this sense 'conservative' determinands which do not interact chemically or biologically within the catchment might be expected to exhibit the strongest mixing characteristics. Although the provenances of quick and slow components of streamflow estimated by IHACRES cannot be ascribed unequivocally, it may be assumed as a working hypothesis that they are the result of processes on (or near) the surface and in deeper layers of the catchment respectively. Furthermore, for many catchments it may be assumed that the water from and near the surface has a distinctly different chemical signature to that from deeper layers within the catchment. With this very simple conceptual model it might be expected that at any time step k there would be a good relationship between the proportion $P_k^{(q)}$ of streamflow which comprises quick flow and the concentration of a conservative determinand c_k. Clearly, however, this approach may not work for catchments, or individual events, wherein the *dominant* storm runoff generation mechanism involves a piston effect such that new rain water pushes out old catchment water (registering by the timing of its appearance as quick flow).

The following section discusses a proposal that IHACRES hydrograph separation might lead to improvements in river mass load estimation in situations when a time series of streamflow is available, matched only with intermittent concentration data.

3.3.1 River mass load estimation

River mass load estimates are required for a range of purposes: to assist with setting and monitoring international agreements on the fluvial discharge of pollutants to lakes, coastal waters, seas and oceans; for river basin management; and for investigations at catchment scale of hydrochemical processes. Many of the issues and practical difficulties concerning river mass load estimation in the United Kingdom have been discussed recently by Littlewood (1992). High-frequency streamflow and concentration time series (with a common sampling interval) should enable good estimates of river loads to be made simply by summing the products of the two variables and applying a factor corresponding to the sampling interval. However, due to the expense of laboratory analyses, concentration data are seldom available as long time series at the same observation frequency as streamflow data. (The use of ion-specific electrodes is improving the situation in this respect for some determinands.) In practice, therefore, recourse is made to a variety of algorithms for combining concentration data with streamflow data according to what data are available.

One popular method is to derive a concentration—discharge relationship from the available observations of streamflow and concentration (y_k, c_k) by

$$c_k = C(y_k + D)^B \tag{31}$$

Eqn (31) can then be used to estimate concentrations from streamflow for all time steps k when observations of concentration are not available. The river mass load is then estimated by summing products and adjusting for the time interval. Values for B and C in eqn (31) can be derived by regressing $\ln(c_k)$ on $\ln(y_k + D)$; D is an optimal 'correction factor' such that the correlation coefficient is a maximum.

In many cases the scatter of points about concentration—discharge curves is largely due, at least in part, to hysteresis (with which the linear regression model is unable to cope). Furthermore, because the regression analysis is undertaken using logarithmically transformed variables, a bias is introduced in estimates made by eqn (31), as discussed by Ferguson (1986). The bias can be large when the scatter is large, a common feature of concentration—discharge relationships. The outcome of using an equation of the form eqn (31) is, therefore, that there are often still large uncertainties in river load estimates.

Use of the transfer function model has been proposed with a view towards possible improvement in river mass load estimation practice (Littlewood, 1992). Inspection of the transfer function model reveals that, unlike the linear regression model, it can cope with hysteretic behaviour. It may be possible, therefore, to account for at least some of the hysteresis often exhibited in scatter-plots of concentration and streamflow by using a transfer function.

Another way of looking at the same problem is to assume that any hysteresis in concentration—streamflow behaviour is due largely to mixing, and to note that $P_k^{(q)}$ (derived using the hydrograph separation facility of IHACRES) plotted against streamflow y_k exhibits similar hysteresis. Making the further assumption that the quick and slow flow components from which $P_k^{(q)}$ is calculated have distinctive concentrations of the determinand in question, it might be expected that there will be a better relationship between $P_k^{(q)}$ and c_k than between y_k and c_k, leading to better estimates of concentration from streamflow at times between samples. While the physical processes which govern streamflow responses to precipitation are much more complex than the assumptions made above allow, there does seem to be some merit in the idea; trials are being undertaken.

3.4 Information transfer to ungauged catchments

Estimates of streamflow behaviour in response to large rainfalls, or during periods of little or no rainfall, are often required at ungauged points on river systems. Several systematic studies have been undertaken (e.g. NERC, 1975; Gustard *et al.*, 1992) to establish statistical relationships between direct runoff UH parameters (or other measures of characteristic streamflow behaviour, e.g. BFI) and physical catchment descriptors (PCDs), e.g. basin area, slope, soil type, etc.

3.4.1 Regionalisation

Once established, the statistical relationships can be employed to describe how particular streamflow characteristics (e.g. for floods or low flows) vary spatially within or between regions. The DRCs from IHACRES $\underline{\tau}$ (τ_w, f, K_2, $\tau^{(q)}$, $\tau^{(s)}$, $\lambda^{(q)}$) provide linkages between meteorological and dynamic hydrological behaviour. Statistical relationships between these DRCs and PCDs would complete the linkage, enabling regionalisation (Jakeman *et al.*, 1992). Relationships between direct runoff UH parameters and PCDs, or between BFI and PCDs, are typically fairly weak, due partly to the rather poor definitions of direct runoff and baseflow involved. With IHACRES, however, problems of flow component definition have been reduced and, importantly, a measure of the covariation between parameters in the linear part of the model is available. We might expect, therefore, to obtain better defined statistical relationships between DRCs and PCDs; investigations are underway (e.g. Sefton *et al.*, 1993).

4 Concluding remarks

A time series analysis methodology has been described for modelling catchment-scale rainfall—runoff dynamic behaviour. With short time-interval data (e.g. hourly for small catchments) either an individual runoff event or a sequence of many events can be modelled in a single operation. For larger catchments, it might be more appropriate to use daily or even monthly data, and periods of several years can be modelled. Because IHACRES models total streamflow, avoiding arbitrary prior hydrograph separation, the resultant UH is arguably superior to a UH derived for direct flow only. In many (if not most) cases, however, the identified structure and parameterisation of the UH for total streamflow facilitates its resolution into separate UHs for quick and slow response components of streamflow. The modelled streamflow hydrograph can then be resolved into its quick and slow flow components by convoluting each of the component UHs with rainfall excess.

Several applications have been discussed of UHs for total streamflow, and separate UHs for dominant quick and slow response components of streamflow: characterisation of floods and low-flows; assessing the sensitivity of change in mean streamflow to scenario changes in rainfall or temperature (or both); assessing the hydrological impact of land-use changes at catchment-scale; estimation of a continually variable mixing ratio to assist with water quality investigations and river mass load estimation; and information transfer from gauged to ungauged catchments using statistical relationships between DRCs and PCDs.

There are other possible applications. The IHACRES computational framework provides a basis with which to assess *at catchment-scale*, and possibly further develop, rainfall excess or evaporation models founded on more detailed physical principles and driven by detailed hydrometeorological measurements. Rainfall excess estimated by evaporation models which use hydrometeorological variables and details of vegetation (type and spatial distribution within a catchment) can be input directly to the linear module of IHACRES (Figure 1). Similarly, where streamflow is usually caused by rainfall but occasionally by snowmelt, the IHACRES computational framework might provide a basis on which to test and further develop appropriate snowmelt modules, e.g. degree-day type models with just one or two additional parameters.

The theme of this chapter has been that the UH model, no longer restricted to analysis of direct runoff on an event basis, has effectively been given a new lease of life. Examples and discussion of applications of IHACRES have been presented in support of this argument. This revitalisation of the structurally simple UH model comes at a time when there is increasing use being made of remotely-sensed or digitised (from maps) environmental variables and geomorphic features in search of better rainfall—runoff models, sometimes within a geographical information system (GIS) framework. In the context of this chapter, examples of this type of work which are of most interest are those which adopt, or have strong similarities with, the geomorphic instantaneous unit hydrograph (GIUH) approach (e.g. Naden, 1992; Calver, 1993; Maidment, 1993). Given that it has been shown that total streamflow for a wide range of catchment types can be modelled well by the IHACRES UH approach, requiring minimal input data (areal rainfall, streamflow and temperature), it seems timely to ask 'What are the benefits of the more data-intensive approaches?' and 'Do the benefits balance the extra costs?' Indeed, the IHACRES approach may provide a bench-mark against which to assess the benefits of other, more costly (more data- or computationally-intensive), rainfall—runoff modelling approaches.

Clearly, the specific objectives of the modelling exercise in question are always of crucial importance. For example, if the objective is simply to provide estimates to fill small gaps in otherwise good streamflow time series then, in principle, any model with the required level of performance will suffice. Similarly, any model can be employed to hindcast a streamflow record for the period before flow measurement began (given suitable rainfall, etc., data). In neither of these cases, though, would it be necessary to use a model which attempts to attain verisimilitude in its representation of the physical processes by which rainfall becomes streamflow.

On the other hand, if the objective is scientific understanding then verisimilitude might be perceived to be important. Here, however, we have a problem. We can describe mathematically the behaviour of water flow over surfaces, in channels and through porous media using well-established physical laws which can be demonstrated in the laboratory. In practice, though, it is difficult to model streamflow from a spatially distributed input of rainfall simply by routing parcels of water through small-scale elements of the catchment for which the physical laws (equations) are assumed to be valid. As a matter of faith we know that physical laws hold true in the catchment just as they do in the laboratory. But for many reasons, verisimilitude in operational catchment rainfall—runoff models is not a practical option, largely because catchments are exceedingly complex heterogeneous physical systems. Our measurements and mathematical representations of processes provide only a crude description of the state and behaviour of environmental systems with respect to the continuum of reality. We might understand how water flows over surfaces, in channels and through porous media but a model which incorporates these and other hydrological processes, representing them at small scale and linking them to obtain a working model for the whole catchment, seems unattainable.

Jakeman and Hornberger (1993) opine that the inclusion of spatial data, e.g. to describe terrain, soils and vegetation, in distributed parameter/physically based models, does not improve our ability to identify more than about half a dozen rainfall—runoff model parameters. They argue that distributed models can be regarded as non-linear versions of the conceptual model given by eqn (6). Precipitation enters the surface component of distributed 'elements' in parallel, and the flow out of these 'storages' may pass to other storages of the same element or of neighbouring elements. The spatial data, presently, permit an attempt to specify the number and configuration of storages. In distributed models, precipitation and streamflow data are still needed to estimate the key parameters of each storage (two in each case if the storage is linear, more if non-linear). But from precipitation and streamflow data we can usually only estimate parameters of two storages, no more. Hence, the over-parameterisation of distributed models with more than two 'storages'. Indeed, in practice, there are many such storages.

We often resort, therefore, to appropriate levels of prescriptive process representation in lumped rainfall—runoff models. The catchment is represented as a series of tanks or reservoirs connected in series or parallel (or both) with simple flow 'splitting' or 'summation' boxes at appropriate junctions. Each tank or reservoir may have a capacity and a leakage rate dependent on its content at a given computational time step. Model parameters which relate to the capacities and leakage rates associated with each tank, the proportions at flow 'splitters,' and any time delays introduced as part of the model structure, are either selected on the basis of experience or calibrated from observations of rainfall input and streamflow output. Such model structures are extremely attractive in the sense that they are easily understood and can give good agreement between observed and modelled streamflow. A problem with such models can be the number of parameters required and their statistical significance. For the practitioner whose interests are restricted to curve-fitting for specific catchments such models may, however, be entirely adequate.

If the objectives of the modelling exercise include efficient *characterisation* of rainfall—runoff behaviour, e.g. to facilitate comparison between catchments (see 3.2.2) or regionalisation (see 3.4.1), then the UH model is a very attractive candidate. Although the UH model is a 'black box' in the sense that it makes no attempt to include any description of physical streamflow generation processes, the IHACRES methodology can often identify separate UHs for dominant quick and slow components of streamflow (the term 'grey box' might be more appropriate, therefore, for this methodology).

Intuitive understanding of a hydrograph being the variable sum of baseflow (most evident in the tails of recessions) and runoff generated by recent rainfall (most evident in hydrograph peaks) is certainly not diminished by IHACRES hydrograph separation analysis (Littlewood and Jakeman, 1991). The larger the hydrograph peak, the more of the streamflow at that instant we might expect to comprise recent rain water, i.e. 'new' water. Some interpretations of tracer measurements in catchments (e.g. Sklash and Farvolden, 1979; Pearce *et al.*, 1986) cast doubt on this relatively straightforward conceptualisation of the dominant mechanism by which rainfall causes variations in streamflow. They indicate that even quite large peak flows in some humid region headwater catchments can comprise mostly 'old' water displaced by 'new' rain water (i.e. a piston effect operates). Unfortunately, it is not possible to ascribe unequivocally the provenances of the quick and slow streamflow components identified by IHACRES solely on the basis of information in rainfall, streamflow and temperature data. It is proposed, however, that the IHACRES rainfall—runoff modelling approach, employing additional information in other time series (e.g. tracer concentrations, groundwater levels and soil moisture storage) offers a way forward.

References

Calver, A. (1993). The time—area runoff formulation revisited. *Proc. Institution of Civil Engineers Water Maritime & Energy*, **101**, 31-36.

Chen, T.H., Hornberger, G.M., Jakeman, A.J. and Swank, W.T. (1993). The performance of different loss models in the simulation of streamflow. *Proc. International Congress on Modelling and Simulation*, M.J. McAleer and A.J. Jakeman (eds.), University of Western Australia, Dec 6-10, 1993.

Ferguson, R.I. (1986). River loads underestimated by rating curves. Water Resources Research, **22**, 74-76.

Gustard, A., Bullock, A. and Dixon, J.M. (1992). Low flow estimation in the United Kingdom. *Institute of Hydrology Report 108.*

Dietrich, C.R., Jakeman, A.J. and Thomas, G.A. (1989). Solute transport in a stream-aquifer system. 1. Derivation of a dynamic model. *Water Resources Research*, **25**(10), 2171-2176.

Jakeman, A.J., Dietrich, C.R. and Thomas, G.A. (1989). Solute transport in a stream-aquifer system. 2. Application of model identification to the River Murray. *Water Resources Research*, **25**(10), 2177-2185.

Jakeman, A.J., Littlewood, I.G. and Whitehead, P.G. (1990). Computation of the instantaneous unit hydrograph and identifiable component flows with application to two small upland catchments. *Journal of Hydrology*, **117**, 275-300.

Jakeman, A.J., Littlewood, I.G. and Symons, H.D. (1991a). Features and applications of IHACRES: A PC program for Identification of unit Hydrographs and Component flows from Rainfall, Evapotranspiration and Streamflow data. *Proc. 13th World IMACS Congress on Computation and Applied Mathematics*, R. Vichnevetsky and J.J.H. Miller (eds.), July 22-26 1991, Trinity College, Dublin.

Jakeman, A.J., Littlewood, I.G. and Whitehead, P.G. (1991b). Catchment-scale rainfall - runoff event modelling and dynamic hydrograph separation using time series analysis techniques. In: D.G. Farmer and M.J. Rycroft (eds.) *Computer Modelling in the Environmental Sciences*. IMA Conference Series, Oxford University Press, 89-101.

Jakeman, A.J., Hornberger, G.M., Littlewood, I.G., Whitehead, P.G., Harvey, J.W. and Bencala, K.E. (1992). A systematic approach to modelling the dynamic linkage of climate, physical catchment descriptors and hydrologic response components. *Mathematics and Computers in Simulation*, **33**, 359-366.

Jakeman, A.J., Littlewood, I.G. and Whitehead, P.G. (1993a). An assessment of the dynamic response characteristics of streamflow in the Balquhidder catchments. *Journal of Hydrology*, **145**, 337-355.

Jakeman, A.J., Chen, T.H., Post, D.A., Hornberger, G.M., Littlewood, I.G. and Whitehead, P.G. (1993b). Assessing uncertainties in hydrological response to climate at large scale. In: W. B. Wilkinson (ed.) *Macroscale Modelling of the Hydrosphere*, IAHS Publ. No. 214, 37-47.

Jakeman, A.J. and Hornberger, G.M. (1993). How much complexity is needed in a rainfall-runoff model? *Water Resources Research*, **29**(8), 2637-2649.

Littlewood, I.G. (1992). Estimating contaminant loads in rivers: a review. *Institute of Hydrology Report No. 117*, 81pp.

Littlewood, I.G. and Jakeman, A.J. (1991). Hydrograph separation into dominant quick and slow flow components. *Proc. Third National Hydrology Symposium*, University of Southampton, 16-18 September 1991, 3.9-3.16.

Littlewood, I.G. and Jakeman, A.J. (1993). Characterisation of quick and slow streamflow components by unit hydrographs for single- and multi-basin studies. Proc. Fourth Conference of the European Network of Experimental and Representative Basins, University of Oxford 29 September - 2 October 1992. In: M. Robinson (ed.) *Methods of hydrological basin comparison*. Institute of Hydrology Report 120, 99-111.

Littlewood, I.G. and Post, D.A. (1993). Preliminary assessment of rainfall excess (loss) models for time series analysis of rainfall—runoff dynamics. *Proc. International Congress on Modelling and Simulation*, M.J. McAleer and A.J. Jakeman (eds.), University of Western Australia, Dec 6-10, 1993.

Maidment, D.R. (1993). Developing a spatially distributed unit hydrograph by using GIS. In: K. Kovar and H.P. Nachtnebel (eds.), *Application of Geographic Information Systems in Hydrology and Water Resources Management*, IAHS Publication No. 211, 181-192.

Naden, P.S. (1992). Spatial variability in flood estimation for large catchments: the exploitation of channel network structure. *Hydrological Sciences Journal*, **37**, 53-71.

NERC, (1975). *Floods Study Report*. 5 volumes. Natural Environment Research Council, UK.

Pearce, A.J., Stewart, M.K. and Sklash, M.G. (1986). Storm runoff generation in humid headwater catchments. I. Where does the water come from? *Water Resources Research*, **22**(8), 1263-1272.

Sefton, C.E.M., Whitehead, P.G., Eatherall, A., Littlewood, I.G. and Jakeman, A.J. (1993). Dynamic response characteristics of the Plynlimon catchments and preliminary analysis of relationships to physical descriptors. *Proc. International Congress on Modelling and Simulation*, M.J. McAleer and A.J. Jakeman (eds.), University of Western Australia, Dec 6-10, 1993.

Sklash, M.G. and Farvolden, R.N. (1979). The role of groundwater in storm runoff. *Journal of Hydrology*, **43**, 45-65.

Wheater, H.S., Jakeman, A.J. and Beven, K.J. (1993). Progress and directions in rainfall - runoff modelling. In: A.J. Jakeman, M.B. Beck and M.J. McAleer (eds), *Modelling Change in Environmental Systems*, John Wiley & Sons Ltd, 99-130.

Whitehead, P.G. (1979). Applications of recursive estimation techniques to time variable hydrological systems. *Journal of Hydrology,* **40**, 1-16.

Whitehead, P.G., Young, P.C. and Hornberger, G.H. (1979). A systems model of stream flow and water quality in the Bedford-Ouse River, I: Streamflow modelling. *Water Research*, **13**, 1155-1169.

Whitehead, P.G., Neal, C., Seden-Periton, S., Christopherson, N. and Langan, S. (1986). A time-series approach to modelling stream acidity. *Journal of Hydrology*, **85**, 281-303.

Young, P.C. (1993). Parallel processes in hydrology and water quality: objective inference from hydrological data. In: R.A. Falconer (ed), *Water quality modelling*, Ashgate, 10-52.

Young, P.C. (1992). Parallel processes in hydrology and water quality: a unified time-series approach. *Journal of Institution of Water and Environmental Management*, **6**, 598-612.

Young, P.C. and Lees, M. (1993). The active mixing volume: a new concept in modelling environmental systems. In: V. Barnett and K.F. Turkman (eds.), *Statistics for the Environment*, John Wiley & Sons Ltd., 3-43.

Young, P.C. (1984). *Recursive estimation and time series analysis*. Springer—Verlag.

Chapter 7

Finite element modeling of the transport of reactive contaminants in variably saturated soils with LEA and non-LEA sorption

G. Gambolati,[a] G. Pini,[a] M. Putti,[a] C. Paniconi[b]

[a] *Department of Mathematical Models, University of Padua, Via Belzoni 7, 35131 Padua, Italy*

[b] *CRS4, Cagliari, Italy*

Abstract

There is a growing need to assess the environmental impact of pollutants in soils and subsurface waters. The origins of this pollution range from everyday activities (disposal of urban sewage and industrial wastes; use of pesticides and fertilizers in agriculture) to accidental releases (spills; leaks; fallout). Once these hazardous substances enter the subsurface, they undergo complex physical and chemical processes that transport and transform the contaminants. Understanding these processes, which include dispersion, diffusion, advection, chemical reaction, adsorption, and decay, is further complicated by the fact that soils and aquifers are highly heterogeneous, and that the processes can occur at different time scales and with intricate feedbacks. Mathematical models describing groundwater flow and reactive transport can be effectively used, in tandem with experimental studies, to study the behavior and effects of contaminants in subsurface waters. The mathematical models are based on the partial differential equations of fluid mass and momemtum balance and solute mass balance, and can be solved numerically by finite element methods. The flow model is developed for the case of variably saturated porous media, applicable to both the unsaturated (soil) zone and the saturated (groundwater) zone. The equation is nonlinear, and is solved iteratively using either Picard or Newton linearization. The transport model is linear and considers advection, dispersion, decay, and sorption. Sorption under equilibrium and nonequilibrium conditions is described, in the latter case using a "dual porosity" concept that subdivides the porous medium into five distinct interacting regions. The coupled model for nonequilibrium sorption is solved using an integro-differential approach. The finite element discretization of the flow and transport equations yields large sparse systems of equations. These systems, which are symmetric for the Picard-linearized flow case and nonsymmetric for the transport and Newton-linearized flow cases, are solved using efficient preconditioned conjugate gradient-like methods.

Key words

Transport, reactive solutes, unsaturated flow, groundwater pollution, finite elements, fly ashes, nonequilibrium sorption.

Introduction

The soil zone acts as a storehouse of nutrients and water for plants and as a buffer and filter between the atmosphere above it and the groundwater zone below it. The pore space in the soil zone is variably saturated, in that it is filled with varying amounts of air and water, while the groundwater zone is fully saturated with water. Pollutants introduced into the unsaturated zone can contaminate not only the soil (thereby damaging vegetation or making crops unfit for consumption), but also the atmosphere (through volatilization), aquifers (through percolation, leaching, and recharge), and streams (through surface and subsurface runoff and seepage). Transported by groundwater, these hazardous substances may also contaminate withdrawal sites at pumping wells, and they may reappear at the surface, emerging from springs and seepage faces. The degradation of soil and water quality which can result from contamination of subsurface water resources can pose a serious risk to public health.

Contamination can occur at point sources (e.g. isolated spills; leaking storage tanks; waste tailings from mining operations; sanitary landfills; septic tanks; radioactive waste disposal) or at nonpoint sources (e.g. herbicides, pesticides, and fertilizers used in agriculture; urban runoff; sewage and waste water; atmospheric deposition such as acid rain). The contaminants can be organics, trace metals, or radionuclides.

An example which illustrates some of the issues surrounding soil and groundwater contamination is that of fly ash wastes originating from thermal combustion processes in coal-fired power plants. Fly ash is composed chiefly of alumina, silica, and iron oxides in an amorphous core. However, trace metal components which are much more toxic are also present, and the extent of their enrichment on fly ash can be large, in some cases one hundred times their concentration in the original coal. These toxic trace metals include arsenic, cadmium, nickel, zinc, lead, manganese, chromium, and copper. Other components, such as boron, sodium, and sulfate have also been found. For a description of the most common chemical constituents of fly ash, their properties, and their potential threat to groundwater quality, see *Theis et al.* [*1978*], *Page et al.* [*1979*], *Simsiman et al.* [*1987*] and *Sharma et al.* [*1989*].

When hazardous wastes such as fly ash are discharged at a disposal site, they may have a significant impact on the quality of subsurface water resources, as described above. These wastes interact with the fluid and solid phases of the porous medium through a wide variety of processes, including chemical diffusion, mechanical dispersion, advection, chemical reaction, decay, and biodegradation. A full understanding of these processes is complicated by numerous factors. For instance, due to heterogeneity and anisotropy, dispersion occurs at very different spatial scales, ranging from a few meters (local scale) to several kilometers (basin or regional scale). Also, chemical reactions between the various contaminant species and between these species and the solid matrix are usually very complex, typically involving highly nonlinear exchange, precipitation/dissolution, and sorption processes, and may occur over time scales different from those characterizing groundwater flow and transport. In addition, there are complicated feedback mechanisms between groundwater flow, transport, and chemical reaction phenomena, and mathematical models of such mechanisms yield coupled and nonlinear equations which cannot be solved analytically. These models, based on the partial differential equations of fluid and solute continuity, describe the migration and fate of contaminants in soils and groundwater.

Depending on the physical and chemical processes of interest, therefore, the governing mathematical equations can exhibit nonlinearities, coupling, and discontinuities. For instance, the transport of ion-exchanging solutes is described by nonlinear equations, and in some cases the concentration profiles may show discontinuities or very sharp fronts. As another example, when the fluid density is affected by the concentration of the dissolved chemicals, the transport process has to be modeled as a system of nonlinearly coupled equations.

To serve as useful water resource management tools, these mathematical models must be able to simulate realistic scenarios (i.e. actual aquifers and catchments at local and regional scales), and to do so in practical (time and cost) terms. Intractable analytically, the models can be solved numerically by finite element and other methods. The advantage of using finite element techniques is that they provide high accuracy and they can be used to simulate irregular three-dimensional domains with complex boundary conditions, such as commonly found in real applications.

The systems of equations resulting from numerical discretization of these models can become very large and complex given the variety of contaminant sources, interaction mechanisms, and large temporal and spatial scales involved. In addition, the equations can be difficult to solve due to strong nonlinearities and coupling, and because of the high degrees of variability in the parameters and boundary conditions required in realistic simulations. This variability is due to heterogeneities inherent in soil and aquifer properties, atmospheric inputs, vegetation, and topography.

One of the most important processes affecting groundwater contaminant transport is sorption onto solid grains. Sorption is frequently simplified by assuming that chemical equilibrium is achieved instantaneously, i.e. with a mass exchange rate equal to infinity. In order for this "local equilibrium assumption" (LEA) to be valid in practical field problems or laboratory experiments, the rate of sorption must be fast relative to other processes (advection, mechanical dispersion) which affect contaminant concentration, so that equilibrium may be established between the sorbent and the solution phase. However, in porous media the time scales associated with attainment of equilibrium conditions may not always be short relative to the time scales associated with changes in pollutant concentration due to macroscopic transport processes. In this case the rate of mass transfer toward equilibrium will influence the space and time concentration of contaminant, yielding asymmetrical or nonsigmoid breakthrough profiles characterized by the so called "tailing" effect [*Giddings, 1963*]. Tailing has been observed under a variety of field conditions including unsaturated flow, flow through complex porous aggregates with dead-end pores, and flow at low seepage velocity [*van Genuchten and Wierenga, 1976*]. For a review of the physical, chemical, and biological processes accounting for nonequilibrium (non-LEA) sorption see *Brusseau and Rao* [*1989*] and *Weber et al.* [*1991*].

One approach to mathematically describe non-LEA sorption is the so-called "dual porosity model", wherein the fluid phase is divided into mobile and immobile regions with a diffusive mass transfer driven by the difference in concentration between the mobile and immobile regions [*Lapidus and Amundson, 1952*; *Coats and Smith, 1964*]. The model can be further enhanced by introducing contaminant decay and instantaneous mass exchange controlled by a linear adsorption isotherm in both the mobile and immobile regions [*van Genuchten and Wierenga, 1976*].

In this report we will present the fundamental equation of nonlinear groundwater flow in variably saturated soils, and develop the basic finite element method for its solution under steady and transient conditions. Two-dimensional porous systems are considered and triangular elements are used. The velocity field is numerically computed from the pressure or potential head solution and is then transferred as an input to the contaminant transport model. This latter model is expressed for both LEA and non-LEA conditions, and is integrated by the finite element method. A constant distribution coefficient is assumed for the instantaneous sorption process and a first order linear kinetics description is postulated for the mass transfer from the mobile water region to the immobile water region. The strong nonlinearity of the finite element flow equation in unsaturated regions is overcome by iterative Picard and Newton methods. The large sparse systems of discrete linear or linearized equations are solved by preconditioned conjugate gradient-like schemes. The linear non-LEA model is solved by an integro-differential approach involving a convolution integral. This approach results in a significant saving of computer storage and CPU time. Finally, the various finite element models are tested on complex sample problems taken from the groundwater literature, and their numerical performance is discussed.

Mathematical Models

Water Flow in Variably Saturated Porous Media

Flow in unsaturated and variably saturated porous media is governed by Richards' equation, which may be written as [*Philip, 1969*]

$$\frac{\partial}{\partial x_i}\left[k_{ij}k_{rw}(S_w)\left(\frac{\partial \psi}{\partial x_j}+\eta_j\right)\right] = \sigma(S_w)\frac{\partial \psi}{\partial t} - q \tag{1}$$

where the indices i and j denote summation over the two coordinate dimensions $(i, j = 1, 2)$ and

x_i is the ith Cartesian coordinate ($i = 1, 2;\ x_2 = z$)

k_{ij} is the saturated hydraulic conductivity tensor [L/T]

k_{rw} is the relative hydraulic conductivity [/]

$S_w = \theta/n$ is the water saturation [/]

θ is the volumetric water content [/]

n is the porosity of the medium [/]

ψ is the pressure head [L]

$\eta_1 = 0,\ \eta_2 = 1$

$\sigma = S_w S_s + n(dS_w/d\psi) \approx d(nS_w)/d\psi$ is the overall storage coefficient [1/L]

S_s is the specific elastic storage of the porous medium [1/L]

t is time [T]

q represents distributed source or sink terms (volumetric flow rate per unit volume) [T^{-1}].

The nonlinear storage and conductivity terms in equation (1) can be modeled using various constitutive or characteristic relations describing the soil hydraulic properties. The characteristic equations used by *Huyakorn et al.* [*1984*] express the water saturation in terms of effective saturation S_e, in the form $S_w(\psi) = (1 - S_{wr})S_e(\psi) + S_{wr}$, where S_{wr} is the residual water saturation. The effective saturation-pressure head relation is then written as

$$\begin{aligned} S_e(\psi) &= [1 + \epsilon^\eta(\psi_a - \psi)^\eta]^{-\gamma} && \psi < \psi_a \\ S_e(\psi) &= 1 && \psi \geq \psi_a \end{aligned}$$

while two possible expressions for the relative conductivity-pressure head relationship are

$$k_{rw}(\psi) = k_{rw}\left(S_e(\psi)\right) = S_e^\mu$$

and

$$k_{rw}(\psi) = 10^{G(S_e(\psi))}$$

where $G(S_e) \equiv aS_e^2 + (b - 2a)S_e + a - b$. In the above expressions ψ_a is the air entry pressure and ϵ, η, γ, μ, a, and b are constants.

The initial and boundary conditions associated with equation (1) can be expressed as

$$\psi(x_i, 0) = \psi_o(x_i) \tag{2a}$$

$$\psi(x_i, t) = \overline{\psi}(x_i, t) \quad \text{on } \Gamma_1 \tag{2b}$$

$$v_i n_i = -q_n(x_i, t) \quad \text{on } \Gamma_2 \tag{2c}$$

where ψ_o is the initial pressure head, $\overline{\psi}$ is the prescribed pressure head on segment Γ_1 of the boundary Γ, and n_i is the direction cosine of the outward normal to the boundary

Γ_2 where the prescribed flux is $-q_n$. We use the sign convention of q_n *positive* for an inward flux and *negative* for an outward flux. Boundary condition (2b) is said to be of the first, principal, or Dirichlet type while (2c) is of the second, natural, or Neumann type. v_i in equation (2c) is the Darcy velocity given by

$$v_i = -k_{ij} k_{rw}(S_w) \left(\frac{\partial \psi}{\partial x_j} + \eta_j \right) \tag{3}$$

Transport Under Equilibrium Conditions (LEA)

The equation describing the transport of a nonreactive contaminant in variably saturated porous media may be written as [*Bear, 1979*; *Nielsen et al., 1986*; *Gambolati et al., 1992*]

$$\frac{\partial}{\partial x_i}\left(D_{ij} \frac{\partial c}{\partial x_j} \right) - \frac{\partial}{\partial x_i}(v_i c) = \frac{\partial (n S_w c)}{\partial t} - q c^* - f \tag{4}$$

where

$D_{ij} = n S_w \tilde{D}_{ij}$

$\tilde{D}_{ij}$ is the dispersion tensor $[L^2/T]$

c is the concentration of the dissolved constituent $[M/L^3]$

c^* is the concentration of the solute injected or withdrawn with the fluid source or sink $[M/L^3]$

f is the distributed mass rate of the solute per unit volume $[M/L^3T]$.

The dispersion tensor for a two-dimensional porous medium is given by [*Bear, 1979*]

$$D_{ij} = n S_w \tilde{D}_{ij} = \alpha_T \mid v \mid \delta_{ij} + (\alpha_L - \alpha_T) \frac{v_i v_j}{\mid v \mid} + n S_w D_0 \tau \delta_{ij} \qquad i,j = 1,2 \tag{5}$$

where

$\mid v \mid = \sqrt{v_1^2 + v_2^2}$

α_L is the longitudinal dispersivity [L]

α_T is the transverse dispersivity [L]

δ_{ij} is the Kronecker delta [/]

D_0 is the molecular diffusion coefficient $[L^2/T]$

τ is the tortuosity ($\tau = 1$ usually assumed) [/].

If the Darcy velocity is parallel to the coordinate axis x_1 equation (5) simplifies to

$$\begin{aligned} D_{11} &= D_L = \alpha_L \mid v_1 \mid + n S_w D_0 \tau \\ D_{22} &= D_T = \alpha_T \mid v_1 \mid + n S_w D_0 \tau \\ D_{12} &= D_{21} = 0 \end{aligned}$$

Applying the Chain Rule to the advective term in equation (4), we obtain

$$\frac{\partial}{\partial x_i}\left(D_{ij}\frac{\partial c}{\partial x_j}\right) - v_i\frac{\partial c}{\partial x_i} - c\frac{\partial v_i}{\partial x_i} = c\frac{\partial (nS_w)}{\partial t} + nS_w\frac{\partial c}{\partial t} - qc^* - f \tag{6}$$

From equations (1) and (3) and the definition of the storage coefficient σ we have

$$\begin{aligned}\frac{\partial v_i}{\partial x_i} &= -\frac{\partial}{\partial x_i}\left[k_{ij}k_{rw}\left(\frac{\partial \psi}{\partial x_j} + \eta_j\right)\right] = -\left(\sigma\frac{\partial \psi}{\partial t} - q\right) \\ &= q - \frac{d(nS_w)}{d\psi}\frac{\partial \psi}{\partial t} = q - \frac{\partial (nS_w)}{\partial t}\end{aligned}$$

and thus equation (6) becomes

$$\frac{\partial}{\partial x_i}\left(D_{ij}\frac{\partial c}{\partial x_j}\right) - v_i\frac{\partial c}{\partial x_i} = nS_w\frac{\partial c}{\partial t} + q(c - c^*) - f \tag{7}$$

If q denotes a sink term then $c = c^*$ and the term $q(c - c^*)$ vanishes. We now introduce two reaction terms to equation (7), decay and equilibrium sorption. If the contaminant is subject to radioactive or biodegradation decay with a half-life constant $T_{1/2}$, equation (7) becomes

$$\frac{\partial}{\partial x_i}\left(D_{ij}\frac{\partial c}{\partial x_j}\right) - v_i\frac{\partial c}{\partial x_i} = nS_w\left(\frac{\partial c}{\partial t} + \lambda c\right) + q(c - c^*) - f$$

where the decay constant $\lambda = \ln 2/T_{1/2}$. If in addition to decay the contaminant also undergoes sorption, the above transport equation is modified to

$$\frac{\partial}{\partial x_i}\left(D_{ij}\frac{\partial c}{\partial x_j}\right) - v_i\frac{\partial c}{\partial x_i} = nS_w\left(\frac{\partial c}{\partial t} + \lambda c\right) + \rho_s\left(\frac{\partial S}{\partial t} + \lambda S\right) + q(c - c^*) - f \tag{8}$$

where S is the concentration of the adsorbed constituent in the solid phase (mass adsorbed per mass of dry sediment), $\rho_s = (1 - n)\gamma_s$ is the bulk soil density, and γ_s is the density of the solid (soil) material (specific weight of dry sediment).

If the contaminant moves slowly enough that chemical equilibrium is achieved and the conditions of the local equilibrium assumption (LEA) hold, then the relationship between S and c can be approximated by the Freundlich isotherm $S = k_d c^\kappa$, where k_d is the distribution coefficient. For the case of a linear sorption isotherm ($\kappa = 1$), we can define the retardation factor $R_d = 1 + (\rho_s k_d)/(nS_w)$ and write equation (8) as

$$\frac{\partial}{\partial x_i}\left(D_{ij}\frac{\partial c}{\partial x_j}\right) - v_i\frac{\partial c}{\partial x_i} = nS_w R_d\left(\frac{\partial c}{\partial t} + \lambda c\right) + q(c - c^*) - f \tag{9}$$

It is well-known that the sorption isotherm may depend on several factors, including surface charge of the sorbing phase, ionic strength, solution pH, competing counter-ions and their concentrations, and the concentration of the adsorbed phase. In some cases the "adsorption" isotherm may be different from the "desorption" isotherm (chemical hysteresis).

The initial and boundary conditions for the transport equation (9) can be expressed as [*Galeati and Gambolati, 1989*]

$$\begin{aligned} c(x_i, 0) &= c_o(x_i) & & \text{(10a)} \\ c(x_i, t) &= \bar{c}(x_i, t) \quad \text{on } \Gamma_3 & & \text{(10b)} \\ D_{ij}\frac{\partial c}{\partial x_j} n_i &= q_c^D(x_i, t) \quad \text{on } \Gamma_4 & & \text{(10c)} \\ \left(D_{ij}\frac{\partial c}{\partial x_j} - v_i c\right) n_i &= q_c^T(x_i, t) \quad \text{on } \Gamma_5 & & \text{(10d)} \end{aligned}$$

where c_o is the initial concentration, $\bar{c}$ is the prescribed concentration on the Dirichlet boundary Γ_3, q_c^D is the prescribed dispersive flux normal to the Neumann boundary Γ_4 (positive outward), and q_c^T is the prescribed total flux of solute across the Cauchy or Rubin boundary Γ_5. Boundary condition (10c) is usually imposed along the outflow boundary with $q_c^D = 0$, that is $D_{ij}(\partial c/\partial x_j)n_i = 0$. Along an impermeable boundary we have $v_i n_i = 0$, and a zero dispersive flux here implies $q_c^T = 0$.

Transport Under Nonequilibrium Conditions (Non-LEA)

Nonequilibrium contaminant transport may be mathematically described using a dual porosity model wherein an unsaturated, aggregated porous medium is subdivided into five regions [*van Genuchten and Wierenga, 1976*]. The five regions are:

1. the air phase;

2. the region containing the *mobile* water phase, located in the largest pores. Fluid flow occurs in this region only, as do the advective and dispersive mechanisms of solute transport. Solute transfer with region 4 occurs by equilibrium sorption, and with region 3 by diffusion;

3. the region containing the *immobile* water phase. There is no fluid flow in this region, and solute transfer occurs by diffusion with region 2, and by equilibrium sorption with region 5;

4. the *dynamic* soil region, located around the mobile water region 2. Solute transfer is by equilibrium sorption with region 2;

5. the *stagnant* soil region, located around the immobile water region 3. Solute transfer is by equilibrium sorption with region 3.

Under these assumptions, the general equations describing solute transport and mass transfer in the mobile and immobile regions are

$$
\begin{aligned}
\frac{\partial}{\partial x_i}\left(\theta_m \tilde{D}_{ij}\frac{\partial c_m}{\partial x_j}\right) - v_i\frac{\partial c_m}{\partial x_i} &= \theta_m\frac{\partial c_m}{\partial t} + \theta_{im}\frac{\partial c_{im}}{\partial t} \\
&+ \rho_s F\frac{\partial S_m}{\partial t} + \rho_s(1-F)\frac{\partial S_{im}}{\partial t} \\
&+ \lambda\left[\theta_m c_m + \theta_{im}c_{im} + \rho_s F S_m + \rho_s(1-F)S_{im}\right] \\
&+ q\left(c_m - c^*\right) - f \qquad (11a)
\end{aligned}
$$

$$
\theta_{im}\frac{\partial c_{im}}{\partial t} + \rho_s(1-F)\frac{\partial S_{im}}{\partial t} = \alpha(c_m - c_{im}) - \lambda\left[\theta_{im}c_{im} + \rho_s(1-F)S_{im}\right] \qquad (11b)
$$

where

$\theta_m = n_m S_{w_m}$ is the volumetric water content in the mobile (“m”) water region [/]

S_{w_m} is the water saturation in the mobile water region [/]

n_m is the porosity of the mobile water region, given as the ratio of the volume of voids in the mobile region to the total volume [/]

$\theta_m \tilde{D}_{ij} = \alpha_T |v| \delta_{ij} + (\alpha_L - \alpha_T) v_i v_j / |v| + n_m S_{w_m} D_0 T \delta_{ij}$ $[L^2/T]$

c_m is the concentration of the dissolved constituent in the mobile region $[M/L^3]$

c_{im} is the concentration of the dissolved constituent in the immobile (“im”) water region $[M/L^3]$

$\theta_{im} = n_{im}$ is the volumetric water content in the immobile water region [/]

n_{im} is the porosity of the immobile water region, given as the ratio of the volume of voids in the immobile region to the total volume [/]

$n = n_m + n_{im}$ is the total porosity [/]

F is the fraction of adsorption sites which are in the dynamic soil region. This parameter describes the amount of adsorption taking place inside the dynamic region of the soil, as a fraction of the total adsorption. F and $1 - F$ characterize the partitioning of the soil matrix into the dynamic (region 4) and stagnant (5) regions [/]

S_m is the concentration of the adsorbed constituent in the dynamic soil region (mass adsorbed per mass of dry sediment) [/]

S_{im} is the concentration of the adsorbed constituent in the stagnant soil region (mass adsorbed per mass of dry sediment) [/]

α is the mass transfer coefficient for the diffusion process between the mobile and immobile water regions [1/T].

We introduce the linear Freundlich isotherm to model the equilibrium (instantaneous) adsorption between the dynamic regions 2 and 4 and between the stagnant regions 3 and 5. For the dynamic regions we have $S_m = k_{d_m} c_m$ while for the stagnant regions $S_{im} = k_{d_{im}} c_{im}$, where k_{d_m} and $k_{d_{im}}$ are distribution coefficients. Substituting the isotherm expressions into equations (11a) and (11b) we obtain

$$\begin{aligned}
\frac{\partial}{\partial x_i}\left(n_m S_{w_m} \tilde{D}_{ij}\frac{\partial c_m}{\partial x_j}\right) - v_i\frac{\partial c_m}{\partial x_i} &= n_m S_{w_m}\frac{\partial c_m}{\partial t} + n_{im}\frac{\partial c_{im}}{\partial t} \\
&+ \rho_s F k_{d_m}\frac{\partial c_m}{\partial t} + \rho_s(1-F)k_{d_{im}}\frac{\partial c_{im}}{\partial t} \\
&+ \lambda\left[n_m S_{w_m} c_m + n_{im}c_{im} + \rho_s F k_{d_m} c_m\right. \\
&\quad \left. + \rho_s(1-F)k_{d_{im}} c_{im}\right] + q(c_m - c^*) - f \\
n_{im}\frac{\partial c_{im}}{\partial t} + \rho_s(1-F)k_{d_{im}}\frac{\partial c_{im}}{\partial t} &= \alpha(c_m - c_{im}) - \lambda\left[n_{im} + \rho_s(1-F)k_{d_{im}}\right]c_{im}
\end{aligned}$$

Collecting terms,

$$\begin{aligned}
\frac{\partial}{\partial x_i}\left(n_m S_{w_m} \tilde{D}_{ij}\frac{\partial c_m}{\partial x_j}\right) - v_i\frac{\partial c_m}{\partial x_i} &= (n_m S_{w_m} + \rho_s F k_{d_m})\frac{\partial c_m}{\partial t} \\
&+ \left[n_{im} + \rho_s(1-F)k_{d_{im}}\right]\frac{\partial c_{im}}{\partial t} \\
&+ \lambda(n_m S_{w_m} + \rho_s F k_{d_m})c_m \\
&+ \lambda\left[n_{im} + \rho_s(1-F)k_{d_{im}}\right]c_{im} + q(c_m - c^*) - f \\
\left[n_{im} + \rho_s(1-F)k_{d_{im}}\right]\frac{\partial c_{im}}{\partial t} &= \alpha(c_m - c_{im}) - \lambda\left[n_{im} + \rho_s(1-F)k_{d_{im}}\right]c_{im}
\end{aligned}$$

Defining the retardation factors $R_m = 1 + (\rho_s F k_{d_m})/(n_m S_{w_m})$ and $R_{im} = 1 + (\rho_s(1 - F)k_{d_{im}})/n_{im}$ and setting $T_m = n_m S_{w_m} R_m$ and $T_{im} = n_{im} R_{im}$, the non-LEA transport equations simplify to

$$\begin{aligned}
\frac{\partial}{\partial x_i}\left(D_{ij}\frac{\partial c_m}{\partial x_j}\right) - v_i\frac{\partial c_m}{\partial x_i} &= T_m\frac{\partial c_m}{\partial t} \\
&+ T_{im}\frac{\partial c_{im}}{\partial t} + \lambda(T_m c_m + T_{im}c_{im}) + q(c_m - c^*) - f \quad (12a) \\
T_{im}\frac{\partial c_{im}}{\partial t} &= \alpha(c_m - c_{im}) - \lambda T_{im} c_{im} \quad (12b)
\end{aligned}$$

To this system we add the flow equation (1) for the partially saturated mobile region and the Darcy equation (3), substituting S_{w_m} for S_w.

Integro-Differential Formulation of Non-LEA Model

Equation (12b) is a linear first order ordinary differential equation which may be integrated analytically. Setting $\beta = \alpha/T_{im}$, a general solution is given by

$$c_{im} = e^{-(\beta+\lambda)t}\left(k + \beta\int_0^t e^{(\beta+\lambda)\tau} c_m d\tau\right)$$

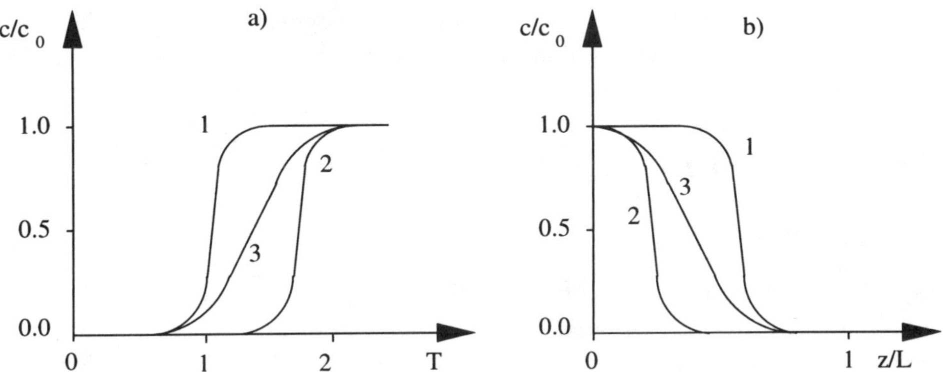

Figure 1. Typical breakthrough curves (a) and concentration front profiles (b) for nonreactive (profile 1), LEA (profile 2), and non-LEA (profile 3) contaminants in a laboratory column experiment.

where k is an integration constant. Assuming $c_{im} = 0$ at $t = 0$, the solution to (12b) becomes

$$c_{im} = \beta e^{-(\beta+\lambda)t} \int_0^t e^{(\beta+\lambda)\tau} c_m \, d\tau \tag{13}$$

Substituting (13) and (12b) into equation (12a), the transport equation for the mobile region becomes

$$\begin{aligned} \frac{\partial}{\partial x_i}\left(D_{ij}\frac{\partial c_m}{\partial x_j}\right) - v_i\frac{\partial c_m}{\partial x_i} &= T_m\frac{\partial c_m}{\partial t} + (\alpha + \lambda T_m + q)\, c_m - (qc^* + f) \\ &\quad - \alpha\beta e^{-(\beta+\lambda)t} \int_0^t e^{(\beta+\lambda)\tau} c_m \, d\tau \end{aligned} \tag{14}$$

Column Experiment Example

Let us examine a typical laboratory experiment involving transport and sorption in a one-dimensional vertical porous column of length L. Define the dimensionless effluent pore volume T as $T = v_z t/(nL)$ where v_z is the vertical Darcy velocity. Assume that at time $t = 0$ the contaminant concentration is instantaneously set equal to a constant value c_o at the top of the column. The "breakthrough curve" is a plot of c/c_o versus T at the bottom of the column ($z = L$). Typical breakthrough curves are shown in Figure 1a. Profile 1 is for a nonreactive contaminant, and the slope of the concentration front is related to the magnitude of dispersion. If the dispersion coefficient is zero, the front maintains a vertical slope while moving through the column. Profile 2 describes a reactive contaminant subject to LEA sorption with a linear isotherm. The difference between profiles 1 and 2 is a simple translation in time (i.e. a time scaling) determined by the retardation factor. Profile 3 is typical of a reactive contaminant for which LEA does not hold. In this case the process is described by linear kinetics tending towards equilibrium. The tailing effect in profile 3 is easily recognized. Qualitatively similar results to those in Figure 1a are obtained when plotting the position of the contaminant front at a fixed time t (Figure 1b).

Solution by Finite Elements

Flow Equation

Equation (1) is solved numerically by finite elements. For a detailed presentation of the finite element method as applied to groundwater flow and transport the reader is referred to *Pinder and Gray* [*1977*], *Huyakorn and Pinder* [*1983*], and *Kinzelbach* [*1986*]. The numerical solution to equation (1) in a flow domain or region R is an approximation $\hat{\psi}(x_1, x_2, t)$ to the exact solution $\psi(x_1, x_2, t)$ obtained by discretizing the region R into p elements and l nodes. The approximation is expressed as

$$\psi \approx \hat{\psi} = \sum_{g=1}^{l} \psi_g(t) N_g(x_1, x_2) \tag{15}$$

where $N_g(x_1, x_2)$ are linear shape or basis functions for two-dimensional triangular finite elements and ψ_g are the unknown nodal pressure head. Recasting equation (1) in operator notation

$$L(\psi) = \frac{\partial}{\partial x_i}\left[k_{ij}k_{rw}\left(\frac{\partial \psi}{\partial x_j} + \eta_j\right)\right] - \sigma\frac{\partial \psi}{\partial t} + q = 0$$

the error, or residual, represented by the finite element approximation (15) is given as $L(\hat{\psi}) - L(\psi)$, or simply $L(\hat{\psi})$. This error is minimized by imposing an orthogonality constraint between the residual and the basis functions, which yields the Galerkin integral

$$\int_R L(\hat{\psi}) N_g(x_1, x_2) dR = 0 \qquad g = 1, \ldots, l \tag{16}$$

We assume for simplicity that the coordinate directions are parallel to the principal directions of hydraulic anisotropy, so that the off-diagonal components of the conductivity tensor k_{ij} are zero. Expanding equation (16) and applying Green's lemma (integration by parts) to the spatial derivative term we get

$$\begin{aligned}
&-\int_R k_{rw}\left[k_{11}\frac{\partial \hat{\psi}}{\partial x_1}\frac{\partial N_g}{\partial x_1} + k_{22}\left(\frac{\partial \hat{\psi}}{\partial x_2} + 1\right)\frac{\partial N_g}{\partial x_2}\right] dR \\
&+\int_\Gamma k_{rw}\left[k_{11}\frac{\partial \hat{\psi}}{\partial x_1}n_1 + k_{22}\left(\frac{\partial \hat{\psi}}{\partial x_2} + 1\right)n_2\right] N_g d\Gamma \\
&-\int_R \sigma\frac{\partial \hat{\psi}}{\partial t} N_g dR + \int_R q N_g dR = 0 \qquad g = 1, \ldots, l
\end{aligned} \tag{17}$$

where n_i, $i = 1, 2$ are the direction cosines of the outer normal to boundary Γ. Substituting equation (15), changing sign, and making use of boundary condition (2c) to replace the boundary integral term above, equation (17) becomes

$$\begin{aligned}
&\sum_{h=1}^{l} \psi_h \left[\sum_{e=1}^{p} \int_{\Delta^e} k_{rw}^e \left(k_{11}^e \frac{\partial N_g^e}{\partial x_1} \frac{\partial N_h^e}{\partial x_1} + k_{22}^e \frac{\partial N_g^e}{\partial x_2} \frac{\partial N_h^e}{\partial x_2} \right) d\Delta^e \right] \\
&+ \sum_{h=1}^{l} \frac{\partial \psi_h}{\partial t} \left[\sum_{e=1}^{p} \int_{\Delta^e} \sigma^e N_g^e N_h^e d\Delta^e \right] + \sum_{e=1}^{p} \int_{\Delta^e} k_{rw}^e k_{22}^e \frac{\partial N_g^e}{\partial x_2} d\Delta^e \\
&- \sum_{e=1}^{p} \int_{\Delta^e} q^e N_g^e d\Delta^e - \sum_{e=1}^{p} \int_{\Gamma_2^e} q_n^e N_g^e d\Gamma^e = 0 \qquad g = 1, \ldots, l
\end{aligned} \tag{18}$$

In equation (18) the integral over domain R has been replaced with the sum of the integrals over each triangular element, where Δ^e is the area of element e. Also, the Dirichlet boundary condition (2b) is imposed after the discretized system has been completely assembled. In matrix form equation (18) can be written as

$$\boldsymbol{H}(\boldsymbol{\psi})\boldsymbol{\psi} + \boldsymbol{P}(\boldsymbol{\psi})\frac{\partial \boldsymbol{\psi}}{\partial t} + \boldsymbol{q}^*(\boldsymbol{\psi}) = \boldsymbol{0} \tag{19}$$

where $\boldsymbol{\psi} = (\psi_1, \psi_2, \ldots, \psi_l)^T$ and

$$\begin{aligned}
H_{gh} &= \sum_{e=1}^{p} \int_{\Delta^e} k_{rw}^e \left(k_{11}^e \frac{\partial N_g^e}{\partial x_1} \frac{\partial N_h^e}{\partial x_1} + k_{22}^e \frac{\partial N_g^e}{\partial x_2} \frac{\partial N_h^e}{\partial x_2} \right) d\Delta^e \\
P_{gh} &= \sum_{e=1}^{p} \int_{\Delta^e} \sigma^e N_g^e N_h^e d\Delta^e \\
q_g^* &= \sum_{e=1}^{p} \left[\int_{\Delta^e} k_{rw}^e k_{22}^e \frac{\partial N_g^e}{\partial x_2} d\Delta^e - \int_{\Delta^e} q^e N_g^e d\Delta^e - \int_{\Gamma_2^e} q_n^e N_g^e d\Gamma^e \right]
\end{aligned} \tag{20}$$

The evaluation of H_{gh}, P_{gh}, and q_g^* for triangular elements is given in Appendix A. Equation (19) can be integrated in time by the weighted finite difference scheme

$$\left(\nu \boldsymbol{H}^{k+\nu} + \frac{1}{\Delta t_k} \boldsymbol{P}^{k+\nu} \right) \boldsymbol{\psi}^{k+1} = \left(\frac{1}{\Delta t_k} \boldsymbol{P}^{k+\nu} - (1-\nu) \boldsymbol{H}^{k+\nu} \right) \boldsymbol{\psi}^k - \boldsymbol{q}^{*^{k+\nu}} \tag{21}$$

where k and $k+1$ denote the previous and current time levels and $\boldsymbol{H}$, $\boldsymbol{P}$, and $\boldsymbol{q}^*$ are evaluated at pressure head $\boldsymbol{\psi}^{k+\nu} = \nu\boldsymbol{\psi}^{k+1} + (1-\nu)\boldsymbol{\psi}^k$. For numerical stability parameter ν must satisfy the condition $1/2 \leq \nu \leq 1$.

Because $\boldsymbol{H}$, $\boldsymbol{P}$, and $\boldsymbol{q}^*$ have coefficients (k_{rw} and σ) which depend on the current solution $\boldsymbol{\psi}^{k+1}$, equation (21) is a nonlinear system. To solve this nonlinear equation, some iterative or other linearization technique is needed. The most common solution approach has been to use either the Picard or Newton iteration methods, with the Picard method being the more popular of the two [*Cooley, 1983*; *Frind and Verge, 1978*; *Huyakorn et al., 1984*; *Neuman, 1973*; *Paniconi et al., 1991*; *Putti and Paniconi, 1992*; *Ross, 1990*]. The Picard scheme has a simple formulation and it preserves symmetry of the finite element matrices. For these reasons it is less costly, on a per iteration basis, than the Newton method, which requires evaluation of Jacobian matrices and yields a nonsymmetric system. The Picard scheme, however, is only linearly convergent compared to quadratic convergence for the Newton method, and therefore it can be expected that for certain problems or under certain accuracy constraints the Newton scheme would exhibit better convergence behavior than Picard.

Denoting by $\boldsymbol{f}\left(\psi^{k+1}\right) = \mathbf{0}$ the discretized equation (21), the Newton scheme can be written as [*Paniconi et al., 1991*]

$$\boldsymbol{f}'(\overset{(m)}{\psi}_{k+1})\mathbf{h} = -\boldsymbol{f}(\overset{(m)}{\psi}_{k+1})$$

where m and $m+1$ denote the previous and current iteration levels,

$$\boldsymbol{h} \equiv \overset{(m+1)}{\psi}_{k+1} - \overset{(m)}{\psi}_{k+1}$$

and

$$f'^{(m)}_{ij} = \nu H^{(m)}_{ij} + \frac{1}{\Delta t_k} P^{(m)}_{ij} + \sum_s \frac{\partial H^{(m)}_{is}}{\partial \psi^{k+1}_j} \overset{(m)}{\psi}{}^{k+\nu}_s + \frac{1}{\Delta t_k} \sum_s \frac{\partial P^{(m)}_{is}}{\partial \psi^{k+1}_j} (\overset{(m)}{\psi}{}^{k+1}_s - \psi^k_s) + \frac{\partial q^{*(m)}_i}{\partial \psi^{k+1}_j}$$

the ijth component of the Jacobian matrix $\boldsymbol{f}'$. The Picard scheme may be written as

$$\left(\nu \overset{(m)}{\boldsymbol{H}}{}^{k+\nu} + \frac{1}{\Delta t_k} \overset{(m)}{\boldsymbol{P}}{}^{k+\nu}\right) \overset{(m+1)}{\psi}{}^{k+1} = \left[\frac{1}{\Delta t_k} \overset{(m)}{\boldsymbol{P}}{}^{k+\nu} - (1-\nu) \overset{(m)}{\boldsymbol{H}}{}^{k+\nu}\right] \psi^k - \overset{(m)}{\boldsymbol{q}}{}^{*k+\nu}$$

or equivalently as

$$\left(\nu \overset{(m)}{\boldsymbol{H}}{}^{k+\nu} + \frac{1}{\Delta t_k} \overset{(m)}{\boldsymbol{P}}{}^{k+\nu}\right) \boldsymbol{h} = -\boldsymbol{f}\left(\overset{(m)}{\psi}{}^{k+1}\right)$$

Since Newton iteration produces a nonsymmetric system while the Picard method preserves symmetry, different solvers have to be used for the resulting linear systems. For instance the Cholesky preconditioned conjugate gradient method can be used for the solution of the symmetric (Picard) system [*Gambolati and Perdon, 1984*]. For the nonsymmetric (Newton) systems, available solvers include the biconjugate gradient method, BI-CGSTAB [*van der Vorst, 1992*], the minimum residual algorithm, GRAMRB, the generalized conjugate residual method, GCRK, and the transpose-free quasi-minimal residual algorithm, TFQMR [*Axelsson, 1980*; *Pini et al., 1989*; *Freund, 1993*].

LEA Transport Equation

For the finite element integration of the transport equation (9) assume, as we did for the flow equation, an approximate solution given in the form of a linear combination of linear basis functions $N_g(x_1, x_2)$:

$$c \approx \hat{c} = \sum_{g=1}^{l} c_g(t) N_g(x_1, x_2) \tag{22}$$

Substituting (22) in equation (9) yields the residual

$$M(\hat{c}) = \frac{\partial}{\partial x_i}\left(D_{ij}\frac{\partial \hat{c}}{\partial x_j}\right) - v_i\frac{\partial \hat{c}}{\partial x_i} - nS_w R_d\left(\frac{\partial \hat{c}}{\partial t} + \lambda\hat{c}\right) - q(\hat{c} - c^*) + f$$

Imposing the condition that the residual $M(\hat{c})$ be orthogonal over R to the l test functions $W_g(x_1, x_2)$, we get the weighted residual equation

$$\int_R M(\hat{c})W_g(x_1,x_2)dR = 0 \qquad g = 1,\ldots,l \tag{23}$$

If $W_g(x_1,x_2) \equiv N_g(x_1,x_2)$, the classical Galerkin method is obtained. If however we use test functions W_g which are different from the basis functions N_g, we obtain a different method, for instance the upwind finite element scheme (e.g., *Huyakorn and Taylor* [*1976*], *Huyakorn et al.* [*1987*], *Pini et al.* [*1989*]) where nonsymmetric test functions are used for the advective component of the transport equation. Upwinding is useful for advection dominated problems. The technique reduces spurious oscillations and gives better resolution at sharp fronts. Other techniques which have been proposed to deal with advection dominated problems include a finite volume upwinding approach [*Putti et al., 1990*] and the Eulerian-Lagrangian method [*Neuman, 1981*; *Sorek, 1988*].

Application of Green's lemma to both the dispersive and advective components of integral (23) would allow us to incorporate Cauchy boundary conditions directly into the integral. However this approach leads to unstable numerical solutions [*Gureghian, 1983*; *Huyakorn and Pinder, 1983*; *Huyakorn et al., 1985*; *Galeati and Gambolati, 1989*], and for this reason we apply Green's lemma to the dispersive component only, obtaining

$$\begin{aligned}
-\int_R\left(D_{ij}\frac{\partial \hat{c}}{\partial x_j}\frac{\partial W_g}{\partial x_i} + v_i\frac{\partial \hat{c}}{\partial x_i}W_g\right)dR + \int_\Gamma\left(D_{ij}\frac{\partial \hat{c}}{\partial x_j}\right)n_i W_g d\Gamma \\
-\int_R nS_w R_d\left(\frac{\partial \hat{c}}{\partial t} + \lambda\hat{c}\right)W_g dR + \int_R[(c^* - \hat{c})q + f]W_g dR \;=\; 0 \qquad g = 1,\ldots,l
\end{aligned}$$

Substituting equation (22), changing sign, and incorporating boundary conditions we obtain

$$\begin{aligned}
&\sum_{h=1}^{l} c_h\left[\sum_{e=1}^{p}\int_{\Delta^e}\left(D_{ij}^e\frac{\partial N_h^e}{\partial x_j}\frac{\partial W_g^e}{\partial x_i} + v_i^e\frac{\partial N_h^e}{\partial x_i}W_g^e + n^e S_w^e R_d^e\lambda^e N_h^e W_g^e\right)d\Delta^e\right] \\
&+\sum_{h=1}^{l}\frac{\partial c_h}{\partial t}\left[\sum_{e=1}^{p}\int_{\Delta^e} n^e S_w^e R_d^e N_h^e W_g^e d\Delta^e\right] + \sum_{h=1}^{l} c_h\left[\sum_{e=1}^{p}\int_{\Delta^e} q^e N_h^e W_g^e d\Delta^e\right] \\
&-\sum_{e=1}^{p}\int_{\Delta^e}(q^e c^{*^e} + f^e)W_g^e d\Delta^e - \sum_{e=1}^{p}\int_{\Gamma_4^e} q_c^{D^e} W_g^e d\Gamma^e - \sum_{e=1}^{p}\int_{\Gamma_5^e} q_c^{T^e} W_g^e d\Gamma^e \\
&-\sum_{h=1}^{l} c_h\left[\sum_{e=1}^{p}\int_{\Gamma_5^e}(v_i^e n_i^e)N_h^e W_g^e d\Gamma^e\right] = 0 \qquad g = 1,\ldots,l
\end{aligned} \tag{24}$$

Equation (24) represents a system of l equations for the unknown nodal concentrations $\boldsymbol{c} = (c_1, c_2, \ldots, c_l)^T$:

$$[\boldsymbol{A} + \boldsymbol{B} + \boldsymbol{E} + \boldsymbol{F}]\,\boldsymbol{c} + \boldsymbol{G}\frac{\partial \boldsymbol{c}}{\partial t} + \boldsymbol{r}^* = 0 \tag{25}$$

where

$$\begin{aligned}
A_{gh} &= \sum_{e=1}^{p} \int_{\Delta^e} D_{ij}^e \frac{\partial N_h^e}{\partial x_j} \frac{\partial W_g^e}{\partial x_i} d\Delta^e \\
B_{gh} &= \sum_{e=1}^{p} \int_{\Delta^e} v_i^e \frac{\partial N_h^e}{\partial x_i} W_g^e d\Delta^e \\
E_{gh} &= \sum_{e=1}^{p} \int_{\Delta^e} n^e S_w^e R_d^e \lambda^e N_h^e W_g^e d\Delta^e \\
F_{gh} &= \sum_{e=1}^{p} \int_{\Delta^e} q^e N_h^e W_g^e d\Delta^e - \sum_{e=1}^{p} \int_{\Gamma_5^e} (v_i^e n_i^e) N_h^e W_g^e d\Gamma^e \\
G_{gh} &= \sum_{e=1}^{p} \int_{\Delta^e} n^e S_w^e R_d^e N_h^e W_g^e d\Delta^e \\
r_g^* &= -\sum_{e=1}^{p} \int_{\Delta^e} (q^e c^{*^e} + f^e) W_g^e d\Delta^e - \sum_{e=1}^{p} \int_{\Gamma_4^e} q_c^{D^e} W_g^e d\Gamma^e - \sum_{e=1}^{p} \int_{\Gamma_5^e} q_c^{T^e} W_g^e d\Gamma^e
\end{aligned} \tag{26}$$

Integration in time of equation (25) is performed using a weighted finite difference scheme:

$$\begin{aligned}
&\left(\nu \left[\boldsymbol{A} + \boldsymbol{B} + \boldsymbol{E} + \boldsymbol{F}\right]^{k+\nu} + \frac{1}{\Delta t_k}\boldsymbol{G}^{k+\nu}\right) \boldsymbol{c}^{k+1} \\
&= \left(\frac{1}{\Delta t_k}\boldsymbol{G}^{k+\nu} - (1-\nu)\left[\boldsymbol{A} + \boldsymbol{B} + \boldsymbol{E} + \boldsymbol{F}\right]^{k+\nu}\right) \boldsymbol{c}^k - \boldsymbol{r}^{*^{k+\nu}}
\end{aligned} \tag{27}$$

Unlike the flow equation, the transport equation is quite sensitive to the value of the weighting parameter ν. A value close to 1/2 leads to accurate but unstable solutions while values close to 1 yield good stability but with large numerical dispersion [*Peyret and Taylor, 1983*]. The evaluation of the matrices in equation (27) is performed using the values of velocity and water saturation at time level $k + \nu$ calculated by the flow model. The coefficients of velocity for triangular elements are given in Appendix B, and the coefficients of $\boldsymbol{A}$, $\boldsymbol{B}$, $\boldsymbol{E}$, $\boldsymbol{F}$, $\boldsymbol{G}$, and $\boldsymbol{r}^*$ are given in Appendix C.

Non-LEA Transport Equation

The integro-differential equation (14) is solved by finite elements together with a scheme to numerically calculate the convolution integral.

Finite Element Discretization

Following the procedure used for the flow and LEA transport equations, we define an approximate solution for c_m as

$$c_m \approx \hat{c} = \sum_{g=1}^{l} c_g(t) N_g(x_1, x_2) \tag{28}$$

Substituting $\hat{c}$ in equation (14) we obtain the residual

$$\begin{aligned} M(\hat{c}) &= \frac{\partial}{\partial x_i}\left(D_{ij}\frac{\partial \hat{c}}{\partial x_j}\right) - v_i \frac{\partial \hat{c}}{\partial x_i} - T_m \frac{\partial \hat{c}}{\partial t} - (\alpha + \lambda T_m + q)\hat{c} + (qc^* + f) \\ &\quad + \alpha\beta e^{-(\beta+\lambda)t} \int_0^t e^{(\beta+\lambda)\tau} \hat{c} d\tau \end{aligned}$$

Prescribing the orthogonality condition between the residual and each test function $W_g(x_1, x_2)$ yields

$$\int_R M(\hat{c}) W_g(x_1, x_2) dR = 0 \qquad g = 1, \ldots, l$$

Applying Green's lemma to the dispersive component, substituting equation (28), changing sign, and incorporating boundary conditions, we obtain

$$\begin{aligned} &\sum_{h=1}^{l} c_h \left[\sum_{e=1}^{p} \int_{\Delta^e} \left(D_{ij}^e \frac{\partial N_h^e}{\partial x_j}\frac{\partial W_g^e}{\partial x_i} + v_i^e \frac{\partial N_h^e}{\partial x_j} W_g^e\right) d\Delta^e\right] \\ &+ \sum_{h=1}^{l} \frac{\partial c_h}{\partial t}\left[\sum_{e=1}^{p}\int_{\Delta^e} T_m^e N_h^e W_g^e d\Delta^e\right] + \sum_{h=1}^{l} c_h \left[\sum_{e=1}^{p}\int_{\Delta^e} (\alpha^e + \lambda^e T_m^e + q^e) N_h^e W_g^e d\Delta^e\right] \\ &- \sum_{h=1}^{l}\left[\sum_{e=1}^{p}\int_{\Delta^e} (\alpha^e\beta^e e^{-(\beta^e+\lambda^e)t} \int_0^t e^{(\beta^e+\lambda^e)\tau} c_h d\tau) N_h^e W_g^e d\Delta^e\right] \\ &- \sum_{e=1}^{p}\int_{\Delta^e} (q^e c^{*^e} + f^e) W_g^e d\Delta^e - \sum_{e=1}^{p}\int_{\Gamma_4^e} q_c^{D^e} W_g^e d\Gamma^e - \sum_{e=1}^{p}\int_{\Gamma_5^e} q_c^{T^e} W_g^e d\Gamma^e \\ &- \sum_{h=1}^{l} c_h \left[\sum_{e=1}^{p}\int_{\Gamma_5^e} (v_i^e n_i^e) N_h^e W_g^e d\Gamma^e\right] = 0 \qquad g = 1, \ldots, l \end{aligned} \tag{29}$$

Equation (29) represents a system of l equations for the unknown mobile region concentrations $\boldsymbol{c} = (c_1, c_2, \ldots, c_l)^T$:

$$[\boldsymbol{A} + \boldsymbol{B} + \tilde{\boldsymbol{E}} + \boldsymbol{F}]\boldsymbol{c} + \tilde{\boldsymbol{G}}\frac{\partial \boldsymbol{c}}{\partial t} + \boldsymbol{r}^* + \boldsymbol{d}^* = \boldsymbol{0}$$

where

$$\begin{aligned} A_{gh} &= \sum_{e=1}^{p}\int_{\Delta^e} D_{ij}^e \frac{\partial N_h^e}{\partial x_j}\frac{\partial W_g^e}{\partial x_i} d\Delta^e \\ B_{gh} &= \sum_{e=1}^{p}\int_{\Delta^e} v_i^e \frac{\partial N_h^e}{\partial x_i} W_g^e d\Delta^e \end{aligned}$$

$$\begin{aligned}
\tilde{E}_{gh} &= \sum_{e=1}^{p} \int_{\Delta^e} (\alpha^e + \lambda^e T_m^e) N_h^e W_g^e d\Delta^e \\
F_{gh} &= \sum_{e=1}^{p} \int_{\Delta^e} q^e N_h^e W_g^e d\Delta^e - \sum_{e=1}^{p} \int_{\Gamma_5^e} (v_i^e n_i^e) N_h^e W_g^e d\Gamma^e \\
\tilde{G}_{gh} &= \sum_{e=1}^{p} \int_{\Delta^e} T_m^e N_h^e W_g^e d\Delta^e \\
d^* &= \sum_{e=1}^{p} M^e s^e \\
M_{gh}^e &= -\int_{\Delta^e} \alpha^e \beta^e e^{-(\beta^e+\lambda^e)t} N_h^e W_g^e d\Delta^e \\
s_h^e &= \int_0^t e^{(\beta^e+\lambda^e)\tau} c_h d\tau \\
r_g^* &= -\sum_{e=1}^{p} \int_{\Delta^e} (q^e c^{*^e} + f^e) W_g^e d\Delta^e - \sum_{e=1}^{p} \int_{\Gamma_4^e} q_c^{D^e} W_g^e d\Gamma^e - \sum_{e=1}^{p} \int_{\Gamma_5^e} q_c^{T^e} W_g^e d\Gamma^e
\end{aligned}$$

Integration in time by a weighted finite difference scheme yields

$$\left[\nu(A + B + \tilde{E} + F)^{k+\nu} + \frac{1}{\Delta t_k}\tilde{G}^{k+\nu}\right] c^{k+1} = \left[\frac{1}{\Delta t_k}\tilde{G}^{k+\nu} - (1-\nu)(A + B + \tilde{E} + F)^{k+\nu}\right] c^k - r^{*^{k+\nu}} - d^{*^{k+\nu}} \quad (30)$$

Calculation of Convolution Integral

The reactive terms which describe the contribution due to mass transfer between the mobile and immobile regions are contained in $\tilde{E}$ and d^* of equation (30). Term d^* contains the convolution integral, which is to be evaluated at time level $k+\nu$: $d^{*^{k+\nu}} = \nu d^{*^{k+1}} + (1-\nu)d^{*^k}$. At time level k we have $d^{*^k} = \sum_{e=1}^{p} M^{e^k} s^{e^k}$ where $s^{e^k} = \int_0^{t^k} e^{(\beta^e+\lambda^e)\tau} c d\tau$. At time level $k+1$, $d^{*^{k+1}} = \sum_{e=1}^{p} M^{e^{k+1}} s^{e^{k+1}}$ where

$$\begin{aligned}
s^{e^{k+1}} &= \int_0^{t^{k+1}} e^{(\beta^e+\lambda^e)\tau} c d\tau = \int_0^{t^k} e^{(\beta^e+\lambda^e)\tau} c d\tau + \int_{t^k}^{t^{k+1}} e^{(\beta^e+\lambda^e)\tau} c d\tau \\
&\equiv s^{e^k} + s_k^{e^{k+1}}
\end{aligned}$$

We can now write

$$d^{*^{k+\nu}} = \nu \sum_{e=1}^{p} M^{e^{k+1}} (s^{e^k} + s_k^{e^{k+1}}) + (1-\nu) \sum_{e=1}^{p} M^{e^k} s^{e^k}$$

The matrix $M^{e^{k+1}}$ can be expressed in terms of M^{e^k} by writing

$$\begin{aligned}
M_{gh}^{e^{k+1}} &= -\int_{\Delta^e} \alpha^e \beta^e e^{-(\beta^e+\lambda^e)t^k} e^{-(\beta^e+\lambda^e)(t^{k+1}-t^k)} N_h^e W_g^e d\Delta^e \\
&= M_{gh}^{e^k} e^{-(\beta^e+\lambda^e)\Delta t_k}
\end{aligned} \quad (31)$$

where $\Delta t_k \equiv t^{k+1} - t^k$. Numerical integration is needed to evaluate the term $s_k^{e^{k+1}}$.

Method Based on Application of the Mean Integral Theorem The ith component of vector $s_k^{e^{k+1}}$ may be written as

$$
\begin{aligned}
(s_k^{e^{k+1}})_i &= \int_{t^k}^{t^{k+1}} e^{(\beta^e+\lambda^e)\tau} c_i d\tau \approx \frac{c_i^{k+1}+c_i^k}{2} \int_{t^k}^{t^{k+1}} e^{(\beta^e+\lambda^e)\tau} d\tau \\
&= \frac{c_i^{k+1}+c_i^k}{2} \frac{e^{(\beta^e+\lambda^e)t^{k+1}} - e^{(\beta^e+\lambda^e)t^k}}{(\beta^e+\lambda^e)} \\
&= \frac{1}{2(\beta^e+\lambda^e)} \left(e^{(\beta^e+\lambda^e)\Delta t_k} - 1\right) e^{(\beta^e+\lambda^e)t^k} \left(c_i^{k+1}+c_i^k\right) \\
&= \frac{1}{2(\beta^e+\lambda^e)} \left(1 - e^{-(\beta^e+\lambda^e)\Delta t_k}\right) e^{(\beta^e+\lambda^e)t^{k+1}} \left(c_i^{k+1}+c_i^k\right)
\end{aligned}
$$

so that $d^{*^{k+1}}$ becomes

$$
\begin{aligned}
\left(d^{*^{k+1}}\right)_g^e &= \sum_{h=1}^{l} M_{gh}^{e^{k+1}} (s^{e^k} + s_k^{e^{k+1}})_h \\
&= -\frac{\alpha^e\beta^e}{2(\beta^e+\lambda^e)} \left(1 - e^{-(\beta^e+\lambda^e)\Delta t_k}\right) \sum_{h=1}^{l} \left(\int_{\Delta^e} N_h^e W_g^e d\Delta^e\right) \left(c_h^{k+1}+c_h^k\right) \\
&+ \sum_{h=1}^{l} M_{gh}^{e^{k+1}} s_h^{e^k}
\end{aligned}
$$

Defining the matrix

$$
Q_{gh}^{e^{k+1}} = -\frac{\alpha^e\beta^e}{2(\beta^e+\lambda^e)} \left(1 - e^{-(\beta^e+\lambda^e)\Delta t_k}\right) \int_{\Delta^e} N_h^e W_g^e d\Delta^e
$$

we obtain

$$
d^{*^{k+1}} = \sum_{e=1}^{p} Q^{e^{k+1}} (c^{k+1} + c^k) + \sum_{e=1}^{p} M^{e^{k+1}} s^{e^k}
$$

Equation (30) can now be written as

$$
\begin{aligned}
&\left\{\nu \left[(A+B+\tilde{E}+F)^{k+\nu} + Q^{k+1}\right] + \frac{1}{\Delta t_k}\tilde{G}^{k+\nu}\right\} c^{k+1} = \\
&\left[\frac{1}{\Delta t_k}\tilde{G}^{k+\nu} - (1-\nu)(A+B+\tilde{E}+F)^{k+\nu} - \nu Q^{k+1}\right] c^k - r^{*^{k+\nu}} - \tilde{q}^{k+\nu}
\end{aligned}
$$

where the vector $\tilde{q}^{k+\nu}$ is given by

$$
\tilde{q}^{k+\nu} = \nu \sum_{e=1}^{p} M^{e^{k+1}} s^{e^k} + (1-\nu) \sum_{e=1}^{p} M^{e^k} s^{e^k}
$$

The updating of matrix M from time level k to $k+1$ is done using the simple recursion relationship (31).

Method Based on the Trapezoidal Rule To validate the results obtained with the previous technique we also include an integration procedure based on the trapezoidal rule:

$$
\begin{aligned}
(s_k^{e^{k+1}})_i = \int_{t^k}^{t^{k+1}} e^{(\beta^e+\lambda^e)\tau} c_i d\tau &\approx \frac{e^{(\beta^e+\lambda^e)t^{k+1}} c_i^{k+1} + e^{(\beta^e+\lambda^e)t^k} c_i^k}{2} \cdot (t^{k+1} - t^k) \\
&= \frac{\Delta t_k}{2} e^{(\beta^e+\lambda^e)t^k} \left(e^{(\beta^e+\lambda^e)\Delta t_k} c_i^{k+1} + c_i^k\right) \\
&= \frac{\Delta t_k}{2} e^{(\beta^e+\lambda^e)t^{k+1}} \left(c_i^{k+1} + e^{-(\beta^e+\lambda^e)\Delta t_k} c_i^k\right)
\end{aligned}
$$

The vector $d^{*^{k+1}}$ may now be expressed as

$$
\begin{aligned}
(d^{*^{k+1}})_g^e &= \sum_{h=1}^{l} M_{gh}^{e^{k+1}} (s^{e^k} + s_k^{e^{k+1}})_h \\
&= -\alpha^e \beta^e \frac{\Delta t_k}{2} \sum_{h=1}^{l} \left(\int_{\Delta^e} N_h^e W_g^e d\Delta^e\right) \left(c_h^{k+1} + e^{-(\beta^e+\lambda^e)\Delta t_k} c_h^k\right) + \sum_{h=1}^{l} M_{gh}^{e^{k+1}} s_h^{e^k}
\end{aligned}
$$

Defining the matrices

$$
\begin{aligned}
(Q_1^e)_{gh}^{k+1} &= -\alpha^e \beta^e \frac{\Delta t_k}{2} \int_{\Delta^e} N_h^e W_g^e d\Delta^e \\
(Q_2^e)_{gh}^{k+1} &= -\alpha^e \beta^e \frac{\Delta t_k}{2} e^{-(\beta^e+\lambda^e)\Delta t_k} \int_{\Delta^e} N_h^e W_g^e d\Delta^e \\
&= (Q_1^e)_{gh}^{k+1} e^{-(\beta^e+\lambda^e)\Delta t_k}
\end{aligned}
$$

we obtain

$$
d^{*^{k+1}} = \sum_{e=1}^{p} \left(Q_1^{e^{k+1}} c^{k+1} + Q_2^{e^{k+1}} c^k\right) + \sum_{e=1}^{p} M^{e^{k+1}} s^{e^k}
$$

Equation (30) can now be written as

$$
\begin{aligned}
&\left\{\nu \left[(A + B + \tilde{E} + F)^{k+\nu} + Q_1^{k+1}\right] + \frac{1}{\Delta t_k} \tilde{G}^{k+\nu}\right\} c^{k+1} = \\
&\quad \left[\frac{1}{\Delta t_k} \tilde{G}^{k+\nu} - (1-\nu)(A + B + \tilde{E} + F)^{k+\nu} - \nu Q_2^{k+1}\right] c^k - r^{*^{k+\nu}} - \tilde{q}^{k+\nu}
\end{aligned}
$$

where as before the vector $\tilde{q}^{k+\nu}$ is given by

$$
\tilde{q}^{k+\nu} = \nu \sum_{e=1}^{p} M^{e^{k+1}} s^{e^k} + (1-\nu) \sum_{e=1}^{p} M^{e^k} s^{e^k}
$$

Numerical Examples

The numerical models are tested by comparing simulation results with solutions reported in the literature.

Table 1. Test Case 1: Initial and Boundary Conditions for the Flow Problem

$$
\begin{aligned}
\psi(x,z,0) &= -90.0 \text{ cm}\\
v_2(x,0,t) &= 0.0\\
v_2(x,10,t) &= 0.0\\
v_1(0,0 \le z < 6,t) &= 0.0\\
\psi(0,6 \le z \le 10,t) &= 6 - z\\
\psi(15,z,t) &= -90.0 \text{ cm}
\end{aligned}
$$

Table 2. Test Case 1: Initial and Boundary Conditions for the Transport Problem

$$
\begin{aligned}
c(x,z,0) &= 0.0\\
\frac{\partial c}{\partial z}(x,0,t) &= 0.0\\
\frac{\partial c}{\partial z}(x,10,t) &= 0.0\\
\frac{\partial c}{\partial x}(0,0 \le z < 6,t) &= 0.0\\
c(0,6 \le z \le 10,t) &= 1.0\\
\frac{\partial c}{\partial x}(15,z,t) &= 0.0
\end{aligned}
$$

Test Case 1

This test case is adapted from *Huyakorn et al.* [*1985*] and concerns the transport of a nonconservative solute in an unsaturated soil. The domain, shown in Figure 2, is initially dry ($\psi = -90.0$ cm) and without solute ($c = 0$). The water and contaminant enter the system through the upper portion of the left-hand boundary ($x = 0$, $6 \le z \le 10$ cm). The initial and boundary conditions for the flow problem are given in Table 1, and in Table 2 for the transport problem. The finite element mesh, shown in Figure 3, is made up of uniform triangles whose geometric characteristics are reported in Table 3. In Table 3 we also give the values used for the various physical parameters of the model.

The solution procedure consisted of first solving the flow problem to determine the Darcy velocities (v_1, v_2) and the water saturation values (S_w) for each triangle. These values were then input to the transport code, which solved for the solute concentrations.

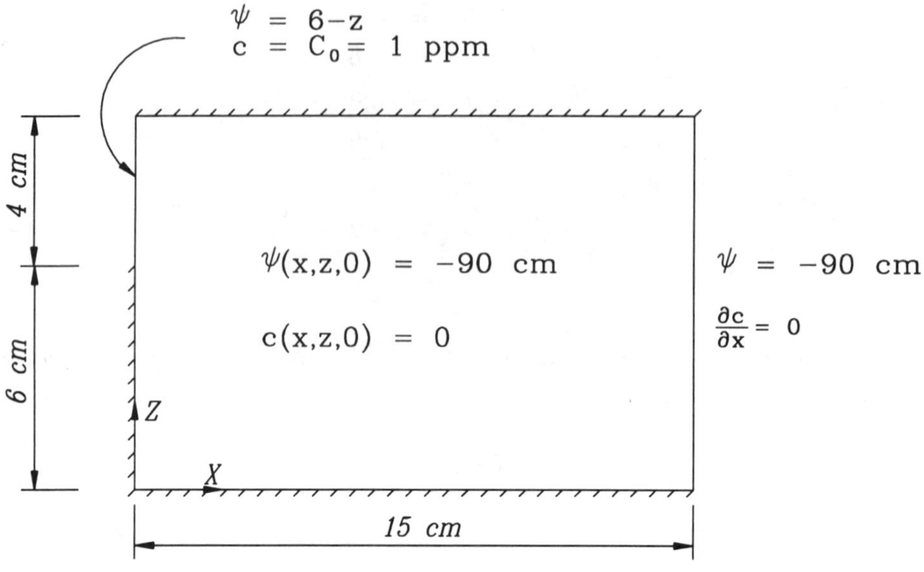

Figure 2. Test case 1: Description of the boundary and initial conditions.

Table 3. Test Case 1: Geometric and Physical Parameter Values

ϵ	η	γ	μ	ψ_a	S_{wr}
−0.01	1.0	−1.0	1.0	0.0 cm	0.333
n	S_s	k_x	k_z	α_L	α_T
0.45	0.0	1.0 cm/d	1.0 cm/d	1.0 cm	0.0 cm
D_o	λ	R_d	t range	x range	z range
0.01 cm^2/d	0.001 d^{-1}	2.0	$[0, 0.508]$ d	$[0, 15]$ cm	$[0, 10]$ cm
# nodes	# triangles	Δt_k (flow)	Δt_k (transport)	Δx	Δz
176	300	$[0.001, 0.05]$ d	$[0.01, 0.05]$ d	1.0 cm	1.0 cm

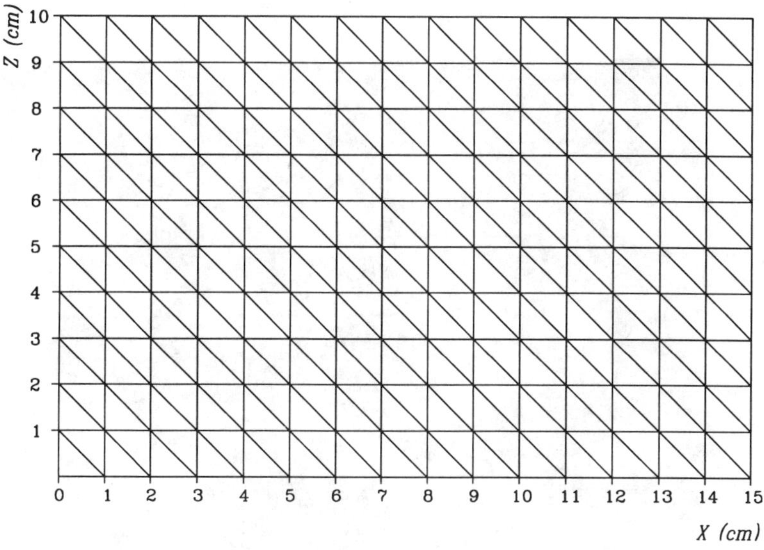

Figure 3. Test case 1: Finite element mesh.

The results of the flow and transport simulations are shown in Figures 4 (pressure head contours and velocity field from the flow model), 5–7 (solute concentration contours at three different times), 8 (vertical concentration profiles at $x = 3$ cm and three different times), and 9 (horizontal concentration profiles at $z = 10$ cm and three different times).

Test Case 2

This test case is adapted from *Gureghian* [*1983*] and concerns flow and transport in a ditch-drained aquifer with incident steady rainfall and trickle infiltration of chloride ion. Figure 10 shows, in vertical cross section, the left half of a ditch-drained aquifer where the boundary $x = 100$ cm (line BC), located midway between two equally spaced ditches, corresponds to a line of symmetry. The soil surface (line FC in Figure 10) is subjected to steady rainfall at a rate $v_2 = 0.1K_s$, where K_s is the saturated hydraulic conductivity of the aquifer, assumed homogeneous and isotropic. A portion of the soil surface (ED) corresponds to a retention pond, and is subjected to trickle type infiltration at a rate $0.05K_s$, in addition to the rainfall. The trickle infiltration on ED carries solute into the aquifer for a period of 15 days. The initial and boundary conditions for the flow problem are given in Table 4, and in Table 5 for the transport problem. The finite element mesh, shown in Figure 11, is made up of 300 nonuniform triangular elements and 176 nodes. Table 6 gives the parameter values used for the model simulations.

The flow problem was solved in steady-state, and the Darcy velocities and water saturation values obtained were then used in a 120-day simulation of the transport

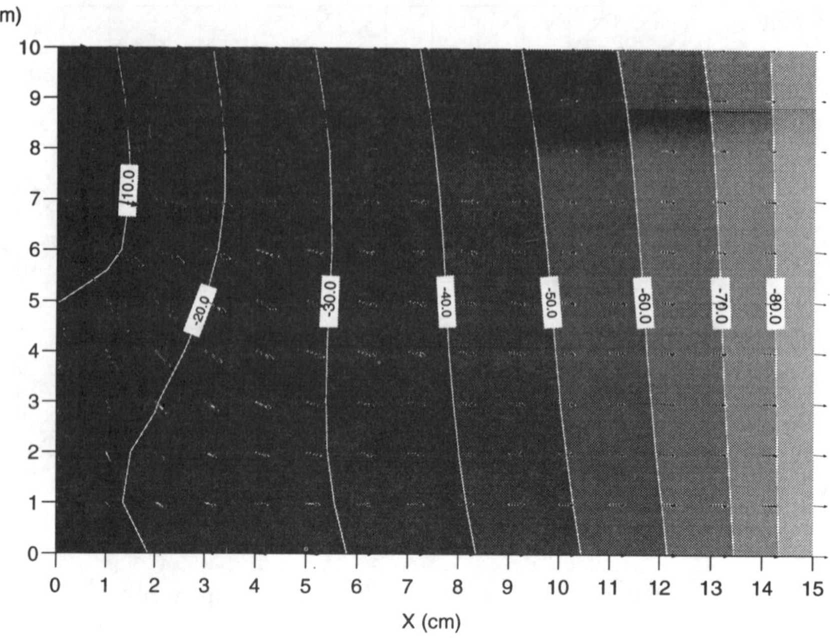

Figure 4. Test case 1: Pressure head contours and velocity field at time 0.508 days.

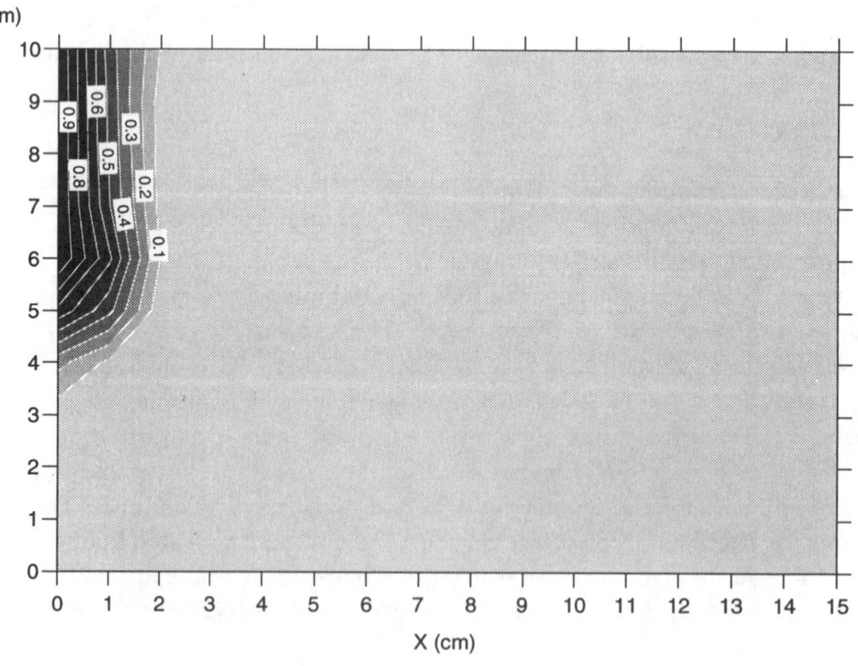

Figure 5. Test case 1: Concentration contours at time 0.053 days.

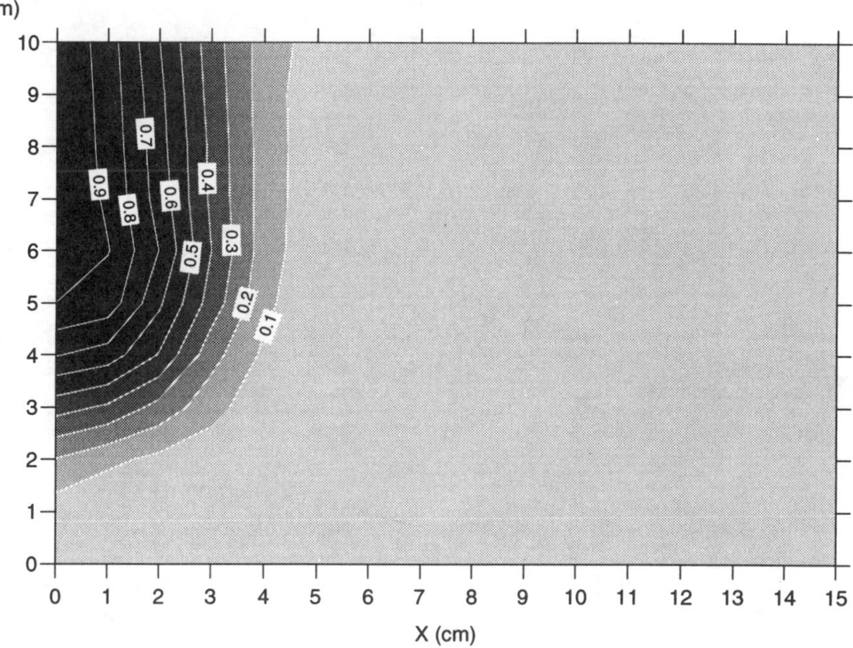

Figure 6. Test case 1: Concentration contours at time 0.165 days.

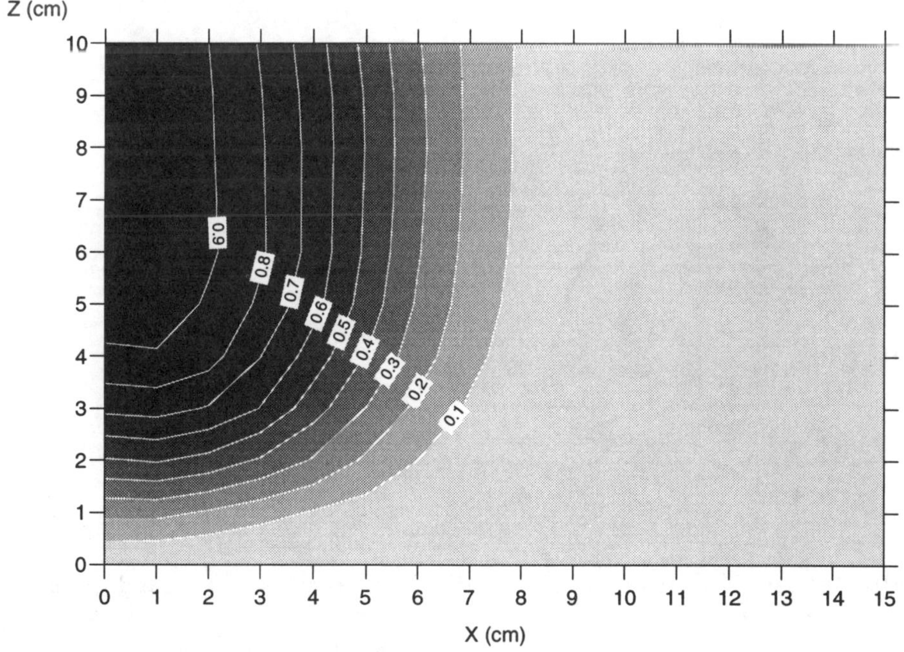

Figure 7. Test case 1: Concentration contours at time 0.508 days.

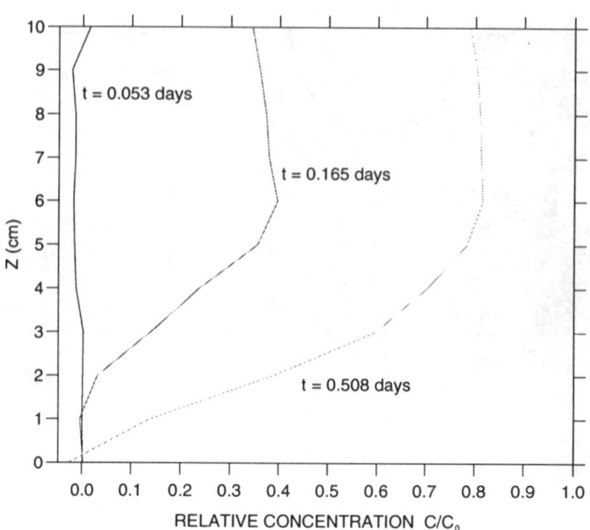

Figure 8. Test case 1: Vertical concentration profiles at $x = 3$ cm and three different times.

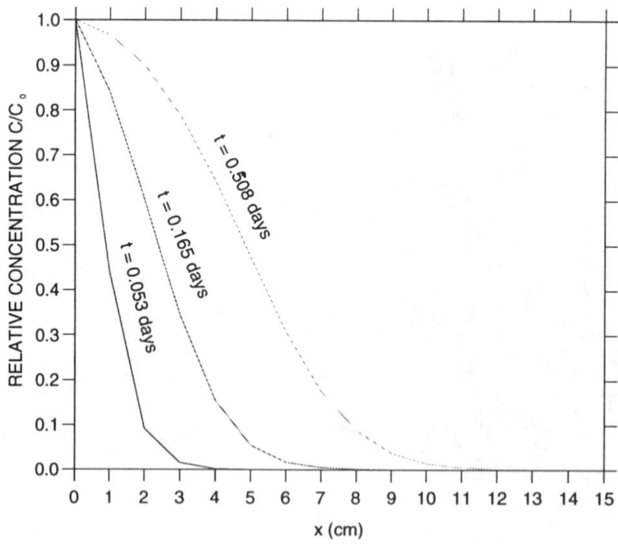

Figure 9. Test case 1: Horizontal concentration profiles at $z = 10$ cm and three different times.

Table 4. Test Case 2: Boundary Conditions for the Flow Problem

ψ	=	0	on AG
v_1	=	0	on FG and BC
v_2	=	0	on AB
v_2	=	$V_r = 0.1K_s$	on EF and CD
v_2	=	$V_r + V_s = 0.15K_s$	on ED

Table 5. Test Case 2: Initial and Boundary Conditions for the Transport Problem

c	=	0	at $t = 0$
c	=	1.0	for $t \leq 15$ days on ED
	=	0.0	for $t > 15$ days on ED
$\partial c/\partial z$	=	0	on AB, EF and CD
$\partial c/\partial x$	=	0	on AF and BC

Table 6. Test Case 2: Physical Parameter Values

ϵ	η	γ	ψ_a	S_{wr}
0.015	2.0	3.0	-10 cm	0.01
n	S_s	K_s	a	b
0.3	0.096 cm^{-1}	40 cm/d	2.0	3.5
α_L	α_T	D_o	λ	k_d
0.5 cm	0.1 cm	10^{-5} cm^2/d	0.0	0.0
t range	x range	z range	Δt_k (transport)	
$[0, 120]$ d	$[0, 100]$ cm	$[0, 50]$ cm	$[0.025, 0.5]$ d	

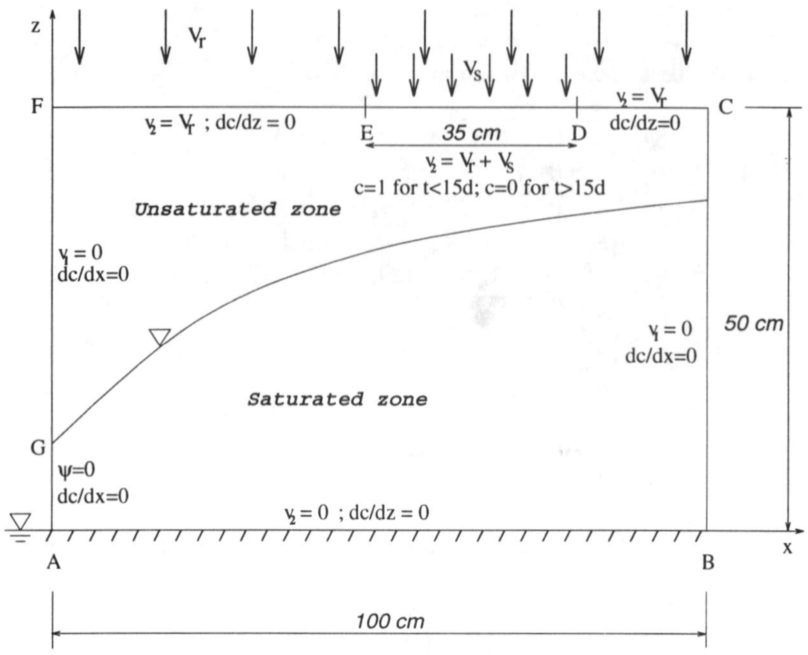

Figure 10. Test case 2: Description of the boundary conditions.

problem. The results of the flow and transport simulations are shown in Figures 12 (pressure head contours and velocity field from the flow model) and 13–16 (solute concentration contours at four different times).

Test Case 3

This test case considers the same problem as test case 2, but we now assume that the solute is reactive and undergoes nonequilibrium sorption. The parameter values are the same as those used for test case 2, with the following additional parameters for the non-LEA model: $n_m = 0.3$; $n_{im} = 0.14$; $k_{d_m} = 0.5$; $k_{d_{im}} = 0.5$; $\gamma_s = 2.0$; $\alpha = 0.1$ d^{-1}; $F = 0.4$.

In Figures 17–20 the solution contours for the mobile region concentrations c_m are shown at four different times. In Figures 21 and 22 some numerical aspects of the flow and transport simulations are illustrated. Figure 21 compares the convergence behavior of various nonlinear iterative schemes used to solve the steady-state flow problem, while Figure 22 shows the effect of time step size on the convergence behavior of the BI-CGSTAB algorithm used to solve the nonsymmetric linear systems arising from the transport problem.

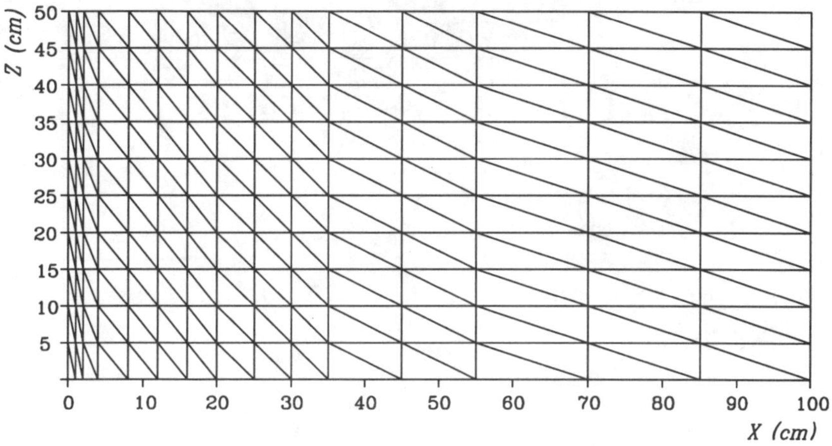

Figure 11. Test case 2: Finite element mesh.

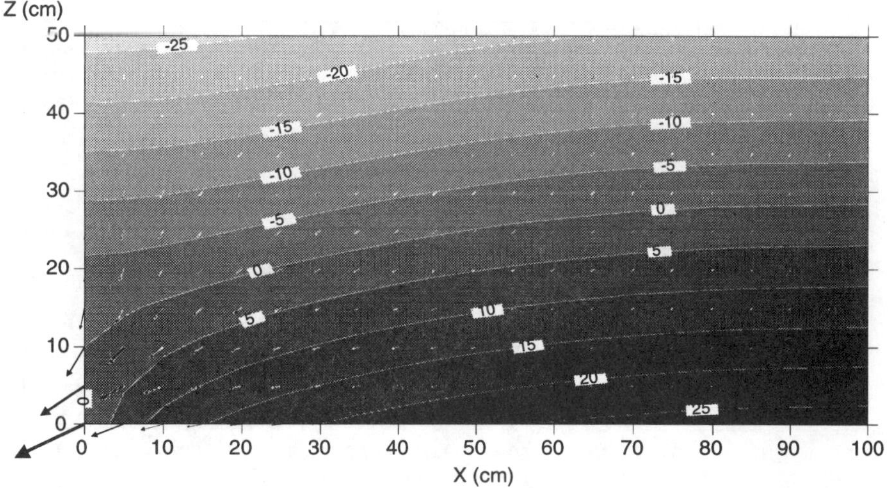

Figure 12. Test case 2: Steady-state pressure head contours and velocity field.

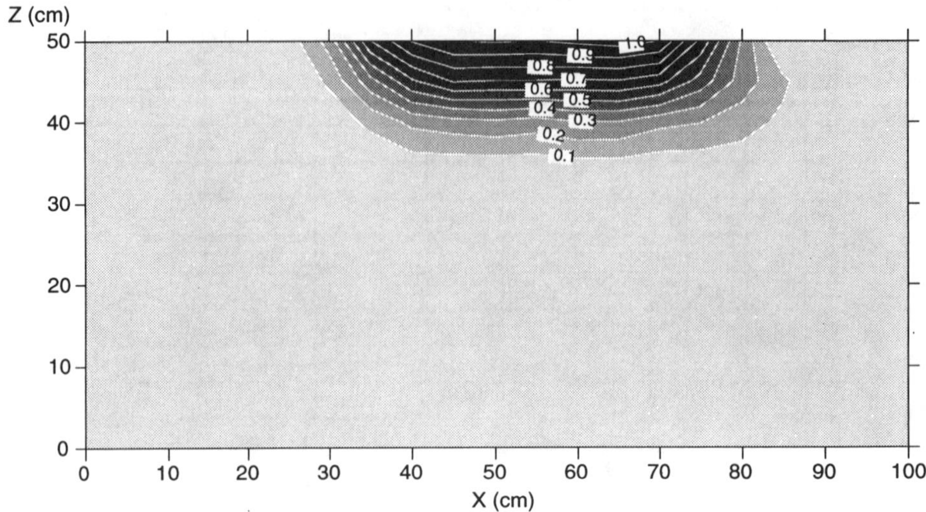

Figure 13. Test case 2: Concentration contours at time 14.775 days.

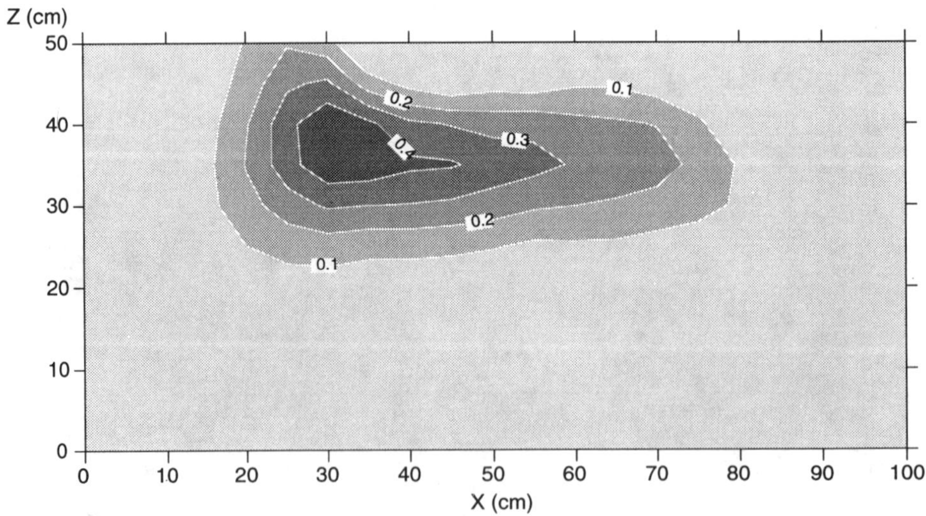

Figure 14. Test case 2: Concentration contours at time 45.275 days.

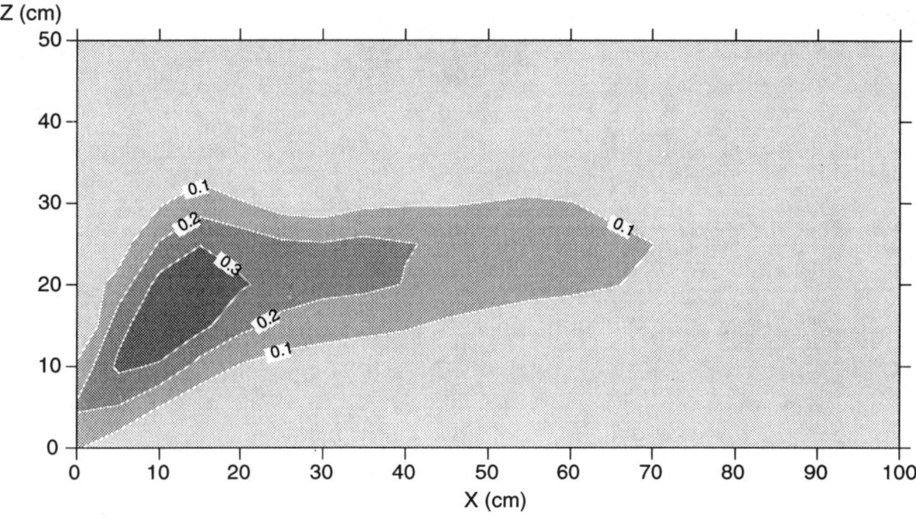

Figure 15. Test case 2: Concentration contours at time 90.275 days.

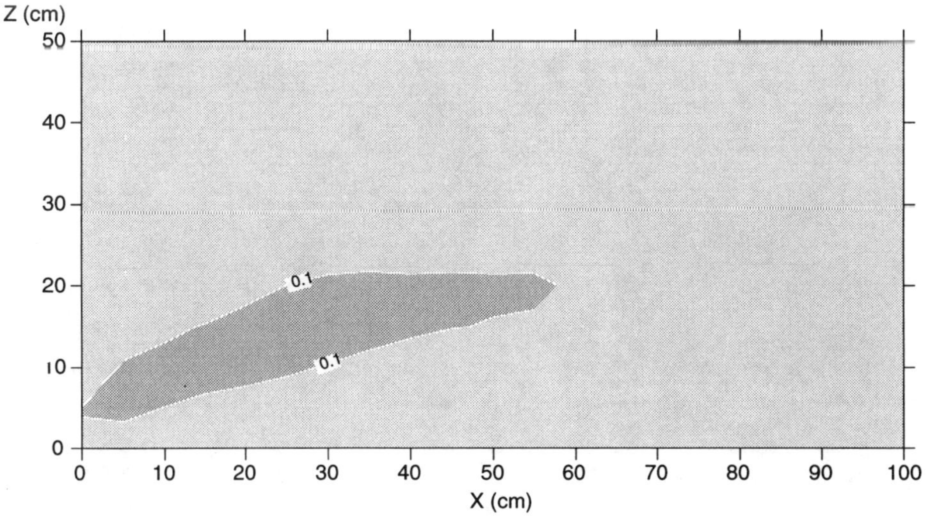

Figure 16. Test case 2: Concentration contours at time 120 days.

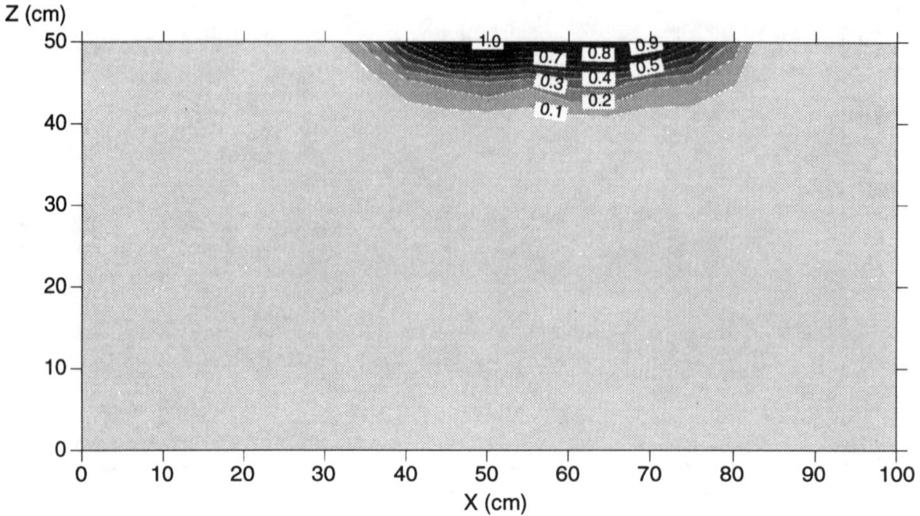

Figure 17. Test case 3: Mobile region concentration contours at time 14.775 days.

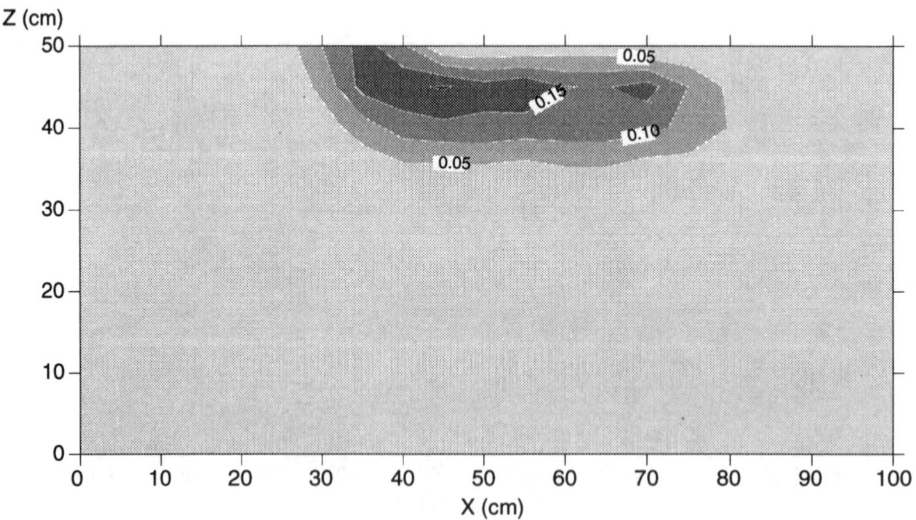

Figure 18. Test case 3: Mobile region concentration contours at time 45.275 days.

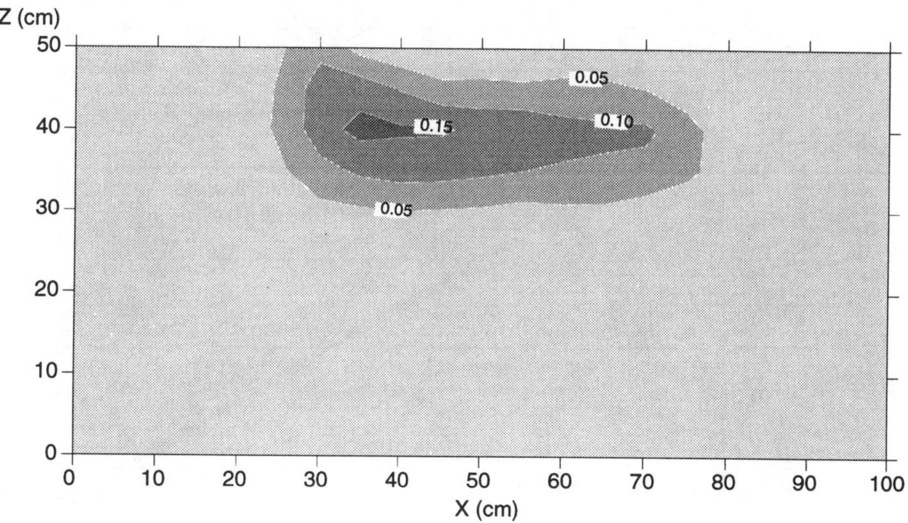

Figure 19. Test case 3: Mobile region concentration contours at time 90.275 days.

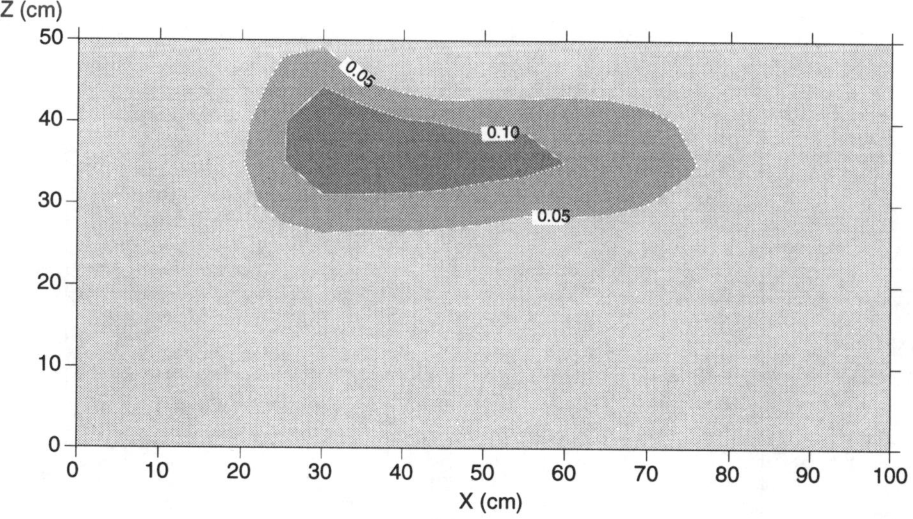

Figure 20. Test case 3: Mobile region concentration contours at time 120 days.

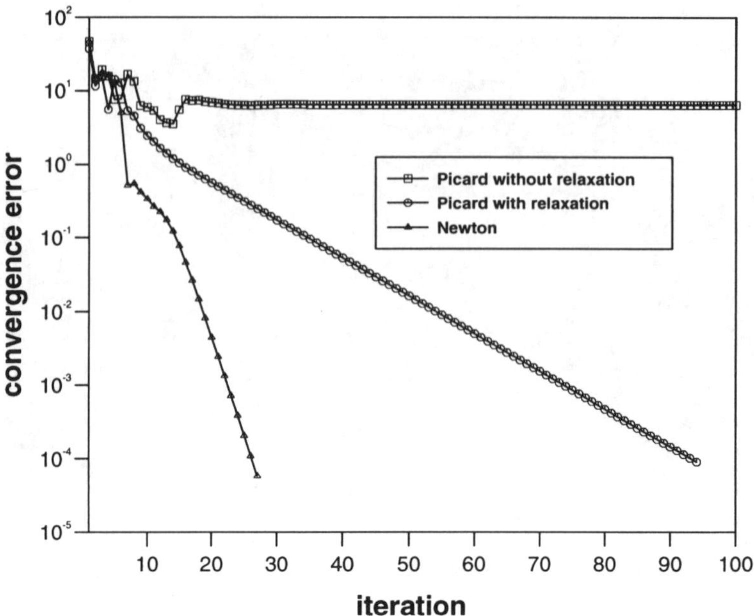

Figure 21. Test case 3: Comparison of Picard, relaxed Picard, and Newton convergence for the nonlinear flow problem.

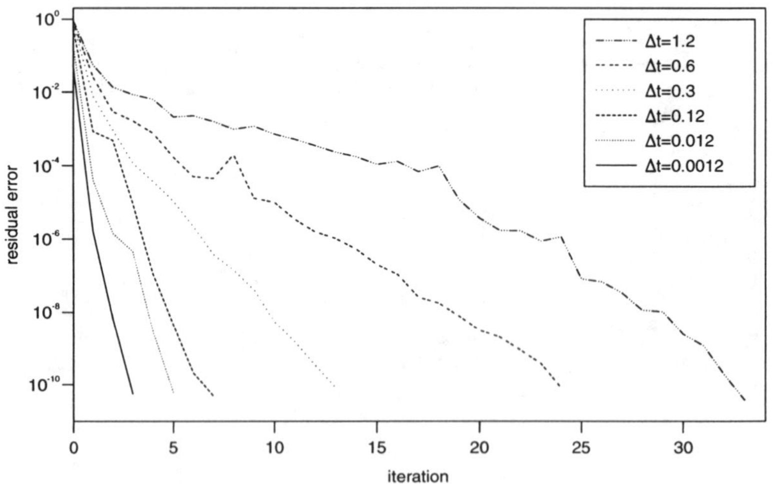

Figure 22. Test case 3: Convergence profiles of the BI-CGSTAB nonsymmetric solver at various time step sizes in the solution of the non-LEA transport problem.

Conclusions

Mathematical models can be effectively used to study the migration and fate of contaminants and their effects on water resources. The models are based on the partial differential equations of fluid and solute continuity, and are solved numerically by finite element methods. Given information on the hydrogeological and physical characteristics of the soil and groundwater zones, including the initial state of the system and the conditions at the boundaries of the region, a numerical simulation will yield information on the fluid pressure and contaminant concentration distributions and on the direction and rate of transport of the pollutants. The numerical procedure consists of first solving the flow equation to obtain the pressure distribution, then applying Darcy's Law to calculate the velocities, and lastly solving the transport equation for the concentration field.

With the aid of Newton or Picard iteration to linearize the unsaturated flow equation, preconditioned conjugate gradient-like schemes to solve large sparse symmetric and nonsymmetric systems, and an integro-differential formulation to deal with the coupling in the non-LEA transport model, the finite element models represent accurate, reliable, and efficient tools for studying flow and reactive transport phenomena in soils and aquifers, with pollutants which undergo equilibrium or nonequilibrium sorption. Running these models on powerful current generation RISC-based workstations allows one to address and solve realistic contamination problems rapidly and inexpensively.

Appendix A: Calculation of Finite Element Matrices for the Flow Equation

In this appendix we evaluate matrices $\boldsymbol{H}$ and $\boldsymbol{P}$ and vector $\boldsymbol{q}^*$ of equation (20). We introduce a slight change in notation, using x, z, k_x, k_z in place of x_1, x_2, k_{11}, k_{22}. With triangular elements and linear basis functions we have, for triangle e with vertices i, j, m, $N_i^e = (\varsigma_i + \zeta_i x + \xi_i z)/2\Delta^e$ where

$$\Delta^e = \frac{1}{2}\begin{vmatrix} 1 & x_i & z_i \\ 1 & x_j & z_j \\ 1 & x_m & z_m \end{vmatrix}$$

is the area of triangle e and $\varsigma_i = x_j z_m - x_m z_j$, $\zeta_i = z_j - z_m$, $\xi_i = x_m - x_j$. The ijth element of matrix $\boldsymbol{H}$ is given by $H_{ij} = \sum_e H_{ij}^e$ where

$$\begin{aligned} H_{ij}^e &= \int_{\Delta^e} k_{rw}^e \left(k_x^e \frac{\partial N_i^e}{\partial x}\frac{\partial N_j^e}{\partial x} + k_z^e \frac{\partial N_i^e}{\partial z}\frac{\partial N_j^e}{\partial z} \right) d\Delta^e \\ &= \int_{\Delta^e} k_{rw}^e \left(k_x^e \frac{\zeta_i}{2\Delta^e}\frac{\zeta_j}{2\Delta^e} + k_z^e \frac{\xi_i}{2\Delta^e}\frac{\xi_j}{2\Delta^e} \right) d\Delta^e \\ &= \frac{k_{rw}^e}{4 \mid \Delta^e \mid}(k_x^e \zeta_i \zeta_j + k_z^e \xi_i \xi_j) \end{aligned}$$

and where the nonlinear coefficient k^e_{rw} is evaluated at the centroid of each element. This treatment of nonlinear coefficients is also used for all the other integral terms. The ijth element of $\boldsymbol{P}$ is $P_{ij} = \sum_e P^e_{ij}$ where

$$P^e_{ij} = \int_{\Delta^e} \sigma^e N^e_i N^e_j d\Delta^e = \sigma^e \frac{|\Delta^e|}{12} \cdot \begin{cases} 2 & \text{if } i = j \\ 1 & \text{if } i \neq j \end{cases}$$

Finally, $q^*_i = \sum_e G^e_i + \sum_e F^e_i + \sum_e L^e_i$ where

$$F^e_i = -\int_{\Gamma^e_2} q^e_n N^e_i d\Gamma^e = -q^e_n \frac{|l^e|}{2}$$

$$L^e_i = -\int_{\Delta^e} q^e N^e_i d\Delta^e = -q^e \frac{|\Delta^e|}{3}$$

$$G^e_i = \int_{\Delta^e} k^e_z k^e_{rw} \frac{\partial N^e_i}{\partial z} d\Delta^e = \frac{|\Delta^e|}{2\Delta^e} k^e_z k^e_{rw} \xi_i$$

In the expression for F^e_i, $|l^e|$ is the length of the boundary segment Γ^e_2.

Appendix B: Velocity Calculation

We compute the Darcy velocity on a triangle with vertices i, j, m. We have, from equation (3),

$$\begin{aligned} v^e_x &= -k^e_x k^e_{rw} \frac{\partial \hat{\psi}^e}{\partial x} \\ v^e_z &= -k^e_z k^e_{rw} \left(\frac{\partial \hat{\psi}^e}{\partial z} + 1 \right) \end{aligned}$$

Since on each triangular element $\hat{\psi}^e = \psi_i N_i + \psi_j N_j + \psi_m N_m$, we get

$$\begin{aligned} v^e_x &= -\frac{k^e_x k^e_{rw}}{2\Delta^e} (\zeta_i \psi_i + \zeta_j \psi_j + \zeta_m \psi_m) \\ v^e_z &= -\frac{k^e_z k^e_{rw}}{2\Delta^e} (\xi_i \psi_i + \xi_j \psi_j + \xi_m \psi_m) - k^e_z k^e_{rw} \end{aligned}$$

where ζ_i and ξ_i are as defined in Appendix A.

Appendix C: Calculation of Finite Element Matrices for the Transport Equation

We evaluate $\boldsymbol{A}$, $\boldsymbol{B}$, $\boldsymbol{E}$, $\boldsymbol{F}$, $\boldsymbol{G}$, and $\boldsymbol{r}^*$ of equation (26). The coefficients of matrix $\boldsymbol{A}$ are computed by performing a rotation of axes so as to align the new coordinate direction x' with the direction of the Darcy velocity. The new coordinate system is defined as $x' = (v_x x + v_z z)/|v|$, $z' = (v_z x - v_x z)/|v|$. Setting $D_{x'} = D_{11}$, $D_{z'} = D_{22}$, we have

$$\begin{aligned} D_{x'} &= \alpha_L |v| + n S_w D_0 \tau \\ D_{z'} &= \alpha_T |v| + n S_w D_0 \tau \end{aligned}$$

We evaluate the finite element matrices and vectors for the Galerkin case ($W_i \equiv N_i$). The ijth element of matrix $\boldsymbol{A}$ is given by $A_{ij} = \sum_e A^e_{ij}$ where

$$
\begin{aligned}
A^e_{ij} &= \int_{\Delta^e} \left(D^e_{x'} \frac{\partial N^e_i}{\partial x'} \frac{\partial N^e_j}{\partial x'} + D^e_{z'} \frac{\partial N^e_i}{\partial z'} \frac{\partial N^e_j}{\partial z'} \right) d\Delta^e \\
&= \int_{\Delta^e} \left(D^e_{x'} \frac{\zeta'_i}{2\Delta^e} \frac{\zeta'_j}{2\Delta^e} + D^e_{z'} \frac{\xi'_i}{2\Delta^e} \frac{\xi'_j}{2\Delta^e} \right) d\Delta^e \\
&= \frac{1}{4 \mid \Delta^e \mid} (D^e_{x'} \zeta'_i \zeta'_j + D^e_{z'} \xi'_i \xi'_j)
\end{aligned}
$$

where ζ', ξ' are computed in the local (rotated) reference frame x', z'. The ijth element of $\boldsymbol{B}$ is $B_{ij} = \sum_e B^e_{ij}$ where

$$
\begin{aligned}
B^e_{ij} &= \int_{\Delta^e} \left(v^e_x \frac{\partial N^e_j}{\partial x} + v^e_z \frac{\partial N^e_j}{\partial z} \right) N^e_i d\Delta^e \\
&= \int_{\Delta^e} \left(v^e_x \frac{\zeta_j}{2\Delta^e} + v^e_z \frac{\xi_j}{2\Delta^e} \right) N^e_i d\Delta^e = \frac{\mid \Delta^e \mid}{6\Delta^e} (v^e_x \zeta_j + v^e_z \xi_j)
\end{aligned}
$$

For $E_{ij} = \sum_e E^e_{ij}$,

$$
E^e_{ij} = \int_{\Delta^e} n^e S^e_w R^e_d \lambda^e N^e_i N^e_j d\Delta^e = n^e S^e_w R^e_d \lambda^e \frac{\mid \Delta^e \mid}{12} \cdot \begin{cases} 2 & \text{se } i = j \\ 1 & \text{se } i \neq j \end{cases}
$$

For $G_{ij} = \sum_e G^e_{ij}$,

$$
G^e_{ij} = \int_{\Delta^e} n^e S^e_w R^e_d N^e_i N^e_j d\Delta^e = n^e S^e_w R^e_d \frac{\mid \Delta^e \mid}{12} \cdot \begin{cases} 2 & \text{se } i = j \\ 1 & \text{se } i \neq j \end{cases}
$$

For $F_{ij} = \sum_e F^e_{ij}$,

$$
\begin{aligned}
F^e_{ij} &= \int_{\Delta^e} q^e N^e_i N^e_j d\Delta^e - \int_{\Gamma^e_5} (v^e_x n_x + v^e_z n_z) N^e_i N^e_j d\Gamma^e \\
&= \left[q^e \frac{\mid \Delta^e \mid}{12} - (v^e_x n_x + v^e_z n_z) \frac{\mid l^e_5 \mid}{6} \right] \cdot \begin{cases} 2 & \text{se } i = j \\ 1 & \text{se } i \neq j \end{cases}
\end{aligned}
$$

If we lump all the extradiagonal contributions onto the main diagonal, we get

$$
\begin{aligned}
F^e_{ii} &= q^e \frac{\mid \Delta^e \mid}{4} - (v^e_x n_x + v^e_z n_z) \frac{\mid l^e_5 \mid}{2} \\
F^e_{ij} &= 0 \quad \text{se } i \neq j
\end{aligned}
$$

Finally, $r^*_i = \sum_e G^{1e}_i + \sum_e G^{2e}_i + \sum_e G^{3e}_i$ where

$$
\begin{aligned}
G^{1e}_i &= - \int_{\Delta^e} (q^e c^{*^e} + f^e) N^e_i d\Delta^e = -(q^e c^{*^e} + f^e) \frac{\mid \Delta^e \mid}{3} \\
G^{2e}_i &= - \int_{\Gamma^e_4} q^{D^e}_c N^e_i d\Gamma^e = -q^{D^e}_c \frac{\mid l^e_4 \mid}{2} \\
G^{3e}_i &= - \int_{\Gamma^e_5} q^{T^e}_c N^e_i d\Gamma^e = -q^{T^e}_c \frac{\mid l^e_5 \mid}{2}
\end{aligned}
$$

In the expressions for F^e_{ij}, G^{2e}_i, and G^{3e}_i, $\mid l^e_4 \mid$ and $\mid l^e_5 \mid$ are the lengths of the boundary segments Γ^e_4 and Γ^e_5, respectively.

Acknowledgments This work has been carried out with the financial support of CNR-ENEL (Project "Interactions of energy systems with human health and environment", Rome, Italy), CNR (Gruppo Nazionale per la Difesa dalle Catastrofi Idrogeologiche, linea di Ricerca n. 4), and the Sardinia Regional Authorities.

References

Axelsson, O., Conjugate gradient type methods for unsymmetric and inconsistent systems of linear equations, *Linear Algebra Appl.* 29, 1–16, 1980.

Bear, J., *Hydraulics of Groundwater.* McGraw-Hill, New York, 1979.

Brusseau, M. L. and P. S. C. Rao, Sorption nonideality during organic contaminant transport in porous media, *Crit. Rev. Env. Contr.* 19(1), 33–99, 1989.

Coats, K. H. and B. D. Smith, Dead-end pore volumes and dispersion in porous media, *Soc. Pet. Eng. J.* 4, 73–84, 1964.

Cooley, R. L., Some new procedures for numerical solution of variably saturated flow problems, *Water Resour. Res.* 19(5), 1271–1285, 1983.

Freund, R. W., A transpose-free quasi-minimal residual algorithm for non-Hermitian linear systems, *SIAM J. Sci. Comput.* 14, 470–482, 1993.

Frind, E. O. and M. J. Verge, Three-dimensional modeling of groundwater flow systems, *Water Resour. Res.* 14(5), 844–856, 1978.

Galeati, G. and G. Gambolati, On boundary conditions and point sources in the finite element integration of the transport equation, *Water Resour. Res.* 25(5), 847–856, 1989.

Gambolati, G. and A. M. Perdon, The conjugate gradients in flow and land subsidence modeling, In: Bear, J. and Y. Corapcioglu (eds.) *Fundamental of Transport Phenomena in Porous Media.* NATO-ASI Series, Applied Sciences 82, Martinus Nijoff B.V., The Hague, pp 953–984, 1984.

Gambolati, G., C. Paniconi and M. Putti, Numerical modeling of contaminant transport in groundwater, In: Petruzzelli, D. and F. G. Helfferich (eds.) *Migration and Fate of Pollutants in Soils and Subsoils.* Springer-Verlag, Berlin. Volume 32 of *NATO ASI Series G: Ecological Sciences*, pp 381–410, 1992.

Giddings, J. C., Kinetic origin of tailing in chromatography, *Anal. Chem.* 35, 1999–2002, 1963.

Gureghian, A. B., TRIPM, a two-dimensional finite element model for the simultaneous transport of water and reacting solutes through saturated and unsaturated porous media. Technical Report ONWI-465, Off. of Nuclear Water Isolation, Columbus, Ohio, 1983.

Huyakorn, P. S. and C. Taylor, Finite element models for coupled groundwater and convective dispersion, In: *Proc. I Int. Conf. on Finite Elements in Water Resources.* Pentech, London, pp 1131–1151, 1976.

Huyakorn, P. S. and G. F. Pinder, *Computational Methods in Subsurface Flow.* Academic Press, London, 1983.

Huyakorn, P. S., S. D. Thomas and B. M. Thompson, Techniques for making finite elements competitive in modeling flow in variably saturated porous media, *Water Resour. Res.* 20(8), 1099–1115, 1984.

Huyakorn, P. S., J. W. Mercer and D. S. Ward, Finite element matrix and mass balance computational schemes for transport in variably saturated porous media, *Water Resour. Res.* 21(3), 346–358, 1985.

Huyakorn, P. S., P. F. Andersen, J. W. Mercer and H. O. White, Saltwater intrusion in aquifers: Development and testing of a three-dimensional finite element model, *Water Resour. Res.* 23(2), 293–312, 1987.

Kinzelbach, W., *Groundwater Modeling.* Elsevier, Amsterdam, 1986.

Lapidus, L. and N. R. Amundson, Mathematics of adsorption in beds, VI. The effect of longitudinal diffusion in ion exchange and chromatographic columns, *J. Phys. Chem.* 56, 984–988, 1952.

Neuman, S. P., Saturated-unsaturated seepage by finite elements, *ASCE, Journal of the Hydraulics Division.* 99, 2233–2250, 1973.

Neuman, S. P., A Eulerian-Lagrangian numerical scheme for the dispersion convection equation using conjugate space time grids, *J. Comp. Phys.* 41, 270–294, 1981.

Nielsen, D. R., M. T. van Genuchten and A. J. W. Biggar, Water flow and solute transport processes in the unsaturated zone, *Water Resour. Res.* 22(9), 89S–108S, 1986.

Page, A. L., A. A. Elseewi and I. R. Straughan, Physical and chemical properties of fly ash from coal-fired power plants with reference to environmental impacts, *Residue Rev.* 7, 83–120, 1979.

Paniconi, C., A. A. Aldama and E. F. Wood, Numerical evaluation of iterative and noniterative methods for the solution of the nonlinear Richards equation, *Water Resour. Res.* 27(6), 1147–1163, 1991.

Peyret, R. and T. D. Taylor, *Computational Methods for Fluid Flow.* Springer-Verlag, New York, 1983.

Philip, J. R., Theory of infiltration, *Adv. Hydrosci.* 5, 215–296, 1969.

Pinder, G. F. and W. G. Gray, *Finite Element Simulation in Surface and Subsurface Hydrology.* Academic Press, San Diego, 1977.

Pini, G., G. Gambolati and G. Galeati, 3-D finite element transport models by upwind preconditioned conjugate gradients, *Adv. Water Resources.* 12, 54–58, 1989.

Putti, M., W. W.-G. Yeh and W. A. Mulder, A triangular finite volume approach with high resolution upwind terms for the solution of groundwater transport equations, *Water Resour. Res.* 26(12), 2865–2880, 1990.

Putti, M. and C. Paniconi, Evaluation of the Picard and Newton iteration schemes for three-dimensional unsaturated flow, In: Russell, T. R., R. E. Ewing, C. A. Brebbia, W. G. Gray and G. F. Pinder (eds.) *Proceedings of the IX International Conference on Computational Methods in Water Resources, Vol. 1, Numerical Methods in Water Resources.* Computational Mechanics and Elsevier Applied Sciences, Southampton, London, 1992.

Ross, P. J., Efficient numerical methods for infiltration using Richards' equation, *Water Resour. Res.* 26(2), 279–290, 1990.

Sharma, S., M. H. Fulekar and C. P. Jayalakshmi, Fly ash dynamics in soil-water systems, *Critical Rev. in Environ. Control.* 19(3), 251–275, 1989.

Simsiman, G. V., G. Chesters and A. W. Andren, Effect of ash disposal ponds on groundwater quality at a coal-fired power plant, *Water Res.* 21(4), 417–426, 1987.

Sorek, S., Eulerian-Lagrangian method for solving transport in aquifers, *Adv. Water Resources.* 11, 67–73, 1988.

Theis, T. L., J. D. Westrick, C. L. Hsu and J. J. Marley, Field investigation of trace metals in groundwater from fly ash disposal, *J. Water Pollut. Control Fed.* 50, 2457–2469, 1978.

van der Vorst, H. A., Bi-CGSTAB: A fast and smoothly converging variant of BI-CG for the solution of nonsymmetric linear systems, *SIAM J. Sci. Stat. Comput.* 13, 631–644, 1992.

van Genuchten, M. T. and P. J. Wierenga, Mass transfer studies in sorbing porous media: 1. Analytical solutions, *Soil Sci. Soc. Amer. J.* 40(4), 473–480, 1976.

Weber Jr., W. J., P. McGinley and L. Katz, Sorption phenomena in subsurface systems: Concepts, models and effects on contaminant fate and transport, *Water Res.* 25(5), 499–528, 1991.

Chapter 8

Mechanisms and models for aggressive permeant interactions with soils

A.A. Jennings,[a] V. Ravi[b]

[a] *Department of Civil Engineering, The Case School of Engineering, Case Western Reserve University, 10900 Euclid Avenue, Cleveland, Ohio 44106-7201, USA*

[b] *Dynamic Corporation, Robert S. Kerr Environmental Research Laboratory, US Environmental Protection Agency, PO Box 1198, Ada, Oklahoma 74820, USA*

Abstract

Aggressive permeants are capable of altering the hydraulic properties of the soils through which they flow. These phenomenon can be significant in several applications, but are probably of most current concern in hazardous waste management situations. The fear is that aggressive permeants will alter the hydraulic conductivity of soil barriers used to contain them, and escape into the environment.

This chapter presents a comprehensive look at the subject of aggressive permeant interactions with soils. The mechanisms thought to be responsible for the most dramatic impact are described and illustrated using laboratory measurements on a wide variety of interaction systems. Strategies for modeling the global impact of aggressive permeants, and for analyzing the preferential flow pathways that can develop are presented. In addition, algorithms are presented for testing alternative aggressive permeant interaction and for extracting the required model coefficients for typical permeameter observations.

Key Words

Aggressive Permanents, Hydraulic Conductivity, Permeameters, Mathematical Modeling, DIP Models, Preferential Flow Models, Parameter Estimation.

1. Introduction

The following sections will present a summary of our current understanding of aggressive permeant interactions with soils. Although these can be of significance in several situations, we will concentrate on those of highest current concern. Wherever possible, we will illustrate phenomenon with actual experimental observations. A considerable database of aggressive permeant observations is being assembled. We have attempted to cite the work of many of the research teams that have reported their experimental observations. We encourage readers to dig farther into this database, and apologize to those researchers who we have inadvertently omitted.

1.1 Hydraulic Conductivity, Intrinsic Permeability and Aggressive Permeants

The saturated hydraulic conductivity (K) of a soil is governed by the physical properties of the soil structure (porosity, pore size distribution, pore surface properties, etc.) as well as properties of the permeating liquid (density ρ, viscosity, μ,etc.) and the gravitational body force due to acceleration g. The intrinsic permeability (k) of a soil is generally defined as that portion of the hydraulic conductivity that is governed by the soil's properties. Often the two are taken to be related as follows.

$$K = \frac{k\rho g}{\mu} \tag{1}$$

Generally the saturated hydraulic conductivity of a soil is measured relative to a base permeant (e.g. water) and base gravitation body force. It is normally assumed to remain constant relative to the base permeant unless there is some "structural" change such as consolidation. Obviously the hydraulic conductivity would vary for fluids significantly different from water (i.e. different ρ, μ values). It is also possible for the intrinsic permeability to vary with different permeants because of the subtle liquid/solid interface interactions that dominate fluid flow in a porous medium. However, until relatively recently, it has almost universally been assumed that the hydraulic permeability and/or intrinsic permeability remains constant for any fixed permeant system. In this presentation we will explore what we now know about permeant systems that violate this assumption.

Aggressive permeants are defined as soil permeants that, during the permeation process, enter into strong physical/chemical interactions with the solid phase, significantly alter the physical or chemical properties of the solid phase, and thereby alter the hydraulic conductivity in ways that cannot be accounted for by changes in liquid density or viscosity. We now know that aggressive permeants can alter the hydraulic conductivity by several orders of magnitude. The induced changes can be reversible, or permanent, but they do occur and can be of great significance to the soil system experiencing the flow.

Consider the following example. Suppose that someone was attempting to design an underdrain system for a solid waste land disposal facility. The facility is (at this point) of indeterminant size, so let's look at a module of 10,000 m^2 (approximately 2.5 acres). Let's further assume that the soil to be used in the facility's underlying hydraulic barrier has a hydraulic conductivity of $1x10^{-9}$ m/sec., that the liner could be subjected to a hydraulic head of as much as a meter, and that the hydraulic barrier will be approximately one meter thick. We could estimate the module's daily leakage as follows using Darcy's law.

$$Q = qA = -K\frac{d\phi}{dx}A \approx -K\frac{\Delta\phi}{\Delta L}A = \left(1.0x10^{-9}\right)\frac{1.0m}{1.0m}(10{,}000m)(86{,}400sec / day) = 0.864m^3 / day \tag{2}$$

Here Q is the volumetric flux, q is the Darcy velocity , A is the area, L is the length, ϕ is the hydrostatic fluid potential and $\Delta\phi$ is the change in hydrostatic fluid potential.

The result would be an expected flow of approximately one cubic meter per day, and this might be used to design a compatible underdrain module. However, what impact would aggressive permeant interactions have on this design? They might, for example alter the soil's hydraulic conductivity by two or three orders of magnitude. Should the hydraulic conductivity decline, there would be little impact on the design. The flow would decline to 0.000864 m^3/day (i.e. approximately one liter per day), but most people would consider this to be a desirable change. In any case, the flow would not overwhelm the underdrain system. However, if the hydraulic conductivity was to increase by three orders of magnitude, the results would be far less benign. In this case, the flow would increase to 864 m^3/day which would clearly overwhelm any underdrain system designed for 1 m^3/day. Increases of this magnitude would cause several additional problems including regulatory agency responses and could lead one to look for a "leak" (i.e. a localized low resistance pathway) that would not exist.

This simple example illustrates one of the possible scenarios in which aggressive permeants can lead to serious, unanticipated difficulties. There are many others. However, before we go on to discuss the mechanisms that can lead to aggressive permeant interactions, and some of the modeling work being done to help predict their consequences, it seems wise to discuss the experimental database from which the concept of "aggressive permeant" has evolved.

1.2 Database of Experimental Observations

It is well-established that a wide variety of chemical interactions can alter the saturated hydraulic conductivity of soils. A great deal of the recent research on this subject has focused on the potential for soil permeability modifications at hazardous waste disposal sites, but the phenomenon is of significance in many other areas. In subsequent sections we will draw from several of these areas to illustrate the magnitude and direction of typical aggressive permeant impacts.

A substantial body of experimental evidence has been assembled on the potential impacts of aggressive permeant interactions that can occur during irrigation (Alperovitch, et al., 1981; Frenkel et al., 1978; McNeal and Coleman, 1966; McNeal et al., 1966; Park and O'Connor, 1980; Quirk and Schofield 1955; Shainberg et al.,1981; Yaron and Thomas, 1968). Problems occur when irrigation waters react with the irrigated soils to decrease hydraulic conductivity. The problems can occur whenever the chemical composition of the irrigation water is significantly different than that of the native pore water as might be the case for irrigation waters imported from other drainage basins. Hydraulic conductivity can apparently decrease significantly leading to crop damage from surface ponding. Similar problems can occur in wastewater land treatment systems (Burton, et al., 1981; Trick et al., 1985) although there are probably mechanistic differences due to increased microbial activity.

Several researchers have demonstrated that aggressive permeant impacts can be significant factors during saltwater intrusion episodes. As with irrigation effects, the impact results from the vast differences between the chemical composition of freshwater and saltwater (Goldenberg et al., 1983; Hardcastel and Mitchell, 1974; Goldenberg et al., 1984; Mehnert and Jennings, 1985). Because of the nature of the mechanisms (probably a flocculation/deflocculation impact which we will discuss in a subsequent section) not all coastal aquifers will experience significant aggressive permeant impacts. However, where they do occur, they can be significant, and may or may not be reversible. In addition, because they alter the rate of fluid flows, "reversible" impacts may be reversed at much more gradual rates than when they were formed.

Aggressive permeant impacts have also been a consideration in petroleum engineering for many years. The potential for encountering "sensitive" aquifers that will dynamically resist secondary or tertiary oil recovery schemes was discussed as early as 1950 (Hewitt, 1963; Land and Baptist, 1965; Smith and Hendrickson, 1965; Neasham, 1977; Shaughnessy and Kunze, 1981). The problem occurs when an oil-bearing formation is flooded with some other solution in an effort to increase the flux of oil to production wells. The idea is that the residual oil can be driven ahead of a new invading solution, thus enhancing the oil recovery efficiency of the formation. However, this does not work very well if the invading fluid greatly decreases the hydraulic conductivity of the formation.

Petroleum engineers also have experience with intentional aggressive permeant interactions. The process of acid "stimulation" of petroleum formations relies on strong dissolution reactions to increase the hydraulic conductivity in the neighborhood of production wells (Smith and Henderickson, 1965 ; Farley et al., 1970; Lund et al., 1976; Shaughnessy and Kunze, 1981). In sandstone formations, concentrated hydrofluoric-hydrochloric (HF-HCl) acid solutions are used to react with the predominantly siliceous granular structure to increase hydraulic conductivity. Hydrofluoric acid is usually strong enough to react with (i.e. dissolve) the solid phase thus increasing the size and connectivity of the pore structure. Although hydrochloric acid is not strong enough to dissolve most sandstones, it helps maintain a low pH as the hydrofluoric acid reacts. This inhibits the formation of undesirable precipitants such as calcium fluoride (CaF_2, etc.) which would be deposited in the renovated pore space and defeat the purpose of the process. A substantial amount of work has been done on the modeling of HF/HCl floods (see Fogler and McCune, 1976; Lund and Fogler, 1976; Schecter and Gidley, 1969, etc.). However, much of this concentrates on the stoichiometry and kinetics of the reactions rather than on the potential physical impacts of the reaction's consequences.

All of the work cited to this point is of considerable significance to the topic of aggressive permeants. However, all of this taken together probably does not rival the volume of work recently assembled on the potential impacts relative to the issues of hazardous waste management. The most obvious concern is that some of the components of hazardous wastes will attack hydraulic barrier systems, significantly alter their permeability, and escape into the environment. This concern was created when some of the initial studies done with hazardous wastes indicated that some chemicals could increase hydraulic conductivity by as much as 5 orders of magnitude. These results created a great deal of professional and regulatory concern (some would say panic) that altered the design and construction of hydraulic barrier systems and led to a more carefully planned search for the chemical constituents and mechanisms responsible for the impacts. Numerous studies have since explored the impacts of concentrated organic permeants such as acetic acid (CH_3COOH), acetone (C_3H_6O), aniline ($C_6H_5NH_2$), benzene (C_6H_6), cyclohexane (C_6H_{12}), carbon tetrachloride (CCl_4), dioxane($C_4H_8O_2$), ethanol (CH_3CH_2OH), ethylene glycol ($C_2H_6O_2$), glycerol ($C_3H_8O_3$), heptane (C_7H_{16}), isopropanol ($(CH_3)_2CH_2OH$), methanol (CH_3OH), nitrobenzene ($C_6H_5NO_2$), phenol (C_6H_5OH), trichloroethylene ($ClCHCCl_2$) and xylene (C_8H_{10}) (see Acar et al., 1985; Anderson et al., 1985; Brown and Anderson, 1981; Brown and Thomas, 1984; Brown et al, 1986; Bowders and Daniel, 1987; Cartwright et al., 1977; Crim et al., 1979; Dunn, 1983; Dunn and Mitchell, 1984; Evans and Kugelman, 1985; Fernandez and Quigley, 1985; Foreman and Daniel, 1986; Green et al., 1981, 1983; Griffin and Shimp, 1978; Mitchell and Madsen, 1987; Schramm et al., 1986, etc.). Similar studies have been conducted on inorganics such as hydrochloric/hydrofluoric acid (HCl/HF), sulfuric acid (H_2SO_4), sodium hydroxide ($NaOH$), and ferric chloride ($FeCl_3$) (see Fogler and McCune, 1976; Lentz et al., 1985; Pierce et al., 1987; Ravi and Jennings, 1990, etc.).

Several of the hazardous-waste-motivated studies have been quite controversial. For example, although it is not difficult to demonstrate dramatic impacts using full-strength organic liquids, researchers have reported that impacts do not persist far below full strength (Acar et al., 1985; Bowders and Daniel, 1987; Evans and Kugelman, 1985; Mitchell and Madsen, 1987). Once the chemicals go into aqueous solution of "realistic" strength, there is much less (or possibly no) impact. Therefore, it is quite commonly held that laboratory measurements of soil/solute domain impacts may overestimate the realistic field impacts. There is also evidence that some laboratory results have been affected by permeameter design. Typically, flexible-wall permeameters do not yield the dramatic responses of fixed-wall permeameters (see Acar et al., 1985; Foreman and Daniel, 1986; Mitchell and Madsen, 1987). This has led some workers to conclude that many of the more dramatic soil/solute interaction data may have been corrupted by "wall effects". There has also been criticism of the high driving gradients used in some laboratory studies. High hydrostatic driving gradients are often used, particularly for testing low permeability soils, because of the extremely long time scales and extremely low flow rates that would be produced at laboratory scale by natural driving gradients. However, it can be convincingly argued that high driving gradients alter the mechanism responsible for aggressive permeant interactions and thus yield a distorted view of the impacts.

In retrospect, it is not difficult to criticize some of the materials and procedures used in aggressive permeant interaction studies. We now know much more about the mechanisms involved and have developed better techniques for studying them. However, the magnitude of literature that has accumulated on the subject makes it perfectly clear that aggressive permeant interactions can be of significance and can be important across a broad range of permeant and porous medium types. In the following section, we will discuss the mechanisms that are now thought to be responsible for many of the observed interactions. We will also use this opportunity to illustrate some of the experimental observations of the studies we have cited in this section.

2. Mechanisms of Aggressive Permeant Interactions

Aggressive permeant impacts actually result from a broad range of phenomena. Generally, one mechanism will dominate, but it is possible for observed results to be a function of several mechanisms that alter hydraulic conductivity to differing degrees and at different times in a dynamic interaction event. Modeling strategies are now evolving that can account for multiple interaction mechanisms, but these are built from single mechanism components. For the moment we will attempt to concentrate on the distinct interaction mechanisms that are currently thought to be most responsible for the observed responses in soils.

2.1 Flocculation Interactions

Although flocculation is most commonly thought of as a phenomenon of dispersed suspensions, it has been observed to yield significant impacts in soil systems if a significant fraction of the solid phase mass is in the size ranges smaller than 2 microns. Flocculation (and its counterpart deflocculation discussed in the next section) result from a change in the ionic strength of the pore water. Both mechanisms result from changes induced in the structure of the electric double layer that develops in the neighborhood of the soil/solution interface in an effort to satisfy charge neutrality.

Soil particle surfaces in aqueous systems typically exhibit electrical charge. This charge can develop in many ways and be either positive or negative, although most soil particles exhibit a net negative charge. Because the total system (solid and liquid phase) must maintain charge neutrality, the net particle surface charge must be counterbalanced by an opposing charge that develops in the solution phase. This is often referred to as the "Diffuse Double Layer". The layer is "diffuse" because the solution ions of opposing charge are attracted to the charged soil surfaces yielding a concentration gradient in the solution phase. This concentration gradient is the driving force for a counter-diffusion of ions back into the bulk solution and thus results in an "opposing" charge that is spread out over a "diffuse" layer for some relatively small distance out into the solution phase (Weber, 1972).

The size and strength of the diffuse double layers that develop around charged solids are of significance when one considers the process of particle agglomeration. If particles move sufficiently close to one another, attractive forces such as Van der Waal's forces and electrostatic attractions will cause them to agglomerate. However, particles in aqueous suspension will resist agglomeration as their (like charged) diffuse double layers repulse one another. Therefore, the size and strength of the diffuse double layer is a significant factor in determining the structure of a soil. There are, of course, other factors such as the Born repulsive forces and hydration forces that come into play, but for the purpose of this discussion, these will be assumed to be relatively minor.

Consider the potential impact of altering the ionic strength of a solution with which a soil system has attained equilibrium. Assume first that one increased the ionic strength of the solution. As the concentration of ions in the solution phase increases, the concentration gradient driving diffusion of the ions opposing the charge of the solid phase is decreased. This lower driving gradient decreases the size of the diffuse double layer, and thereby allows soil particles to approach one another more closely than before. This is illustrated in Figure 1. Obviously, not all soil particles will be mobilized in response to this reduced constraint on motion, but if there is a net mobility, it will probably be characterized by the movement of smaller, more mobile solids toward the larger, more immobile solids. This is because the small solids are less likely to be trapped in place by the structural loads (such as overburden) distributed throughout the solid phase. The net result should be a decrease in the size of the small pores separating small particles from the larger particles, and an increase in the size of the larger pores as the small solids retreat from them. These rearrangements will not alter the solid phase porosity, but will alter how the available pore space is arranged. The ensemble result should be an increase in the soil's hydraulic conductivity.

Figure 2 illustrates data of Pierce et al., (1987) for the transient hydraulic conductivity response of a clay soil accompanying an increase in permeant ionic strength. The data are for the dynamic impact of a solution of ferric chloride (500 mg/l $FeCl_3$) displacing a low ionic strength calcium sulfate solution (0.01 N $CaSO_4$). The test was conducted in a fixed-wall permeameter cell using a hydraulic gradient in the range of 65 to 190. The soil used was the Faceville clay of Claredon County, South Carolina which is primarily composed of kaolinite. For the moment, consider only the actual experimental data indicated by the solid square points. The lines and coefficients also indicated on this figure will be discussed in the subsequent section on aggressive permeant modeling.

Note that the data for relative (nondimensional) hydraulic conductivity change (K(t)/K(t=0)) have been plotted versus a nondimensional time measured in pore volumes. The value K(t=0) is the hydraulic conductivity at time zero with respect to a "base" fluid. The nondimensional time scale in pore volumes is commonly used because it provides a better sense of how much contact the soil has actually had with the aggressive permeant. One pore volume corresponds to the time required for the flow to displace the permeameter cell's pore volume exactly once. After one pore volume, it is generally assumed that the whole sample is in contact with aggressive permeant. There will, of course, be some of the original "base" solvent trapped in dead-end pores, chemically bound to surfaces, etc., but this is assumed to be a relatively small fraction of the pore volume.

Note that the data of Fig. 2 depicts a relatively gradual increase in hydraulic conductivity. The reactions responsible for the impact are certainly not instantaneous with the soil's first exposure to the aggressive permeant. The experiment was apparently terminated before the

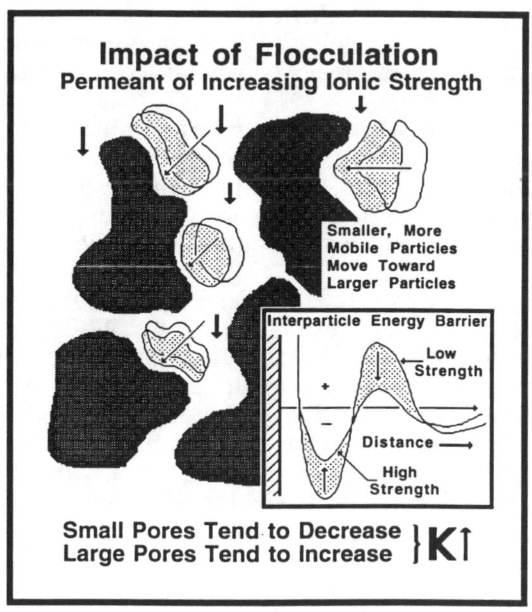

Fig. 1 - Aggressive permeant flocculation interactions.

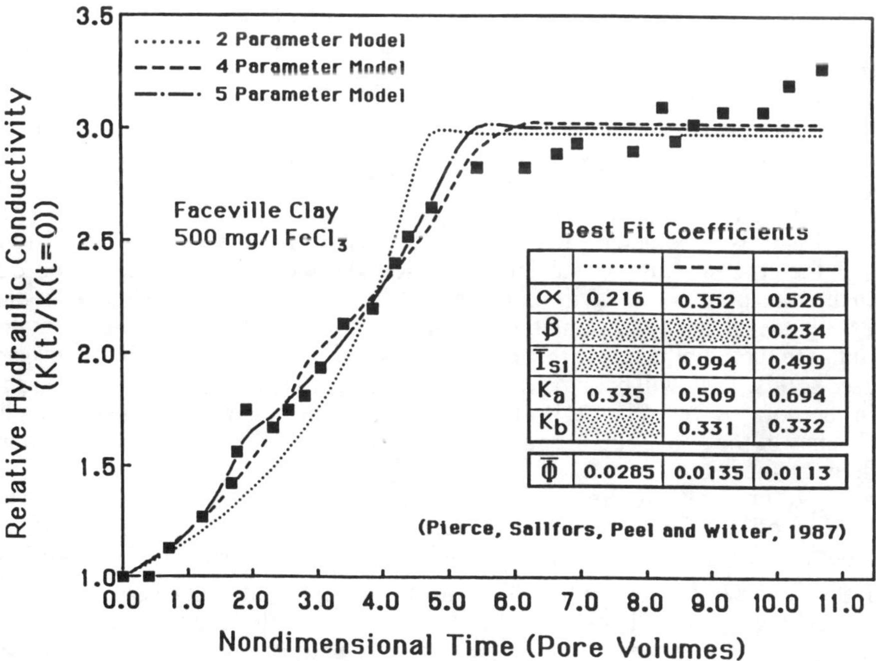

Fig. 2 - Aggressive permeant interaction (Faceville clay - $FeCl_3$ solution) characteristic of flocculation (Pierce et al., 1987).

ultimate equilibrium condition was reached. The soil was apparently still reacting after a nondimensional time of eleven pore volumes. Longer-term experiments indicated that the soil ultimately reached an equilibrium relative hydraulic conductivity of approximately 3.9. (Pierce et al.,1987)

Also note that Fig. 2 shows what appears to be a distinct change in the impact rate (at approximately 5.5 pore volumes) during the course of the experiment. This probably indicates that more than one mechanism was probably responsible for the total impact. However, these data are characteristic of what one should expect from a soil experiencing flocculation that was gradually shifting the pore size distribution in favor of the larger pores.

For comparison, Fig. 3 shows results of Pierce et al., (1987) for the same experiment repeated on a different soil. These results are for the White Store clay from Durham County, North Carolina which is predominantly montmorillonite. Note that the resulting impact is of approximately the same magnitude, but that the soil appears to attain equilibrium more quickly. There also does not seem to be evidence of a second, more gradual mechanism. The differences in the two responses could probably be attributed to subtle differences in the pore size and grain size distributions. What is most significant for this discussion, however, is that both soils yield similar responses that are characteristic of a flocculation impact.

2.2 Deflocculation Interactions

Deflocculation (also often referred to as dispersion) is the reverse of flocculation. If the soil system is in equilibrium with a solution of relatively high ionic strength, the diffuse double layers will be "compressed" by the presence of so many ions in solution. If this solution is then displaced with one of lower ionic strength the diffuse double layers will expand, and will increase the probability of repulsive forces causing nearby particles to migrate away from one another. The processes of deflocculation are illustrated in Fig. 4.

Deflocculation can have one of at least three impacts on the soil structure. If the net separation between soil particles is increased, but the "deflocculated" solids do not become entrained in the flow field, one should expect a modest decrease in hydraulic conductivity. The reason for this is that, although there will be no net change in the volume of the pore space, there will be a modest shift in the pore size distribution toward the smaller pore sizes as the smaller, more mobile particles move away from the larger solids. However, if the "deflocculated" solids become entrained in the flow field, the results are normally more dramatic. If the pore structure is sufficient to allow the mobilized particles to be carried away, then the total pore volume will increase, and the increase will favor the formation of large pores. Under these conditions one should expect a dramatic increase in hydraulic conductivity. On the other hand, if the pore space is not sufficient to allow the passage of the mobilized solids, they will become lodged in the constrictions of pore channels and act as plugs in the pore space. When this occurs there is no net increase in the pore volume, but the efficiency of the pore structure can be greatly reduced. The result is often a dramatic decrease in hydraulic conductivity.

Figure 5 presents data of Frenkel et al. (1978) that are characteristic of the dynamic domain impact of soil deflocculation. The data are for kaolinitic (10.4%) and montmorillonitic (10%) soils originally in equilibrium with a 1N $NaCl$-$CaCl_2$ solution and then permeated with a low strength (< 0.025 M) salt solution. The data were measured in 5 cm dia., 30 cm long, fixed-wall, constant-head permeameter cells. Note that the data exhibit a rapid decline in hydraulic conductivity of nearly an order of magnitude and the majority of the impact is exerted relatively quickly (< 3 pore volumes). These results are characteristic of deflocculation followed by mobilization and subsequent entrainment. Several other workers have reported similar, or more dramatic results (see Goldenberg et al., 1983; Goldenberg et al. 1984; Mehnert and Jennings, 1985; Shainberg et al., 1981 a,b; Yaron

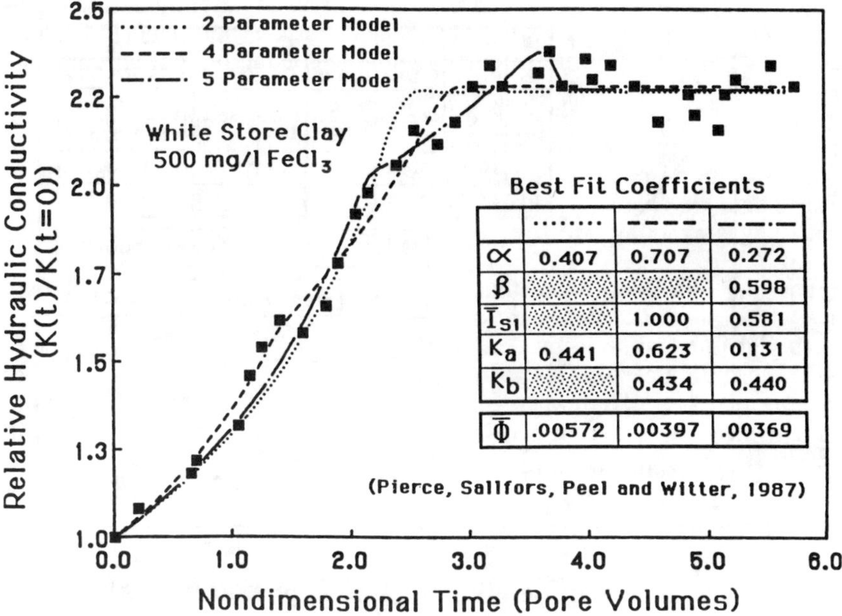

Fig. 3 - Aggressive permeant interaction (White Store clay - $FeCl_3$ solution) characteristic of flocculation (Pierce et al., 1987).

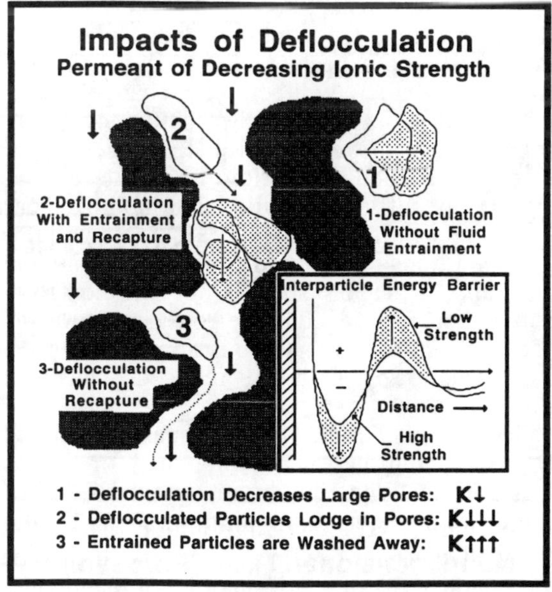

Fig. 4 - Aggressive permeant deflocculation interactions.

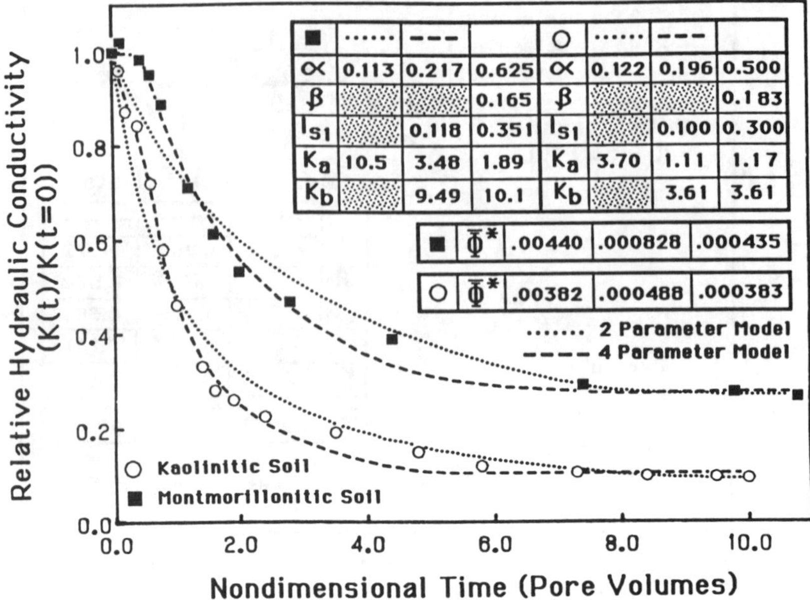

Fig. 5 - Aggressive permeant interaction (clay soils - low ionic strength solution) characteristic of delocculation (Frenkel et al., 1978).

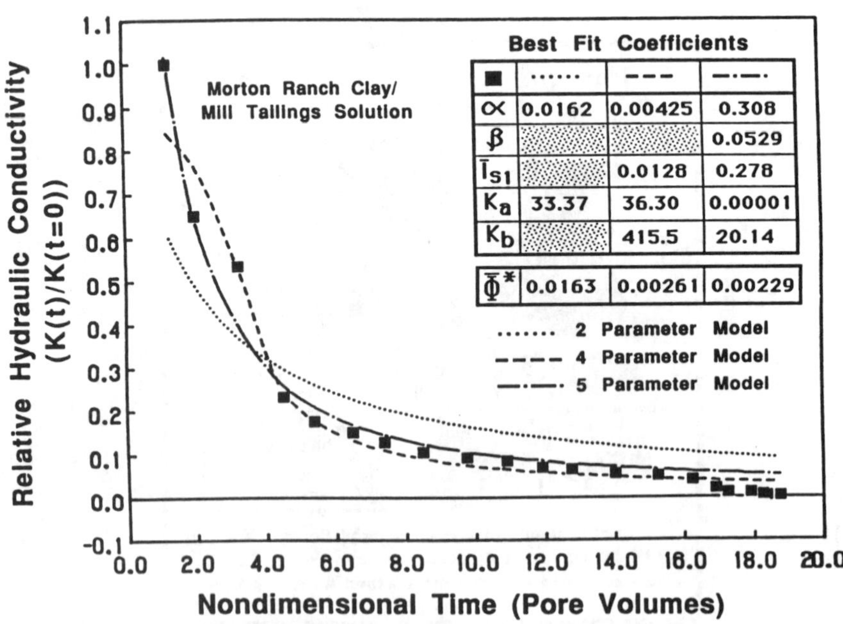

Fig. 6 - Aggressive permeant interaction (Morton Ranch clay - acidic mill tailings solution) characteristic of deflocculation (Peterson and Gee, 1985).

and Thomas, 1968) There has been some debate about whether the impacts are the result of deflocculation or in-place swelling (see Section 2.3). Although both can occur and both can have nearly identical impacts, sufficient direct measurements have now been made to confirm that particle migration is often the cause of the domain impact in "non-swelling" soils. It is relatively easy to distinguish deflocculation from swelling by looking for mobilized solids in the permeameter test cell discharge.

Figures 6 and 7 present results reported by Peterson and Gee (1985) on the impact between clay liner materials and acidic uranium mill tailings solutions. The tests were conducted to evaluate the long-term impacts that these acidic permeants would have on clay liners. The particular results illustrated here are for the Morton Ranch clay (actually a clayey silt) permeated with a real uranium mill tailings solution (pH= 1.8) and a synthetic tailings solution (pH=2.0). The data were generated in constant head permeameter cells operated at modest differential heads (1400 cm). The interactions are of interest because they were attributed to both deflocculation and precipitation (to be discussed in Section 2.4). Generally at these low pH one would not expect precipitation interactions. If anything, the reactions should be in the other direction. However, the Morton Ranch soil has a relatively high acid neutralizing capacity. Therefore, as the acid is neutralized the pH increases and previously dissolved solutes may precipitate. Nevertheless, it is most likely that the results of Figs 6 and 7 were dominated by deflocculation interactions motivated by the increased ionic strength of the "neutralized" permeant. This would explain why the permeability did not recover after the neutralizing capacity was exhausted.

2.3 Shrinking/Swelling Interactions

A solid phase volume change (either shrinking or swelling) can also exert a considerable impact on a soil's hydraulic conductivity. When soil particles swell, the volume of the available pore space is generally reduced. This is because the confining forces often restrict global deformation. When a soil swells the hydraulic conductivity generally declines. The reverse is often true when soils shrink. Although global deformations are more likely, when soils shrink the volume of the pore space often increases, and hydraulic conductivity generally increase. This is illustrated schematically in Fig. 8.

Shrinking/ swelling interactions are most often associated with organic permeants coming in contact with clays. Although there is still some debate about the fundamental mechanisms involved, it is generally held that shrinking/swelling reactions of clays are due primarily to the hydration of their layer surfaces. Double-layer particle interactions are generally believed to play only minor roles (Chen et al., 1987).

Clay particles in water swell due to the uptake of water. The amount of swelling depends on the clay mineralogy. Clays such as montmorillonite can uptake (in interlayer and intercrystalline spaces) up to 10 g of water per gram of "dry" clay mass. However, clay particles in pure organic solvents generally uptake far less solvent mass. Therefore, when a water-saturated clay is exposed to a organic solution, the displacement of water associated with the solid phase can lead to a significant decrease in the total solid phase volume. This decreased degree of swelling will be perceived as shrinkage, and often leads to dramatic increases in hydraulic conductivity. The degree to which this shrinkage will occur depends on physical properties of both the clay and organic. Generally, organic compounds that ionize in water have the greatest impact.

Figure 9 presents data of Acar et al. (1985) for kaolinite saturated with a 0.01N $CaSO_4$ and then permeated with full strength acetone. The test was conducted in a fixed-wall permeameter cell using a hydraulic gradient of 100. It appears that shrinkage caused by the displacement of water from the solid phase results is a rapid and dramatic hydraulic conductivity increase of nearly two orders of magnitude. The reader is cautioned, however, that not all clays react this dramatically with acetone. Figure 10 presents an example where the hydraulic conductivity of a kaolinite actually declined when permeated with a 1000 mg/l

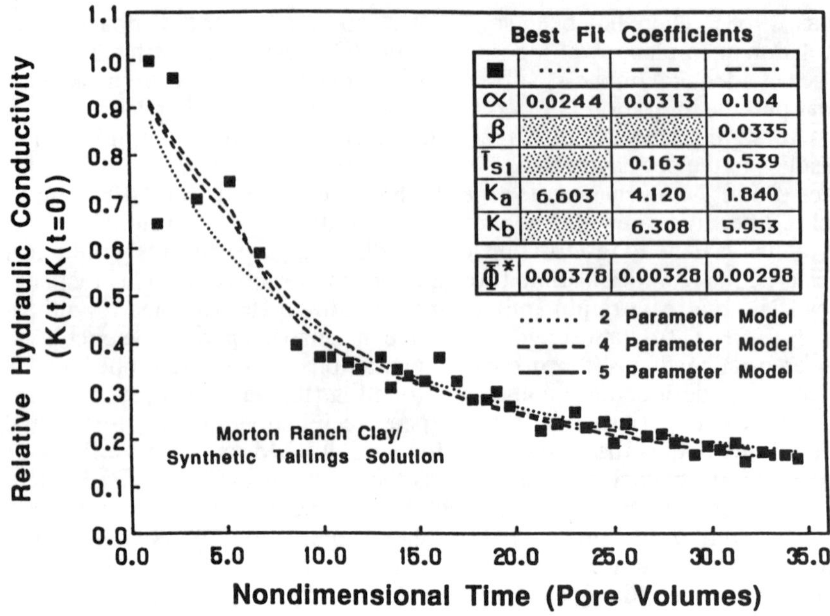

Fig. 7 - Aggressive permeant interaction (Morton Ranch clay - synthetic mill tailings solution) characteristic of deflocculation (Peterson and Gee, 1985).

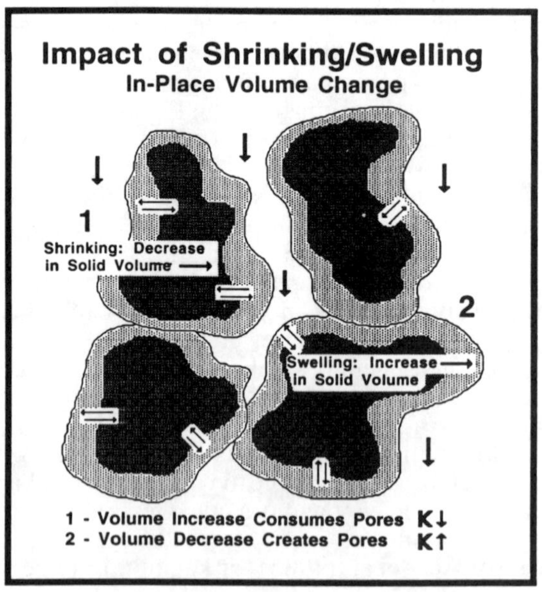

Fig. 8 - Aggressive permeant shrink/swell interactions.

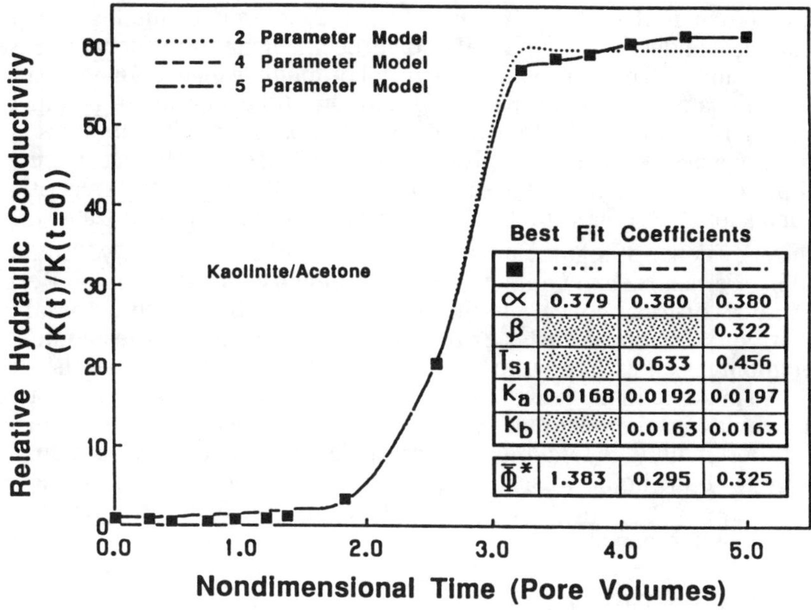

Fig. 9 - Aggressive permeant interaction (kaolinite - acetone) characteristic of shrinking (Acar et al., 1985).

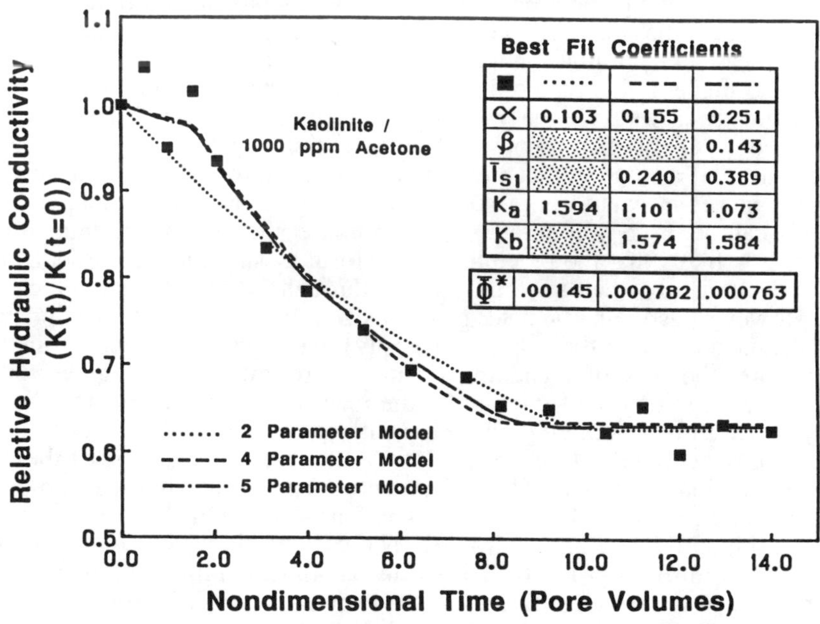

Fig. 10 - Aggressive permeant interaction (kaolinite - acetone solution) characteristic of swelling (Acar et al., 1985).

acteone solution (Acar et al., 1985). Figures 11 and 12 present examples of experiments where the hydraulic conductivity of the soil initially declined, but ultimately increased when permeated with acetone. This is probably the result of multiple interaction mechanisms that function at different rates, but the mechanistic origins are still under investigation.

Figures 13, 14, and 15 illustrate the potential impacts of other organics (Aniline, Methanol, and Ethylene Gycol respectively) on various clay soils. The data were generated in fixed wall permeameters run at hydraulic gradients of 361.6 (for the smectite) or 61.1 (for the illite and kaolinite) after saturation with a base permeant of 0.01M $CaSO_4$. Note that all of these show evidence of significant increase in hydraulic conductivity. Similar results may be found in Fernandez and Quigley, 1985 and Green et al., 1983, etc. Results such as these were (are?) of great concern to the solid and hazardous waste disposal industry. The fear is that liquid chemical wastes or the leachates from solid wastes could attack the clay soils commonly used as hydraulic barriers, and substantially increase their hydraulic conductivity. An increase in the hydraulic conductivity of a clay liner of one or two orders of magnitude would certainly degrade its role as an effective barrier to the migration of liquid wastes. Fortunately, regulations now preclude most instances where direct contact would be likely for bulk liquid wastes, and research has indicated that organics in dilute aqueous solution (as in a modest strength leachate) are not likely to have the dramatic impacts of full strength organics.

2.4 Precipitation Interactions

Precipitation can also alter the hydraulic conductivity of soils by one of at least two distinct mechanisms (see Fig. 16). Generally these interactions consume pore volume in the domain and thus decrease the soil's hydraulic conductivity. If the precipitant is formed in a reaction between a mobile (i.e. solution phase) species and some chemical component of the solid phase, the resulting precipitation will probably evolve as an addition to an existing immobile soil surface. The result will be a gradual consumption of the available pore space and a modest decrease in hydraulic conductivity. The formation of metal oxide coatings on soil grains is an example of this type of interaction. Often the reaction is self-governing in that it stops or greatly slows once the initial surface coating has formed. However, if the precipitation reaction is between previously soluble solution components, the impact can be much more dramatic. If a solution that previously satisfied all solubility product constraints on solubility at one pH is swept into a domain with a significantly different pH, the once soluble species can begin to precipitate. If solid phase species do not enter into these reactions, then it is likely that the precipitants will form as suspended colloidal solids. If these solids are sufficiently small to be transported through the pore structure there will be little impact. However, as the particulates grow in size (due to continued precipitation and colloid flocculation induced by the domain tortuosity), they can become lodged in the pore space. Therefore, the precipitants both consume pore volume, and greatly alter the efficiency of the pore structure. This may be the mechanism responsible for the data of Figs 6 and 7. Such situations often produce dramatic impacts on hydraulic conductivity.

Figure 17 presents data of Lentz et al. (1985) that are characteristic of the physical impacts of precipitation reactions. These data were measured during the permeation of kaolinite, magnesium montmorillonite, and a kaolinite (85%) plus bentonite (15%) soils mixture with a high pH sodium hydroxide (NaOH) solution. The data are for soil specimens of 3.56 cm dia. by 1.3 to 2.5 cm in length tested in a flexible-wall triaxial permeameter operated under a confining pressure of 34.5 kPa (5 psi) and a potential gradient of 400 to 500. The soils were tested with NaOH solutions at pH values of 9, 11, and 13, but little impact was observed below a pH of 13. The impacts at pH 13 were attributed to the precipitation of previously soluble calcium carbonate ($CaCO_3$) and magnesium hydroxide ($Mg(OH)_2$) that formed and then plugged the soil's pore space.

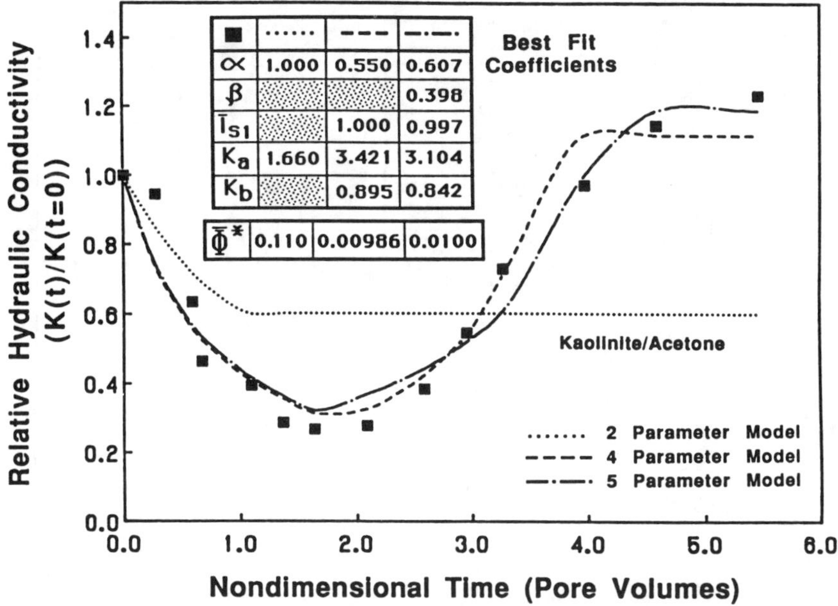

Fig. 11 - Aggressive permeant interaction (kaolinite - acetone) characteristic of a compound interaction with little ultimate impact (Acar et al., 1985).

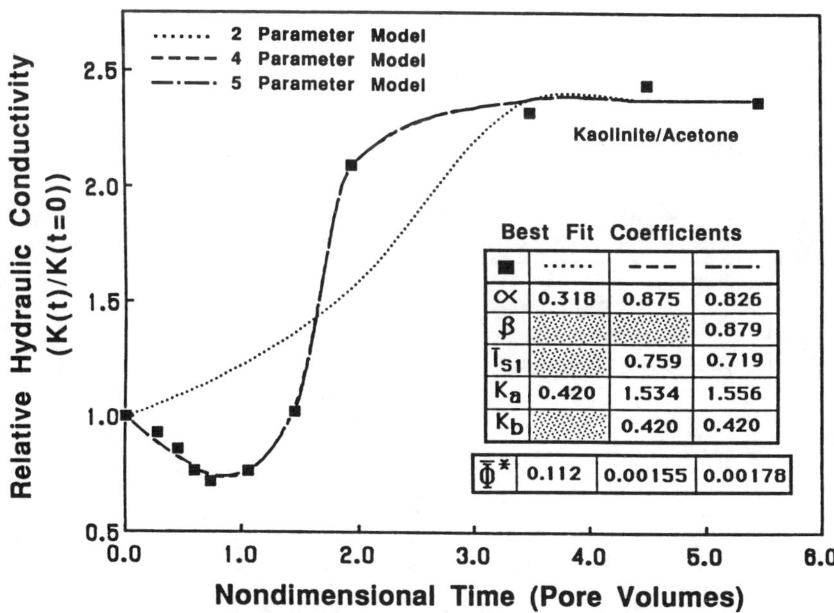

Fig. 12 - Aggressive permeant interaction (kaolinite - acetone) characteristic of a compound interaction with a modest ultimate impact (Acar et al., 1985).

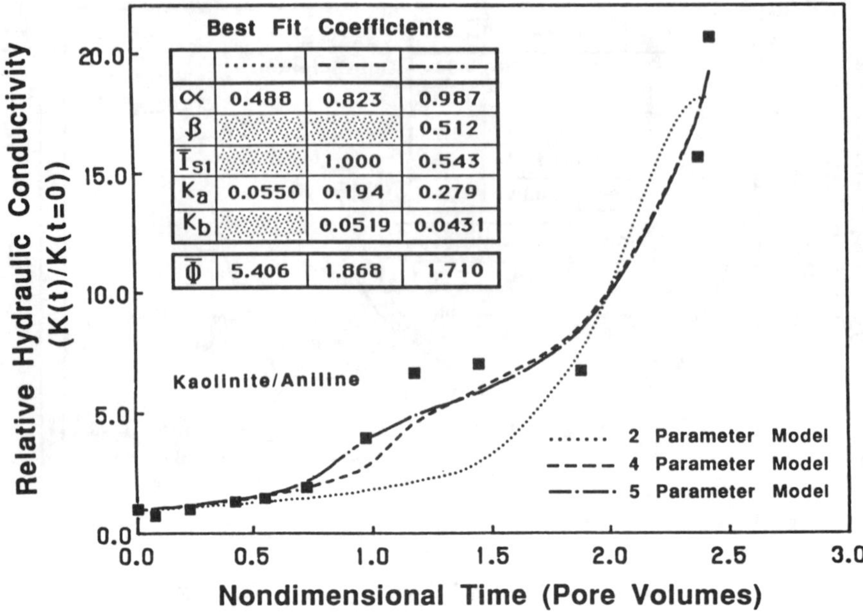

	....	– – – –	—·—·—
α	0.488	0.823	0.987
β			0.512
$\bar{I}_{S1}$		1.000	0.543
K_a	0.0550	0.194	0.279
K_b		0.0519	0.0431
$\bar{\Phi}$	5.406	1.868	1.710

Fig. 13 - Aggressive permeant interaction (kaolinite - aniline) characteristic of a significant volume change interaction (Anderson et al., 1985).

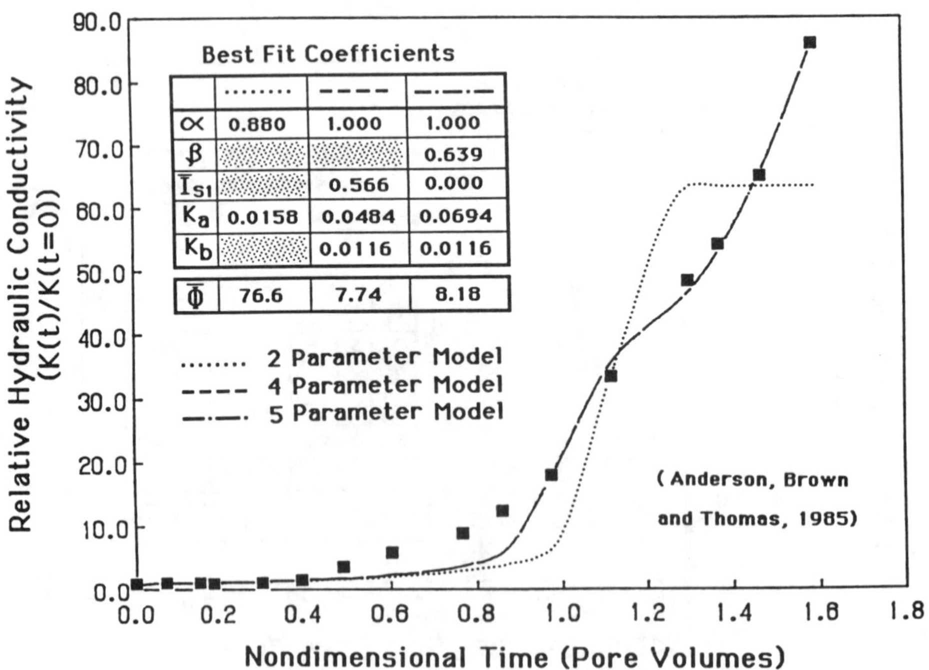

	....	– – – –	—·—·—
α	0.880	1.000	1.000
β			0.639
$\bar{I}_{S1}$		0.566	0.000
K_a	0.0158	0.0484	0.0694
K_b		0.0116	0.0116
$\bar{\Phi}$	76.6	7.74	8.18

Fig. 14 - Aggressive permeant interaction (illite - methanol) characteristic of a significant volume change interaction (Anderson et al., 1985).

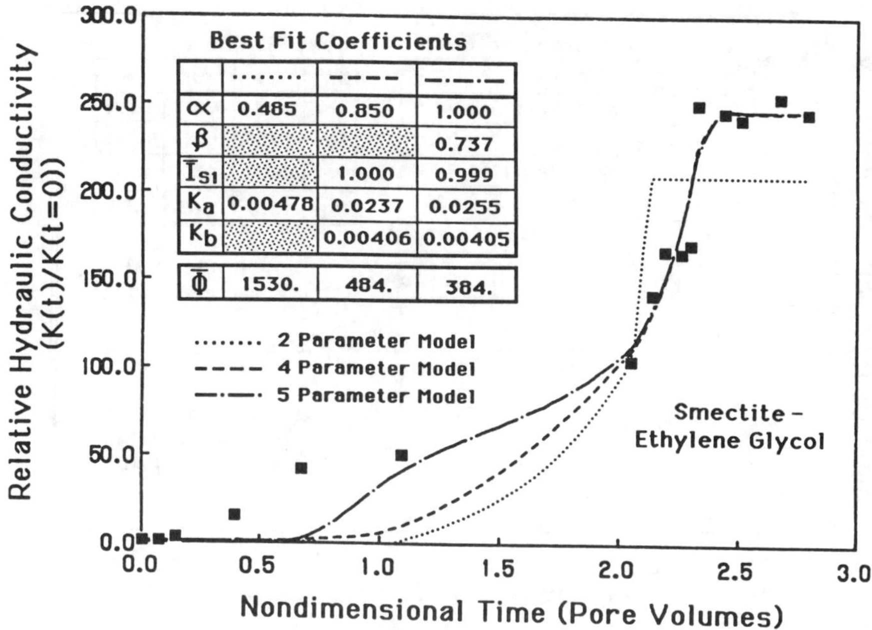

Fig. 15 - Aggressive permeant interaction (smectite - ethylene glycol) characteristic of a significant volume change interaction (Anderson et al., 1985).

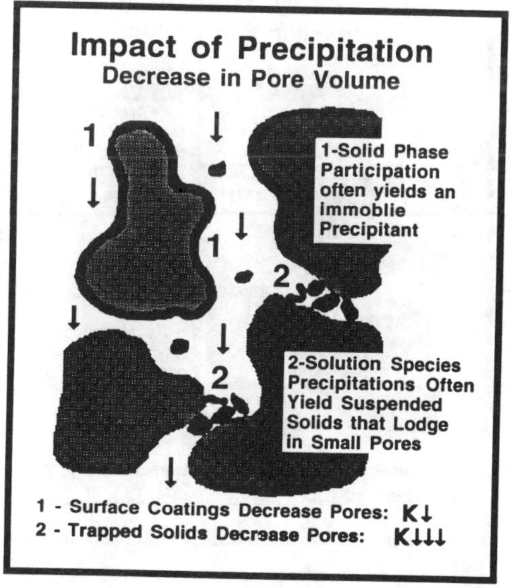

Fig. 16 - Aggressive permeant precipitation interactions.

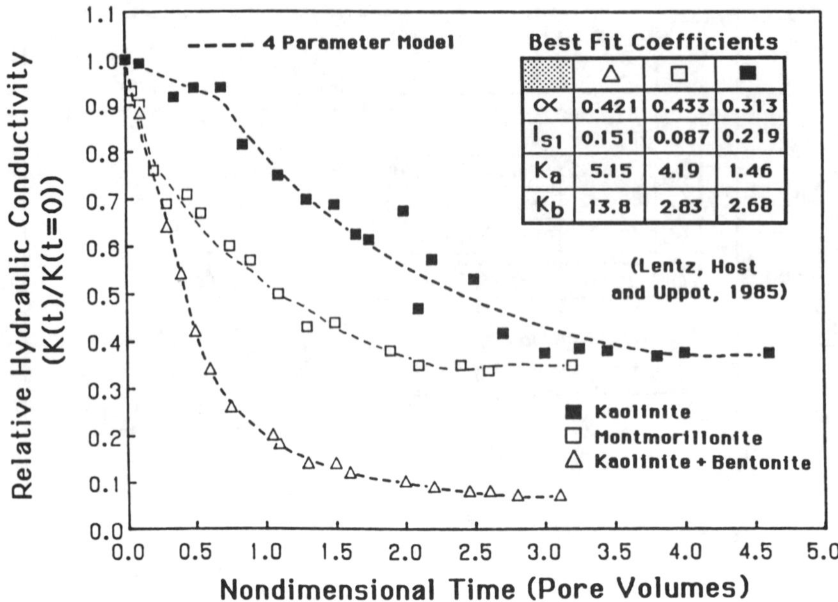

	△	□	■
α	0.421	0.433	0.313
I_{s1}	0.151	0.087	0.219
K_a	5.15	4.19	1.46
K_b	13.8	2.83	2.68

Fig. 17 - Aggressive permeant interaction (clays - sodium hydroxide) characteristic of a precipitation interaction (Lentz et al., 1985).

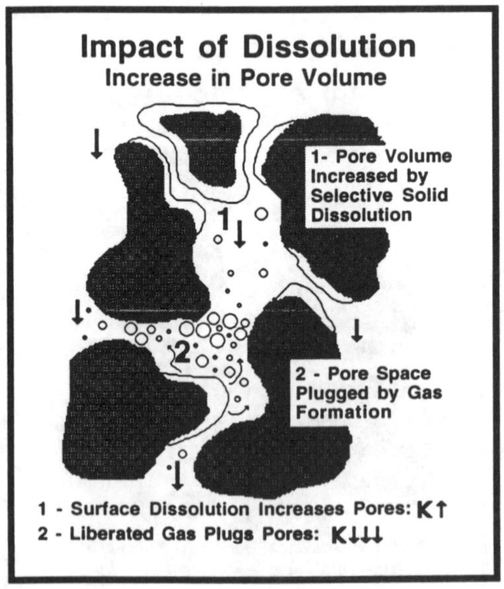

Fig. 18 - Aggressive permeant dissolution interactions.

2.5 Dissolution Interactions

Dissolution reactions are the reverse of precipitation reactions. Precipitation involves the formation of new solid phase mass either by direct precipitation onto immobile surfaces or by the trapping of colloidal participants in the pore space. Dissolution involves the transfer of immobile (solid phase) mass to the liquid phase in response to an unsatisfied soluble species solubility constraint. Since the result should be an increase in the volume of the pore space, dissolution reactions are often expected to increase hydraulic conductivity. This is not always the case (see Figure 18). If the species entering solution remain in solution, then one should expect the pore space to increase. This generally yields a relatively modest increase in hydraulic conductivity. The increase is often modest because, the rates of dissolution reactions are effected by the solution phase species concentrations and will often be greatest in the zones of highest solution velocity. These are the locations where the products of the dissolution reaction are most rapidly swept away from the reaction site thus supplying the highest possible concentration driving gradient. However, since the zones of highest solution velocity generally correspond to the largest domain pores, the impacts of the increased pore volume are less significant than they might otherwise be.

Figure 19 presents result characteristic of hydraulic conductivity increases resulting from dissolution reactions (Fogler and McCune, 1976). These data are for sandstone cores permeated with strong HF/HCl solutions in a constant-flux, fixed wall permeameter cell using cores of 2.5 to 7 cm in length and applied permeant fluxes of 5 to 20 ml/min. Note the initial delay in time (probably attributed to the kinetics of the reactions) followed by the more dramatic increase in hydraulic conductivity.

It is interesting to note that dissolutions do not always yield the expected result. If solid mass is removed, the porosity increases and the Hydraulic Conductivity should increase. If the solution is quite aggressive, a great deal of the solid phase may attempt to dissolve. However, there is no guarantee that all of the products that dissolve will remain in solution. For example, carbonate mineral may be readily dissolved by aggressive acid permeant, but as this occurs it is possible for the result to exceed carbon dioxide (CO_2) solubility. If this occurs, the excess CO_2 may evolve into a distinct gas phase (see Fig. 18) that can plug the pore space and thus decrease the hydraulic conductivity. Although this would probably be a temporary condition, "temporary" can be a very long time because the liquid flux required to carry off the gas declines with the hydraulic conductivity.

Figure 20 presents hydraulic conductivity data of Ravi and Jennings (1990) for an aggressive permeant dissolution interaction that was not constrained by the formation of a gas phase. These data were measured in a 2.5 cm dia. by 8.5 cm long fixed-wall permeameter under a constant potential driving gradient of 10. The soil was taken from the calcareous Aquia Greensand formation of Maryland and was permeated with a 0.005 M sulfuric acid (H_2SO_4) solution. Note that the impact is rather gradual, but ultimately the hydraulic conductivity is substantially increased. Dissolved solids testing of the test cell effluent confirmed that previously immobile mass was being transported out of the domain.

Figure 21 presents data for this same soil permeated with a more concentrated (0.025 M) sulfuric acid solution. Logic suggests that the dissolution would be accomplished more rapidly, so the response should occur at small values of (pore volume) time, and that the hydraulic conductivity should increase to an even higher value since a higher degree of reaction completion should be attained. However,contrary to this expectation, the hydraulic conductivity actually declined. The reason is that the increased strength of the permeant was sufficiently aggressive to dissolve more calcium carbonate that could remain in solution. A CO_2 gas phase formed within the domain, and this effectively plugged the pore space.

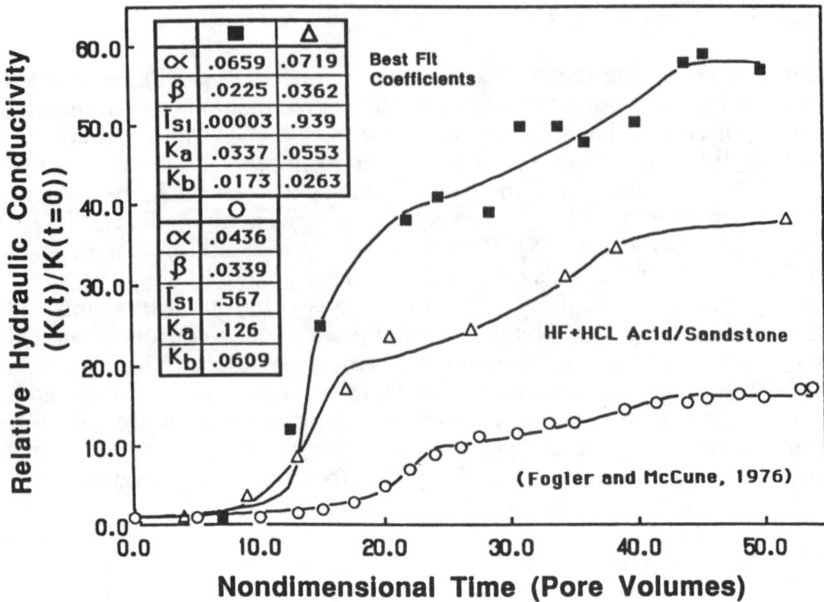

	■	Δ
α	.0659	.0719
β	.0225	.0362
$\bar{T}_{S1}$	.00003	.939
K_a	.0337	.0553
K_b	.0173	.0263

	○
α	.0436
β	.0339
$\bar{T}_{S1}$	.567
K_a	.126
K_b	.0609

Fig. 19 - Aggressive permeant interaction (sandstone - hydrofluoric/hydrochloric) characteristic of a concentrated acid dissolution (Fogler and McCune, 1976).

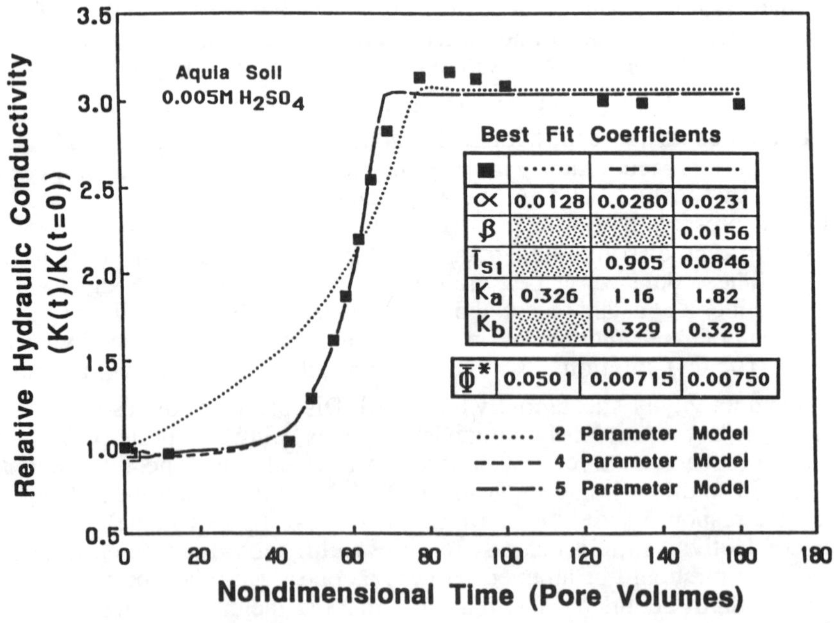

■	......	– – –	–·–·
α	0.0128	0.0280	0.0231
β			0.0156
$\bar{T}_{S1}$		0.905	0.0846
K_a	0.326	1.16	1.82
K_b		0.329	0.329
$\bar{\Phi}^*$	0.0501	0.00715	0.00750

Fig. 20 - Aggressive permeant interaction (Aquia soil - sulfuric acid solution) characteristic of a dilute acid dissolution (Jennings and Ravi, 1990).

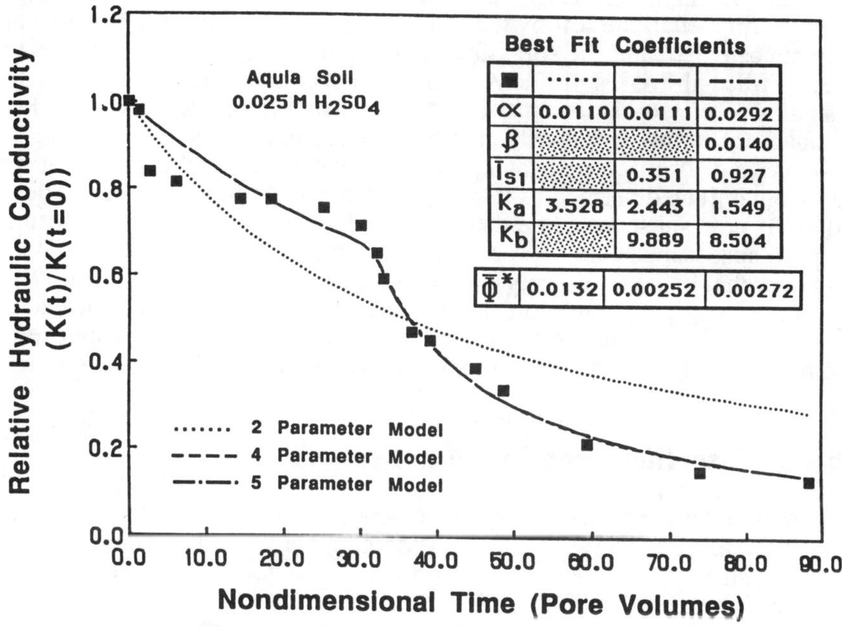

Fig. 21 - Aggressive permeant interaction (Aquia soil - sulfuric acid solution) characteristic of a dilute acid dissolution plus gas phase evolution (Jennings and Ravi, 1990).

2.6 Additional Aggressive Permeant Interaction Mechanisms

The five classes of interaction mechanisms discussed in Sections 2.1 through 2.5 are certainly not exhaustive. There are several possible chemical interactions (ion exchanges-sorptions, surface complexations, chelations,etc.) that can alter the surface properties of the solid phase and thus alter the particle-to-particle structure of the domain. Under field conditions many permeants also carry suspended solids (often of very fine colloidal matter) that can become trapped in the soil by mechanisms such as filtration. The physical state of the soil domain can also alter the soil's hydraulic conductivity. For example, as the waste disposal cell is filled above a soil liner system, the added load induces additional stress in the soil, leading to consolidation that decreases the hydraulic conductivity. Finally, the potential impacts of biological activity cannot be overlooked. The introduction of a new substrate source (e.g. from groundwater pollution) may accelerate the growth of native soil organisms. Contaminated permeants may also seed the domain with microbes that are acclimated to the new substrate. In either case accelerated biological growth can consume (plug) the pore space and lead to a substantial decline in hydraulic conductivity. Such conditions are probably temporary since as the hydraulic conductivity declines, the flux of substrate to the biological population also declines and the increased biomass cannot be sustained. However, as in the problem with gas phases evolving in response to strong dissolution reactions, "temporary" conditions can exist for a surprisingly long time.

3. Discrete Interface Penetration Models

It has long been known that soil/solute interactions such as those illustrated in Section 2 can alter the hydraulic conductivity of soils. Methods for actually quantifying the domain impact are more difficult to come by. Modeling attempts have been made (see Fogler and McCune, 1976; Guin et al. 1971), but these tend to be very specialized. The Discrete Interface Penetration (DIP) strategy for formulating domain impact models is one exception to this generalization. The DIP modeling strategy yields relatively simple analytical solutions for the domain impacts that will result from dynamic soil/solute interactions. The advantage of this approach is that analytical solutions may be derived for the transient saturated hydraulic conductivity of a soil experiencing aggressive permeant interactions K(t), for the space- and time-dependent potential field within the domain during the interaction $\phi(x,t)$, and for the transient volumetric flux Q(t) discharged from the domain. The analytical solutions are nonlinear, piecewise differentiable functions. In some cases, these properties make them challenging to calibrate and use. Model calibration will be discussed in Section 5. Nevertheless, DIP models represent the state-of-the art in general modeling of aggressive permeant interactions. The next generation of models will undoubtedly be much more mechanism specific, and applicable under much more restrictive conditions.

3.1 The DIP Modeling Strategy

Several DIP models have been formulated. Otte and Jennings, (1984) describe a single interface model based on a "unreacted" zone of unaltered soil and a "reacted" zone of altered soil properties. Mundell and Jennings (1987) extended this concept to include a "reacting" zone of fixed length where the soil's properties could take on values other than their equilibrium values. Jennings and Ravi (1990) extended the three zone DIP model to allow the "reacting" zone to vary in size as the interactions develop in time. Jennings and Ravi (1990) also present solutions in a nondimensional form that are independent of the boundary conditions applied. Therefore, the solutions may be applied to a wide variety of permeameter designs (constant head, constant flux, falling head, etc.) to evaluate the required domain coefficients.

The fundamental concept of the DIP modeling approach is illustrated in Fig. 22. Basically, it is assumed that the domain may be divided into distinct zones separated by traveling interfaces. The zones define areas where soil/solute interactions are occurring. Distinct physical properties may be specified for each zone. The zones are separated by discrete interfaces that are propagated through the domain at velocities dictated by the rates of soil/solute interaction. Note that Phases I through III of Fig. 22 represent one continuous event that is only separated into "phases" for analytical convenience. These are described as follows:

Phase I - At time zero(t_0) an aggressive permeant begins to infiltrate a soil of "unreacted", homogeneous saturated hydraulic conductivity K_1. This permeant begins to interact with the soil to yield the "reacting" hydraulic conductivity value K_2. The reacting hydraulic conductivity represents an intermediate stage in the transition from the original value of K_1 to some final value K_3 that is achieved after all interactions have attained equilibrium. The interface separating K_1 and K_2 is assumed to propagate through the domain at a time dependent velocity $v_1(t)$ that is proportional to the permeant's interstitial velocity $q(t)/\varepsilon$ where $q(t)$ is the Darcy velocity and ε is the domain porosity. This is equivalent to assuming that the interface propagation rate is proportional to the transport of reactant mass to the leading interface. The proportionality would depend on the capacity of the reaction. For example, Otte and Jennings (1984) illustrate that this is essentially the soil's acid neutralizing capacity (ANC) for dilute acid penetration problems. Phase I continues until an initial interface separation depth of I_{s1} is achieved (at $t=t_1$), and soil at $x = 0$ attains the equilibrium value K_1.

Phase II - The "reacting" zone propagates through the soil leaving behind material characterized by a final equilibrium saturated hydraulic conductivity K_3. The interface separating K_2 and K_3 propagates through the domain at time dependent velocity $v_2(t)$ which is also proportional to the permeant's interstitial velocity. Note that there is no requirement for equal interface propagation rates but the second interface may not surpass the first. Therefore, the interface separation $I_s(t)$ must be non-negative.

Phase III - At time t_2 the interface separating K_1 and K_2 encounters the domain boundary at $x = L$. Phase III continues until the second interface encounters the boundary at time t_3 leaving behind a homogeneous domain of significantly altered hydraulic conductivity K_3.

Although this may be a simplistic phenomenological description for what is often mechanistically complex, it has compelling attractions. First, given that there is a method of predicting the rate at which the interfaces will be propagated, time shift analytical solutions can be obtained for $K(t)$, $\phi(x,t)$, and $Q(t)$. Mundell and Jennings (1987) discuss numerous interaction mechanisms that may be accommodated by this approach, and suggest methods for quantifying interface penetration rates. Secondly, although the model maintains little of the original mechanistic formulation, it preserves the physical soil response to the interaction. Therefore, it becomes possible to predict the consequences of events such as hazardous waste penetration through clay soil barriers. Finally, the model is a very attractive partner for permeameter testing. It presents a method for extrapolating from laboratory to field scale and offers hope of generalizing permeameter results to other soils and/or permeants.

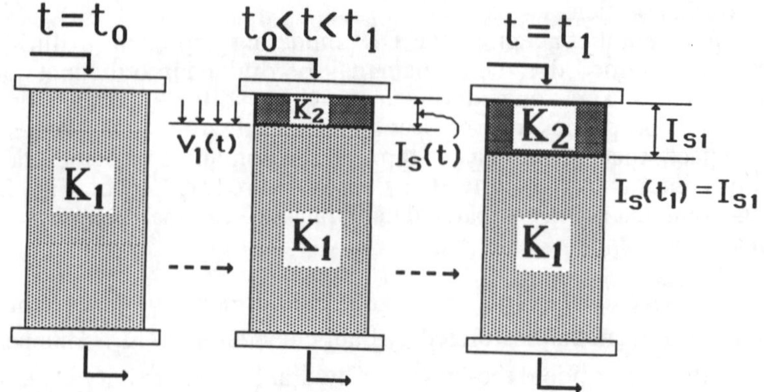

(a) Phase I – Reacting Zone Development

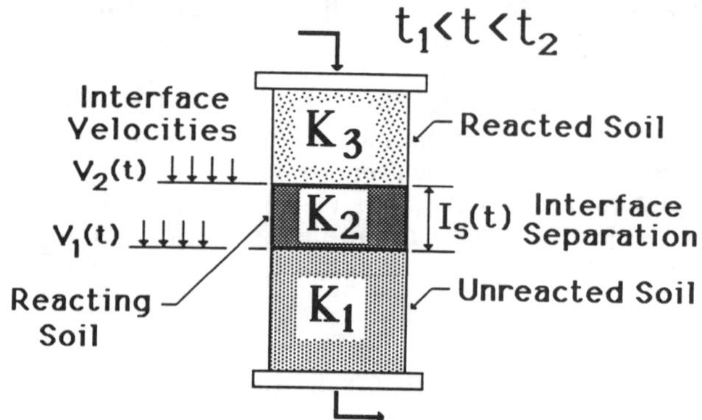

(b) Phase II – Dual Interface Propagation

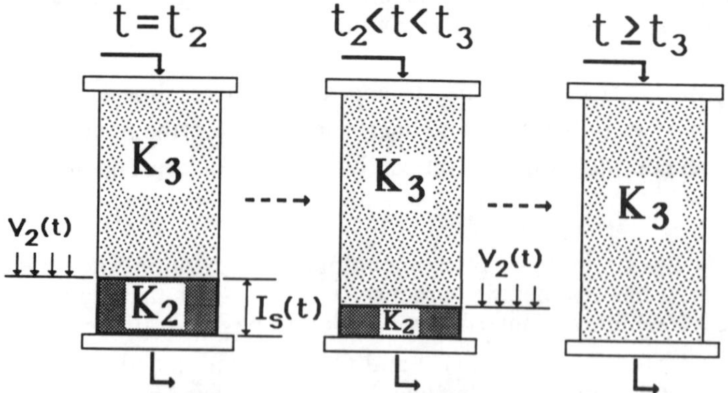

(c) Phase III – Interaction Completion

Fig. 22 - Schematic illustration of the analytically-distinct phases of a Discrete Interface Penetration (DIP) model (Jennings and Ravi, 1990)

An example dual interface penetration model derivation will be presented for the constant head permeameter design illustrated in Fig. 23. Constant head tests are often applied to relatively permeable soils or (at high driving gradients) to low permeability soils. Jennings and Ravi (1990) also present derivations for the falling head and constant flux configurations illustrated on Fig.23. Falling head tests are applied to low permeability soils at potential differences resembling natural driving gradients (Head,1982 ; Klute,1965). Constant flux experiments are probably least frequently done, but they do offer the advantage of transferring the transient response to the potential field where it may be more easily measured. Unfortunately, constant flux experiments preserve the disadvantages of experimentation at unnaturally high driving gradients.

Figure 24 is provided to assist in the definition of the notation to be used in the following derivation. The derivation will also be presented in nondimensional form. It proves convenient to present solutions in nondimensional form. However, although nondimensionalization of most quantities is relatively straightforward, one must be cautious about nondimensional time. There are two alternatives (identified here as τ and θ) for nondimensionalizing time. Equation (1) yields a linear, nondimensional time scale τ relative to the initial domain flux. Equation (2) yields nondimensional time θ corresponding to the number of pore volumes discharged. Although θ is most commonly reported, it is

$$\tau = \frac{q(t_0)t}{\varepsilon L} = \frac{K_1(\phi_1 - \phi_2)t}{\varepsilon L^2} \tag{3}$$

$$\theta(t) = \int_{t_0}^{t} \frac{q(t)}{\varepsilon L}\, dt \tag{4}$$

important to note that, since flux is inherently transient, θ yields a non-linear time scale.

The transient bulk saturated hydraulic conductivity K(t) is defined as the apparent domain resistance at any time and may be expressed directly from Darcy's law using superficial permeameter data for any time scale (t, τ, θ). This value is most conveniently nondimensionalized by dividing by the initial soil hydraulic conductivity.

$$K(t) = \frac{q(t)L}{(\phi_1 - \phi_2)} \tag{5}$$

$$\hat{K}(t) = \frac{K(t)}{K_1} \tag{6}$$

Two nondimensional hydraulic conductivities are sufficient to quantify the dual-interface penetration problem. These will be defined as $K_a = K_1/K_2$ and $K_b = K_1/K_3$. The interface penetration depths will be nondimensional as follows.

$$\hat{X}_1(t) = \frac{X_1(t)}{L} \tag{7}$$

$$\hat{X}_2(t) = \frac{X_2(t)}{L} \tag{8}$$

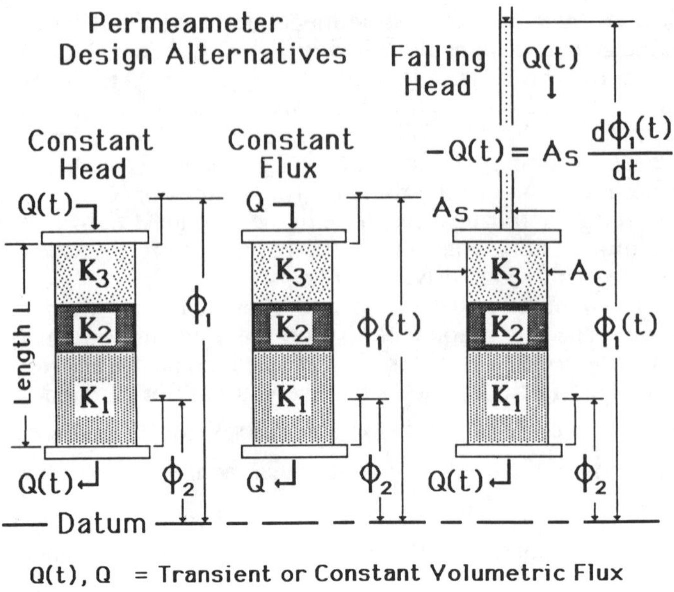

Fig. 23 - Typical laboratory permeameter boundary condition arrangements (Jennings and Ravi, 1990)

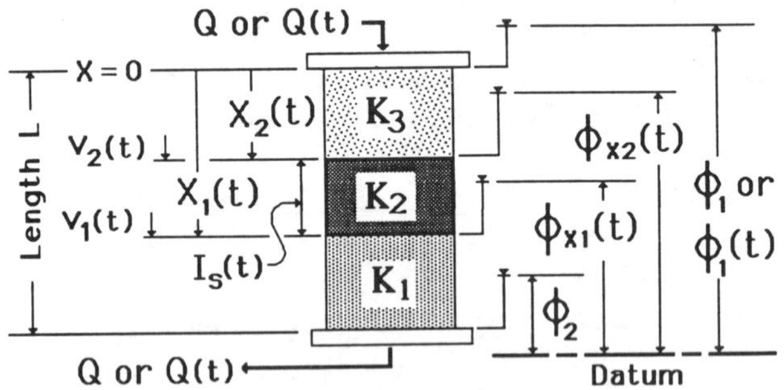

Fig. 24 - Schematic illustration of the notation of Discrete Interface Penetration (DIP) analysis (Jennings and Ravi, 1990)

It is also convenient to define the nondimensional interface separation function as follows.

$$\hat{I}_s(t) = \frac{I_s(t)}{L} = \hat{X}_1(t) - \hat{X}_2(t) \tag{9}$$

3.2 Example DIP Model Derivation - Constant Head Permeation

The following solution derivation is for the constant head permeation example illustrated in Fig. 23. Each of the phases are analytically distinct, but the flow problems that exist during each phase may be solved separately. The overall solution is then spliced together at the phase initiation times. The solution presented here follows that of Jennings and Ravi (1990).

Phase I ($t_0 \leq t < t_1, \quad \tau_0 \leq \tau < \tau_1, \quad 0 \leq \theta < \theta_1$)

The instantaneous bulk hydraulic conductivity of a horizontally stratified domain may be expressed as follows by equating the volumetric flux through each of the strata.

$$K(t) = \frac{\sum d_i(t)}{\sum (d_i(t) / K_i} \tag{10}$$

Here $d_i(t)$ represent the thickness of the ith strata at time t and K_i is the hydraulic conductivity of the ith strata. In nondimensional form and applied to the geometry of Fig. 24, this yields the following relationship.

$$\hat{K}(t) = \frac{1}{1 + (K_a - 1)\hat{X}_1(t)} \tag{11}$$

Note that the location of the interface must be known. This location may be determined given that the rate of interface penetration v_1 is proportional to the interstitial velocity $q(t)/\varepsilon$ where α is the required penetration constant.

$$v_1(t) = \frac{dX_1(t)}{dt} = \frac{\alpha q(t)}{\varepsilon} \tag{12}$$

Applying the chain rule, Eq. (12) and the definitions Eq. (3) and (7) yield the following.

$$\frac{d\hat{X}_1(t)}{d\tau} = \frac{d\hat{X}_1(t)}{dX_1(t)}\frac{dX_1(t)}{dt}\frac{dt}{d\tau} = \frac{\alpha q(t) L}{K_1(\phi_1 - \phi_2)} \tag{13}$$

or, by Eqs (5), (6) and (11),

$$\frac{d\hat{X}_1(t)}{d\tau} = \alpha\hat{K}(t) = \frac{\alpha}{1+(K_a-1)\hat{X}_1(t)} \tag{14}$$

which may be integrated from $(0,\tau_0)$ to $(\hat{X}_1(t),\tau)$ to yield a quadratic polynomial in $\hat{X}_1(t)$.

$$0.5(K_a-1)(\hat{X}_1(t))^2 + \hat{X}_1(t) - \alpha(\tau-\tau_0) = 0 \tag{15}$$

Alternatively, Eq. (12) may be integrated from $(0,\tau_0)$ to $(X_1(t),\tau)$ to express the interface location as a function of pore volumes.

$$\hat{X}_1(t) = \alpha\theta(t) \tag{16}$$

Equations (11) and (15) or (16) yield a system that may be solved sequentially to yield both the bulk hydraulic conductivity and the first interface location at any time in Phase I. The transient potential corresponding to this interface location may also be computed from the continuity of volumetric flux.

$$q(t) = \frac{K_1(\phi_{x1}(t)-\phi_2)}{L-X_1(t)} = \frac{K_1K_2(\phi_1-\phi_2)}{K_1X_1(t)+K_2(L-X_1(t))} \tag{17}$$

$$\hat{\phi}_{x1}(t) = \frac{1-\hat{X}_1(t)}{1+(K_a-1)\hat{X}_1(t)} \tag{18}$$

Here $\phi_{x1}(t) = (\phi_{x1}(t) - \phi_2)/(\phi_1 - \phi_2)$. The whole transient potential field $\phi(x,t)$ is then defined by a piecewise linear function passing through the points ϕ_1, $\phi_{x1}(t)$, and ϕ_2. The time boundary between Phases I and II is given by $\hat{X}_1(t_1) = \hat{I}_{s1} = I_{s1}/L$ in Eq. (15) or (16)

Phase II ($t_1 \le t < t_2$; $\tau_1 \le \tau < \tau_2$; $\theta_1 \le \theta < \theta_2$)

Phase II begins when the second interface denoting the zone where the soil interactions have attained equilibrium begins to enter the domain. Mundell and Jennings (1987) assumed equal rates of penetration (α=β) for the fronts of Phase II. Jennings and Ravi (1990) present a more general analysis that will be followed here. The only restriction that will be placed on interface penetration is the relatively mild restriction that the second interface will not be allowed to surpass the first.

Similar to Eqs (12) and (14), the rate of penetration of this second interface may be expressed as follows where β is the second required penetration constant.

$$v_2(t) = \frac{dX_2(t)}{dt} = \frac{\beta q(t)}{\varepsilon} \tag{19}$$

or, similar to Eq. (13),

$$\frac{d\hat{X}_2(t)}{d\tau} = \beta\hat{K}(t) \tag{20}$$

where the hydraulic conductivity may be expressed as follows.

$$\hat{K}(t) = \frac{1}{1+(K_a-1)\hat{I}_s(t)+(K_b-1)\hat{X}_2(t)} \tag{21}$$

Note that because $\alpha \neq \beta$, the interface separation distance is a function of time and the bulk hydraulic conductivity is a function of two unknowns. However, the interface separation may be expressed in terms of $X_2(t)$ by equating the bulk hydraulic conductivity in Eq. (14) and Eq. (20) and integrating the result from the time boundary of Phase II.

$$\frac{1}{L}\int_{\hat{I}_{s1}}^{\hat{X}_1(t)} dX_1(t) = \left(\frac{\alpha}{\beta L}\right)\int_0^{\hat{X}_2(t)} dX_2(t) \tag{22}$$

$$\hat{X}_1(t) = \left(\frac{\alpha\hat{X}_2(t)}{\beta}\right)+\hat{I}_{s1} \tag{23}$$

Substitution into the definition of the interface separation function (Eq. (9)) yields the following.

$$\hat{I}_s(t) = \left(\frac{\alpha}{\beta}-1\right)\hat{X}_2(t)+\hat{I}_{s1} \tag{24}$$

Eq. (21) and (24) may now be substituted into Eq. (20) and integrated from τ_1 to τ yield a quadratic equation for the location of the second interface.

$$0.5\left\{(K_b-1)+\left(\frac{\alpha}{\beta}-1\right)(K_a-1)\right\}(\hat{X}_2(t))^2+\left\{1+(K_a-1)\hat{I}_{s1}\right\}\hat{X}_2(t)-\beta(\tau-\tau_1)=0 \tag{25}$$

Alternatively, the second interface location and interface separation may be expressed in terms of the pore volume time scale.

$$\hat{X}_2(t) = \beta\{\theta(t)-\theta_1\} = \beta\left\{\theta(t)-\frac{\hat{I}_{s1}}{\alpha}\right\} \tag{26}$$

$$\hat{I}_s(t) = \{\alpha-\beta\}\theta(t)+\frac{\beta\hat{I}_{s1}}{\alpha} \tag{27}$$

Similar to Phase I, the bulk hydraulic conductivity may be evaluated by sequential solution of Eqs. (25), (24) (or (26), (27)) and (21). To avoid conceptual difficulties associated with the trailing front overtaking the leading front, it is also necessary to impose the following coefficient constraint.

$$\alpha > \beta(1 - \hat{I}_{s1})$$

Expressions for the potentials associated with the traveling interfaces may be derived by equating the transient volumetric flux through the whole domain to that through the "reacted" and "unreacted" domain partitions. Note that $X_2(t)$ has been eliminated by observing that $X_2(t) + I_s(t) = X_1(t)$.

$$\frac{\phi_1 - \phi_2}{\frac{X_1(t) - I_s(t)}{K_3} + \frac{I_s(t)}{K_2} + \frac{L - X_1(t)}{K1}} = \frac{K_1(\phi_{x1}(t) - \phi_2)}{L - X_1(t)} = \frac{K_3(\phi_1 - \phi_{x2}(t))}{X_1(t) - I_s(t)}$$

Rearrangement and simplification yields the desired interface potentials.

$$\hat{\phi}_{x1}(t) = \frac{1 - \hat{X}_1(t)}{1 + (K_b - 1)\hat{X}_1(t) + (K_a - K_b)\hat{I}_s(t)} \tag{28}$$

$$\hat{\phi}_{x2}(t) = \frac{1 - \hat{X}_1(t) + K_a\hat{I}_s(t)}{1 + (K_b - 1)\hat{X}_1(t) + (K_a - K_b)\hat{I}_s(t)} \tag{29}$$

The complete potential field is then given by a piecewise linear function passing through the values ϕ_1, $\phi_{x1}(t)$, $\phi_{x2}(t)$, and ϕ_2. The boundaries in time between Phase II and Phase III (i.e. time t_2) may be located as follows.

$$\hat{X}_1(t_2) = 1 \tag{30}$$

$$\hat{X}_2(t_2) = \frac{\beta(1 - \hat{I}_{s1})}{\alpha} \tag{31}$$

Phase III ($t_2 \le t < t_3$; $\tau_2 \le \tau < \tau_3$; $\theta_2 \le \theta < \theta_3$)

Phase III is similar to Phase I in that the domain response is governed by the location of only one interface. As before, the bulk hydraulic conductivity may be expressed by appropriate substitutions into Eq. (10).

$$\hat{K}(t) = \frac{1}{K_a + (K_b - K_a)\hat{X}_2(t)} \tag{32}$$

Substitution of this into Eq. (20) and integration yields a polynomial for the nondimensional interface penetration depth.

$$\frac{(K_b - K_a)(\hat{X}_2(t))^2}{2} + K_a\hat{X}_2(t) - \frac{\beta K_a\left(1-\hat{I}_{s1}\right)}{\alpha} - \frac{\beta^2(K_b - K_a)\left(1-\hat{I}_{s1}\right)^2}{2\alpha^2} - \beta(\tau - \tau_2) = 0 \tag{33}$$

Therefore the transient bulk hydraulic conductivity may be evaluated by sequential solution of Eqs (33) and (32). The interface potential may also be determined as in Phase II by equating the flux through the domain to that through the remainder of the "reacting" domain partition.

$$\hat{\phi}_{x2}(t) = \frac{K_a\left(1-\hat{X}_2(t)\right)}{K_a + (K_b - K_a)\hat{X}_2(t)} \tag{34}$$

The potential field solution is again defined as a piecewise linear function of space passing through ϕ_1, $\phi_{x2}(t)$, ϕ_2. The solution terminates at time t_3 when the second interface reaches the domain boundary.

$$\hat{X}_2(t_3) = 1 \tag{35}$$

It is interesting to note that since Eq. (10) is independent of domain boundary conditions, the nondimensional bulk hydraulic conductivity and interface penetration functions for Phases I, II, and III are independent of permeameter designs that only differ by boundary conditions.

Several other DIP model formulations have been suggested, and analytical solutions for many of these have been derived. Generally, the function for the transient bulk hydraulic conductivity response is the most useful of the possible dependent variables because this is generally what is measured in a dynamic permeability experiment. Tables 1and 2 presents a summary of formulations available for two-, three- and five-parameter models. Jennings and Ravi (1990) have demonstrated that with the correct definition of nondimensional parameters these solutions are valid for any of the permeameter configurations illustrated in Fig. 23. The necessary parameter nondimensionalizations are summarized in Table 3.

DIP model formulations are not capable of capturing all of the behavior that has been observed in laboratory. Often this is because the observed response is not the result of a global soil property change. Rather, it is possible that under some conditions the majority of the impact results from the formation of preferential flow pathways. This will be discussed in greater detail in Section 4. It is also true that the process of DIP model calibration can be quite challenging. Procedures for using DIP model and laboratory observations to quantify essential parameters will be discussed in Section 5. However, DIP modes of the form summarized in Table 1 have been rather successful in reproducing the kind of behavior that has been observed in most dynamic permeability experiments. All of the model fits that are illustrated with the figures of Section 3 are the equations of Table 1. Not all models fit all data, but generally there is a DIP model simulation of the measured data that most people would accept as a reasonably good fit.

4. Preferential Flow Pathway Models

There has been a considerable amount of interest in the chemistry and physical consequences of soil interactions with aggressive permeants. There has also been debate regarding the permeameter cell designs, unnaturally high driving gradients and unrealistic

Table 1 - Discrete Interface Penetration model solutions for bulk hydraulic conductivity

Penetration Phase	Bulk Hydraulic Conductivity K(t)	Interface Location Functions
2-Parameter Model (Otte and Jennings, 1984)		
$0 \le t \le t_1$	$[1+(K_a-1)\hat{X}_1(t)]^{-1}$	$\hat{X}_1(t) = \alpha\theta$
4-Parameter Model (Mundell and Jennings, 1987)		
$0 \le t \le t_1$	$[1+(K_a-1)\hat{X}_1(t)]^{-1}$	$\hat{X}_1(t) = \alpha\theta$
$t_1 < t \le t_2$	$[1+(K_a-1)\hat{I}_{S1}+(K_b-1)\hat{X}_2(t)]^{-1}$	$\hat{X}_2(t) = \alpha\theta - \hat{I}_{S1}$
$t_2 < t \le t_3$	$[K_a+(K_a-K_b)\hat{X}_2(t)]^{-1}$	$\hat{X}_2(t) = \alpha\theta - \hat{I}_{S1}$
5-Parameter Model (Jennings and Ravi, 1990)		
$0 \le t \le t_1$	$[1+(K_a-1)\hat{X}_1(t)]^{-1}$	$\hat{X}_1(t) = \alpha\theta$
$t_1 < t \le t_2$	$[1+(K_a-1)\hat{I}_S(t)+(K_b-1)\hat{X}_2(t)]^{-1}$	$\hat{X}_2(t)=\beta(\theta-(\hat{I}_{S1}/\alpha))$ $\hat{I}_S(t)=(\alpha-\beta)\theta+(\beta\hat{I}_{S1}/\alpha)$
$t_2 < t \le t_3$	$[K_a+(K_a-K_b)\hat{X}_2(t)]^{-1}$	$\hat{X}_2(t)=\beta(\theta-(\hat{I}_{S1}/\alpha))$

Nondimensional Quantities

Saturated Hydraulic Conductivities: $K_a = \frac{K_1}{K_2}$; $K_b = \frac{K_1}{K_3}$

Interface Spatial Relationships:

$$\hat{X}_1(t) = \frac{X_1(t)}{L} \; ; \; \hat{X}_2(t) = \frac{X_2(t)}{L} \; ; \; \hat{I}_S(t) = \hat{X}_1(t) - \hat{X}_2(t) \; ; \; \hat{I}_{S1} = \hat{X}_1(t_1)$$

Nondimensional Time (Pore Volumes): $\theta(t) = \int_0^t q(t)dt/\epsilon L$

Table 2 - Parameter constraints for DIP models

2-Parameter Model	4-Parameter Model	5-Parameter Model
1. $\alpha > 0$	1. $\alpha > 0$	1. $\alpha > 0$
2. $1-\alpha \ge 0$	2. $1-\alpha \ge 0$	2. $1-\alpha \ge 0$
3. $K_a > 0$	3. $\hat{I}_{S1} \ge 0$	3. $\beta > 0$
	4. $1-\hat{I}_{S1} \ge 0$	4. $1-\beta \ge 0$
	5. $K_a > 0$	5. $\hat{I}_{S1} \ge 0$
	6. $K_b > 0$	6. $1-\hat{I}_{S1} \ge 0$
		7. $\alpha-\beta(1-\hat{I}_{S1}) > 0$
		8. $K_a > 0$
		9. $K_b > 0$

Table 3 - Permeameter design-dependent nondimensionalizations for Discrete Interface Penetration analysis design-independent analytical solutions

Nondimensional Parameter	Permeameter Design Alternatives		
	Constant Head	Constant Flux	Falling Head
Time $\tau =$	$q(t_0)t/\epsilon L$	$qt/\epsilon L$	$q(t_0)t/\epsilon L$
Time $\theta =$	$\int_{t_0}^{t} q(t)dt/\epsilon L$	$qt/\epsilon L$	$A_S(\phi_1(t_0)-\phi_1(t))/A_C\epsilon L$
Interface Potential $\hat{\phi}_{x1}(t) =$	$\dfrac{\phi_{x1}(t)-\phi_2}{\phi_1-\phi_2}$	$\dfrac{\phi_{x1}(t)-\phi_2}{qL/K_1}$	$\dfrac{\phi_{x1}(t)-\phi_2}{\phi_1(t)-\phi_2}$
Interface Potential $\hat{\phi}_{x2}(t) =$	$\dfrac{\phi_1}{L}$	$\dfrac{\phi_{x2}(t)-\phi_2}{qL/K_1}$	$\dfrac{\phi_1(t)}{L}$
Boundary Potential $\hat{\phi}_1(t) =$	$\dfrac{\phi_2}{L}$	$\dfrac{\phi_1(t)-\phi_2}{qL/K_1}$	$\dfrac{\phi_2(t)}{L}$

Where: t = dimensional time, t_0 = dimensional initial time,
ϵ = domain porosity, L = domain length, q = Darcy velocity,
ϕ_1 = boundary potential at x = 0, ϕ_2 = boundary potential at x = L,
ϕ_{x1} = potential associated with the leading interface,
ϕ_{x2} = potential associated with the trailing interface,
K_1 = initial "unreacted" saturated hydraulic conductivity, and
(t) implies a function of dimensional time t.

contaminant concentrations used in many aggressive permeant studies (Acar and D'Hollosy, 1987; Fogler and McCune, 1976; Foreman and Daniel, 1986; Mitchell and Madsen, 1987). Despite these criticisms, there remains general agreement that under some conditions aggressive permeants can alter soil hydraulic conductivity by several orders of magnitude.

Efforts to model interactions between soils and aggressive permeants have provided interesting results. The Discrete Interface Penetration (DIP) modeling approach described in Section 4 have been proposed as a general, phenomenological approach to the modeling problem. DIP model predictions have also been shown to capture many of the characteristic behavior patterns of soil-permeant interactions and fit many observed results quite well. However, calibration studies have also indicated that there are soil-permeant systems that cannot be modeled well with the existing DIP formulations. For example, Fig. 15 illustrates data for which DIP modeling is not particularly successful. Recall that these data are for a non-calcareous smectite permeated with full strength ethylene glycol in a fixed wall permeameter using a hydraulic gradient of 361.6 (Anderson et al., 1985). Although the DIP models are successful in approximating the ultimate behavior of the system, they do not capture the behavior of the bulk hydraulic conductivity early in nondimensional time. In this case, the reason appears to be that the soil's behavior (particularly at early time) was not the consequence of a "global" change in bulk soil properties. Rather, it appears that the response was influenced by the development of a preferential flow pathway. Anderson et al. (1985) reported the presence of observable changes in the samples after the experiment and concluded that ". . . . the formation and subsequent flow through cracks and macropores is the probable mechanism for increased permeability". Although it is not known whether wall separations provided the most significant preferential flow pathways, it seems reasonable to assume that volume changes in the soil near the top of the permeameter resulted in a wall separation that rapidly (relative to the pore volume time scale) propagated down the length of the permeameter cell wall. The presence of such a preferential flow pathway would allow a significant fraction of the test cell inflow to bypass the soil and rapidly appear as cell discharge. This would account for the very early "apparent" domain response, the ultimate permeability increase of over two orders of magnitude, and the DIP strategy's inability to capture the dynamic behavior because it did not occur within the porous media. It will subsequently be demonstrated that, although the behavior illustrated in Fig. 15 cannot be described well by domain impact models, the description can be improved by adding a preferential pathway flow component.

Results of the type illustrated in Fig. 15 can be very important. If, for example, the test was being conducted to examine the integrity of a proposed hydraulic barrier for a hazardous waste site, one might conclude that the soil was unacceptable. Obviously, if the soil's hydraulic conductivity would increase by orders of magnitude when it came in contact with the waste, then the soil could not used to protect the underlying environment. However, if this response is only an artifact of the experiment and does not reflect the soil's actual propensity to interact with the permeant, this incorrect conclusion might eliminate an otherwise viable and cost-effective design alternative. Additional insight might be gained by repeating the test, but it is possible that the phenomenon yielding the preferential flow is reproducible. It is also true that permeameter tests on aggressive permeants and low permeability soils are difficult, time consuming, and expensive to replicate. It would be desirable to have analytical tools to help determine if a permeameter experiment have been impacted by the formation of preferential flows.

In this section strategies for preferential flow analysis will be explored. These will be divided into wall separation models and internal domain models. Wall separation models are based on the assumption that the preferential flow pathway will develop along the interface between the soil sample and the permeameter cell wall. Internal domain models are based on the assumption the pathway evolves within the soil, away from the wall. The latter allows for greater geometric flexibility and a more realistic treatment of preexisting conditions such as "fossil" cracks that are closed but not sealed.

4.1 Wall Separation Preferential Flow Pathway Models

Wall separation preferential flows develop between the wall of a permeameter cell and the soil sample being tested. This type of preferential flow is more characteristic of fixed-wall permeameters than of flexible wall permeameters, although wall effects have been observed in both. Often these are thought of as resulting from some fault or flaw in the experimental technique. This is certainly possible. It the pathway exists at time zero, then the test cell was not properly loaded and the experiment should be terminated. However, it is possible for a wall separation to develop sometime after time zero in a properly loaded cell. When this occurs, the domain impact results will be corrupted, but this is not an experimental flaw.

It should be noted that permeameter cell modifications have been proposed to reduce the potential for wall influences, or to separate wall-influenced flow from the total cell discharge (Hill and King ,1982 ; McNeal and Reeve,1964 ; Daniel et al.,1985; Anderson et al., 1985). Although these can be effective for many conditions, it has been our experience that they are not commonly used. This is probably because "split-flow" test cells are not generally available from equipment suppliers, and the required modifications of conventional test cells are not discussed in procedure guidance references such as Methods of Soil Analysis (ASA-ASTM, 1965) or in ASTM Standards (e.g. D 2434, and D 5084), (ASTM,1990).

Figure 25 illustrates a permeameter cell being influenced by the formation of a "wall separation" preferential flow pathway (Jennings and Ravi, 1992). Note that the pathway is associated with the boundary at $r=R$ over some significant portion of the surface $2\pi RL$. This pathway yields a fluid velocity significantly different than the Darcy velocity. Figure 25 also illustrates the basic modeling strategy proposed to analyze the effect of the wall separation. It is assumed that after time zero ($t=o$) when the permeameter influent is switched from the base saturating solution to an aggressive permeant, a wall separation of average radial extent $(R - \delta/2)\theta$ and width δ forms at the top of the test cell around a θ radians section of the circumference. It is assumed that flow is not governed by Darcy's law within the wall separation (since there is no porous medium occupying this volume) and that this zone offers significantly less resistance to flow than the soil. Therefore, once formed, the vertical extent of the wall separation denoted here as $X_w(t)$ continues to grow and the zone's propagation velocity $dX_w(t)/dt$ will accelerate as it experiences a decreasing depth of resisting soil.

If a wall separation is taking place within the test cell of a typical permeameter experiment, only "superficial" observations are available. Data may be available for the potential at either end of the cell, and/or for the cell's total discharge, but no data will be available from within the permeameter cell. All inferences must be made from boundary information. What is desired is a method of detecting when the fraction of cell discharge that may be attributed to wall separations (i.e. $Q_w(t)$ of Fig. 25) is a significant component of the total flow $Q_T(t)$.

It should be noted that there may always be some degree of wall impacted flow due to the interface between the permeameter cell wall and the porous medium structure. This is thought to be the result of the altered grain structure in the neighborhood of the cell wall. The rigid wall may force a higher degree of order in the packing arrangement of soil grains immediately adjacent to it. This component of "wall-influenced " flow is generally negligible whenever the permeameter cell diameter exceeds the mean particle size by a factor of about 50 (Dudgeon, 1967 ; Franzinz, 1956 ; Somerton and Wood, 1988). It should also

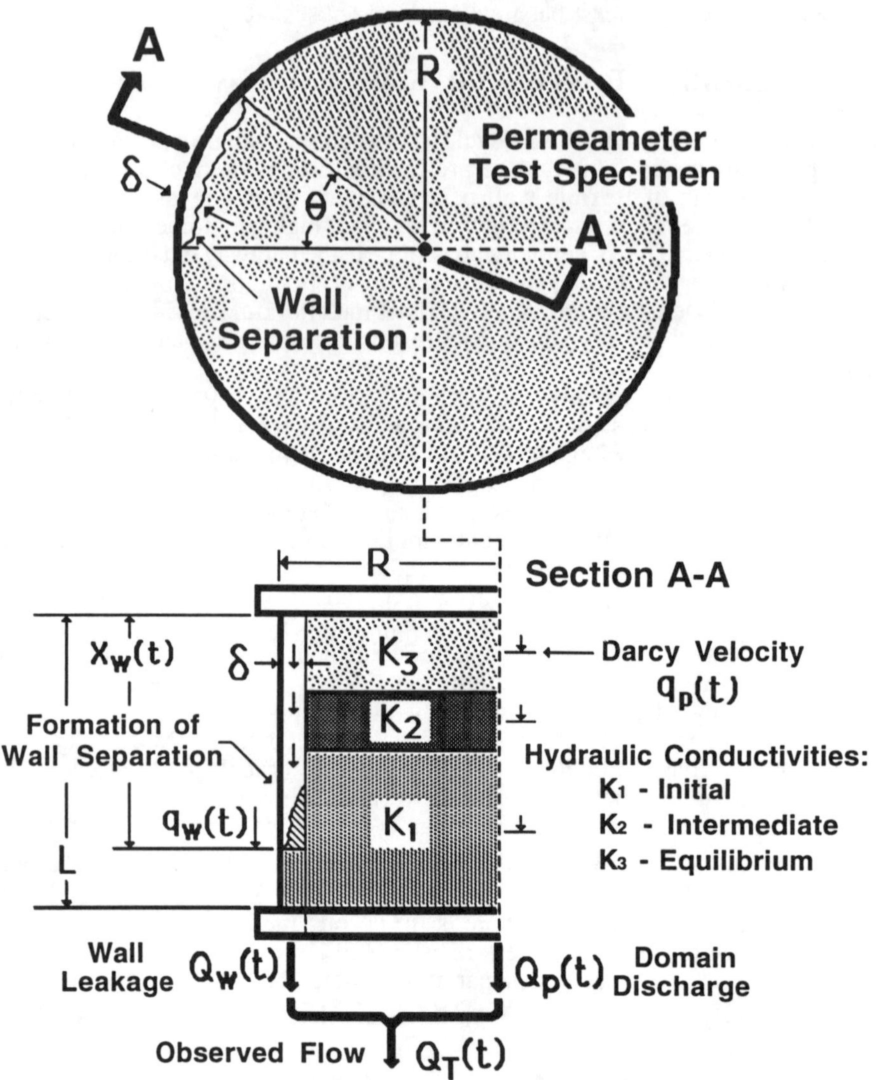

Fig. 25 - Schematic illustration of the development of a wall separation preferential flow pathway (Jennings and Ravi, 1992)

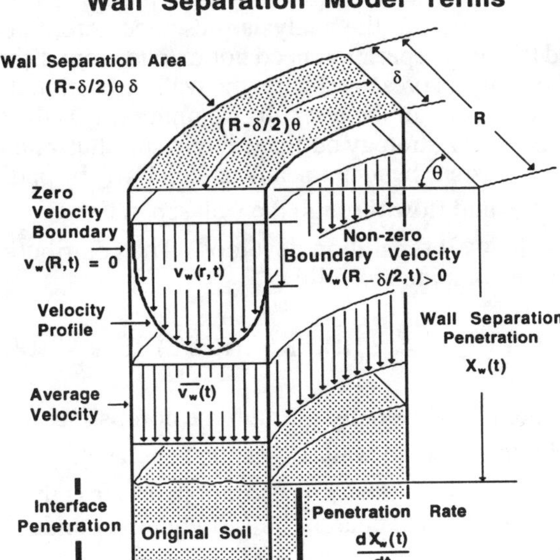

Fig. 26 - Schematic illustration of the significant terms of wall separation preferential flow analysis (Jennings and Ravi, 1992).

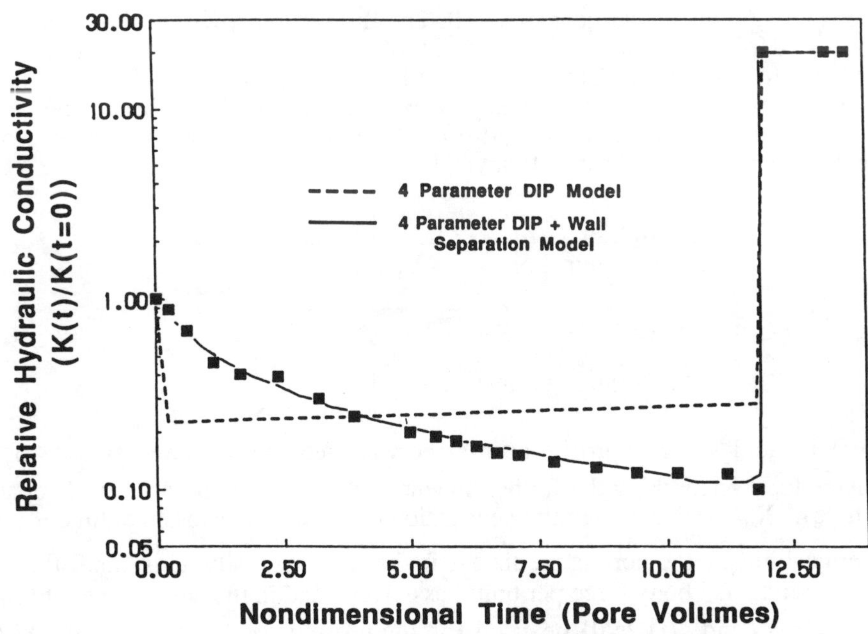

Fig. 27 - Aggressive permeant interaction (illite/chlorite soil - heptane) characteristic of domain interaction response subsequently corrupted by preferential pathway flow (Bowders, 1985 ; Jennings and Ravi, 1992).

be noted that the concept described in Fig. 25 and Fig. 26 is similar to the work of Tokunaga (1988). However, in the analysis presented here, steady-state conditions will not be imposed and the wall separation need not extend around the whole circumference of the cell. Also, in the analysis presented here, the wall separation does not exist at time zero. It develops over some fraction of the test cell circumference as the test proceeds.

Given the basic modeling strategy outlined above and illustrated in Fig. 25 and Fig. 26, let the total flow of a constant head permeameter cell $Q_T(t)$ be partitioned into flow through the porous media $Q_p(t)$ and flow through the wall separation $Q_w(t)$. Both of these fractions may vary in time as the wall separation develops. The contribution of each fraction to the total discharge may be computed as follows,

$$Q_T(t) = Q_p(t) + Q_w(t) = q_p(t)[\pi R^2 - (R-\delta/2)\,\theta\delta] + v_w(t)[(R-\delta/2)\theta\delta] \tag{36}$$

where $q_p(t)$ is the Darcy fluid velocity through the porous media at any time t, $v_w(t)$ is the average fluid velocity through the wall separation zone at any time t, R is the permeameter cell radius, δ is the thickness of the wall separation, and θ is the angular extent of the wall separation (see Fig. 26). The apparent bulk hydraulic conductivity K(t) (as defined by Darcy's Law) is given by:

$$K(t) = \frac{LQ_T(t)}{\pi R^2(\phi_1 - \phi_2)} \tag{37}$$

where ϕ_1 is the permeameter boundary potential at x= 0, ϕ_2 is the boundary potential at x=L, and L is length of the permeameter cell. The fluid velocity within the porous media may be evaluated from either a simple statement of Darcy's law or a DIP model analysis. The latter leads to a transient velocity that can be a strong function of time. The fluid velocity within the wall separation zone may be evaluated at any time t by analogy to flow through a circular annulus (Bird et al.,1960). At time t, a momentum balance on the Newtonian fluid within the wall separation yields:

$$\frac{d}{dr}\left\{-r\mu\frac{dv_w(r,t)}{dr}\right\} = \frac{\rho g(\phi_1 - \phi_w(t))}{X_w(t)} \tag{38}$$

Subject to $$v_w(R,t) = 0 \tag{39}$$

and $$v_w(R-\delta,t) = k\rho g(\phi_1 - \phi_2)/\mu L \tag{40}$$

Here $v_w(r,t)$ is fluid velocity profile across the wall separation at any time t, $\phi_w(t)$ is potential at the location marking the farthest extent of the penetration of the wall-separation regime at time t, $X_w(t)$ is the maximum penetration of the wall separation at time t, k is the intrinsic permeability of the porous media, μ is the fluid viscosity, ρ is the fluid density and g is the gravitational body force per unit mass. Here, and in the development to follow, the notation "(r,t)" and "(t)" will be used for the terms $v_w(r,t)$, $\phi_w(t)$, and $X_w(t)$ to emphasize that these variables are functions of space and/or time.

The boundary condition of Eq. (39) requires that the bulk fluid velocity be zero on the interface between the fluid and the test cell wall. Boundary condition of Eq. (40) requires that the fluid velocity on the interface between the fluid and the soil sample to be equal to the initial Darcy flow velocity within the soil. When a domain impact model is applied to the soil, this should actually be treated as a transient condition because the domain's intrinsic permeability would be a function of time. This was not done for this presentation. The solution presented here is intended to be applied when the volume flowing through the wall separation is large compared to the total flow. This is unlikely to be the case if the domain flow is changing dramatically. Since the impact of the boundary condition itself is relatively modest, it was assumed that the additional complexity was not required.

An alternative approach for boundary condition (40) is to treat the vertical plane at $r=R-\delta$ as a non-slip plane and specify a zero velocity on the boundary. This does not imply that there is no flow within the soil, but simply that the two flow regimes are hydraulically isolated from one another. This approach is generally acceptable whenever the velocity in the wall separation is substantially greater than the Darcy velocity in the bulk soil. The advantage of specifying a zero velocity at $r= R-\delta$ is that it simplifies the solution.

The solution of Eq. (38) subject to boundary conditions (39) and (40) is as follows,

$$v_w(r,t) = \frac{\rho g}{4\mu}\left(\frac{\phi_1 - \phi_w(t)}{X_w(t)}\right)(R^2 - r^2) + \frac{\rho g \ln(R/r)}{\mu \ln(\eta)}\left\{k\left(\frac{\phi_2 - \phi_1}{L}\right) + \frac{\phi_1 - \phi_w(t)}{X_w(t)}\left(\frac{R^2}{4}\right)(1-\eta^2)\right\} \tag{41}$$

where η is a non-dimensional parameter quantifying the fraction of the domain radius not occupied by the wall separation.

$$\eta = (R-\delta)/R \tag{42}$$

The average vertical flow velocity down the wall separation $\bar{v}_w(t)$ may be computed from Eq. (41) by averaging $v_w(r,t)$ across the width of the wall separation.

$$\bar{v}_w(t) = \frac{\int_{R-\delta}^{R} r v_w(r,t)\,dr}{\int_{R-\delta}^{R} r\,dr} \tag{43}$$

Substitution of Eq. (41) into Eq. (43) and integration yields the required average.

$$\bar{v}_w(t) = \frac{k\rho g\Omega}{\mu}\left(\frac{\phi_1 - \phi_w(t)}{X_w(t)}\right) + \frac{k\rho g\Psi}{\mu}\left(\frac{\phi_2 - \phi_1}{L}\right) \tag{44}$$

Here Ω and Ψ are nondimensional functions of η defined as follows.

$$\Omega = \frac{R^2}{8k}\left[1 + \eta^2 + \frac{1-\eta^2}{\ln(\eta)}\right] \tag{45}$$

$$\Psi = \frac{1}{2\ln(\eta)} + \frac{\eta^2}{1-\eta^2} \tag{46}$$

The coefficient group Ω can be thought of as the relative intrinsic permeability of the wall separation. The value of Ψ is negative and determines the relative significance of the non-zero boundary velocity's contribution to the average wall separation zone velocity.

Although Eq. (44) appears to offer the solution to the problem, it contains two functions that are as yet unquantified. The vertical penetration of the wall separation $X_w(t)$ and the potential at the leading edge of the separation $\phi_w(t)$ are not yet known. These may be determined by equating the flow through the wall separation to the flow through the soil directly below the developing separation. Since flow through the unaltered soil is governed by Darcy's law, the following may also be written.

$$q_w(t) = \frac{k\rho g}{\mu}\left(\frac{\phi_w(t) - \phi_2}{L - X_w(t)}\right) \tag{47}$$

Since these two velocities must be equal, Eq. (44) and (47) may be equated and solved for the potential difference driving flow through the unaltered soil.

$$[\phi_w(t) - \phi_2] = \frac{(L - X_w(t))(\phi_1 - \phi_2)\left[\Omega - \frac{X_w(t)\Psi}{L}\right]}{X_w(t) \ + \ (L - X_w(t))\Omega} \tag{48}$$

Flow through the soil below the separation may be expressed as follows.

$$q_w(t) = \frac{L\,\Omega - X_w(t)\Psi}{X_w(t) \ + (L - X_w(t))\Omega}\left(\frac{k\rho g(\phi_1 - \phi_2)}{\mu L}\right) \tag{49}$$

However, the depth of the developing wall separation $X_w(t)$ remains unknown. To evaluate this, a constitutive relationship must be specified to quantify how the wall separation propagates along the permeameter cell wall. For the development presented here, it will be assumed that the rate of wall separation propagation is proportional to the average fluid velocity within the zone already impacted. This provides for an acceleration of the propagating front in response to the increasing fraction of the potential gradient available to act on the unaltered soil depth. This approach is consistent with the strategy applied in DIP models, and is equivalent to specifying that the rate of prorogation is proportional to the rate of aggressive permeant mass transport to the leading edge of the separation zone. By this approach, the rate of wall separation propagation may be expressed as follows,

$$\frac{dX_w(t)}{dt} = \alpha_w \overline{v_w}(t) = \alpha_w q_w(t) \tag{50}$$

where α_w is the wall separation propagation rate constant. Substitution of Eq. (49) into Eq. (50) yields an initial value problem that may be solved subject to the initial condition ($X_w(0) = 0$) for the transient depth of the wall effect zone. The result is an implicit function of $X_w(t)$ that may be evaluated at any time t.

$$\frac{X_w(t)}{\Psi}(1-\Omega) + \frac{L\Omega}{\Psi^2}(1+\Psi-\Omega)\ln\left[1-\frac{\Psi X_w(t)}{L\Omega}\right] + \frac{\alpha_w k\rho g}{\mu}\left(\frac{\phi_1-\phi_2}{L}\right)t = 0 \qquad (51)$$

Equations (44), (48) and (51) supply all the information required to evaluate the unknown velocity in Eq. (36) and complete the solution for the observed permeameter discharge $Q_T(t)$ and the bulk hydraulic conductivity.

The duration of the wall separation development event (T_w) is the time required for the wall separation to reach the bottom of the permeameter cell. This may be evaluated from Eq. (51) by observing that $X_w(T_w) = L$. Substitution yields the following result.

$$T_w = \frac{-\mu L}{\alpha_w k\rho g(\phi_1-\phi_2)}\left\{(1-\Omega)\left(\frac{L}{\Psi}\right) + \frac{L\Omega}{\Psi^2}(1+\Psi-\Omega)\ln\left[1-\frac{\Psi}{\Omega}\right]\right\} \qquad (52)$$

If this solution is rederived using the following "no-slip" boundary condition to replace Eq. (40),

$$v_w(R-\delta,t) = 0 \qquad (53)$$

the resulting solution is as follows.

$$v_w(r,t) = \frac{\rho g(\phi_1-\phi_w(t))}{4\mu X_w(t)}\left\{R^2 - r^2 + \left(2R\delta-\delta^2\right)\left(\frac{\ln[R/r]}{\ln[\eta]}\right)\right\} \qquad (54)$$

$$\overline{v_w}(t) = \frac{k\rho g\,\Omega(\phi_1-\phi_w(t))}{\mu X_w(t)} \qquad (55)$$

$$\phi_w(t)-\phi_2 = \frac{\Omega(L-X_w(t))(\phi_1-\phi_2)}{\Omega(L-X_w(t))+X_w(t)} \qquad (56)$$

$$q_w(t) = \frac{k\rho g\,\Omega(\phi_1-\phi_2)}{\mu\,\Omega(L-X_w(t))+\mu X_w(t)} \qquad (57)$$

$$X_w(t) = \frac{-\Omega L + \left\{(\Omega L)^2 + \dfrac{2\alpha_w k\rho g\Omega}{\mu}(1-\Omega)(\phi_1-\phi_2)t\right\}^{1/2}}{(1-\Omega)} \qquad (58)$$

$$T_w = \frac{\mu L^2(1+\Omega)}{2\alpha_w k\rho g(\phi_1-\phi_2)\Omega} \qquad (59)$$

The fundamental difference between these equations and the solutions previously presented can be seen most clearly by comparing Eq. (55) with Eq. (44). The second term in the velocity of Eq. (44) represents the contribution of the non-zero boundary condition, and guarantees that the Eq. (44) will yield a value greater than that of Eq (55). This, in turn, ensures that the breakthrough time predicted by Eq. (52) will be sooner than that of Eq. (59).

The relative importance of modeling the velocity on the soil/wall interface can be determined by comparing the magnitude of the two terms in Eq. (44). For cases where the second term is negligible, the simpler formulation based on the non-slip boundary condition is recommended.

One less obvious advantage of this simpler formulation is that the transient interface location $X_w(t)$ can be solved for explicitly using Eq. (58). This avoids what can be substantial numerical difficulties associated with solving the implicit non-linear relationship of Eq. (51) (see Jennings and Ravi, 1992).

Considering a wall-separation component of observed aggressive permeant interaction data can greatly increase the ability of models to reproduce observed behavior. As an example, Fig. 27 presents experimental results and DIP model predictions with and without a wall separation flow component. This example was selected because the presence of a preferential flow was actually observed (albeit post hoc). Had the preferential flow not been observed, one might consider that these data were valid for the domain impact of the soil/solvent interaction tested.

Figure 27 illustrates data of Bowders (1985) measured for an illite-chlorite soil permeated with full strength Heptane in a flexible wall permeameter using a hydraulic gradient of 200. The initial decrease in the hydraulic conductivity was attributed to shrinkage of the soil. In a fixed wall permeameter this would probably lead to an increase in permeability, but in a flexible wall test cell, the confining pressure forces the soil to consolidate as it shrinks and the bulk domain hydraulic conductivity declines. In the work of Bowders (1985), permeability declined until there was a "heptane blowout". This was observed as an increase in flow and a corresponding increase in heptane effluent concentration. The cause of this was attributed to a preferential flow pathway that opened between the soil sample and the flexible wall membrane.

Figure 27 illustrates DIP model fit to the data if the impact is assumed to be a bulk domain phenomenon and the fit of a DIP plus wall separation model. The coefficient values used to simulate these results may be found in Jennings and Ravi (1992), but are not particularly significant to the intent of the example. What is significant is the fact that standard DIP modeling procedures can simulate the soil's behavior if data beyond the "blowout" are ignored, but cannot provide a reasonable representation if the data on the rapid hydraulic conductivity increase are included. The optimum 4-parameter DIP model fit had a least squares regression analysis "fit" residual of 0.0485 . This residual was reduced by more than an order of magnitude (0.00380) when a wall separation flow component is added.

This illustrates the potential value of wall separation modeling. For this case, had it not actually been observed, one would not know that the experiment had experienced a preferential flow. The pattern of initial hydraulic conductivity decline followed by an eventual increase is not particularly unusual in the data of aggressive permeant interactions. Acar et al., (1985) and Acar and D'Hollosy (1987) present data on several acetone/soil interactions that yield a similar pattern. What is significant is that, in this case, the observed behavior cannot be simulated with a domain impact model, but can be captured very well with a model including a preferential flow. Given this, it would be prudent to conclude that there had been a preferential flow pathway even if it had not actually been observed.

4.2 Pore Penetration Preferential Flow Pathway Models

The modeling strategy of the previous section is appropriate to analyze the possibility that a preferential flow pathway opened along the wall of the permeameter cell. "Wall effects" are actually a special case of preferential pathway development, but are of particular interest because they are often observed in permeameter experiments. The goal of the analysis presented in this section is to extend this same basic impact modeling strategy to predict the consequences of preferential flow pathways that open within the soil domain at some time after the experiment begins.

Figure 28 illustrates two proposed analysis strategies. In the first, a circular preferential flow pathway of radius R_p is assumed to open at the top of the test cell sometime after time zero. This pathway then penetrates in the direction of the Darcy velocity at some finite rate governed by the chemistry and physics of the interaction phenomenon. Since flow within the pathway is not governed by Darcy's law, the pathway offers significantly less resistance to flow. Therefore, once the pathway opens it will continue to "grow" (penetrate) as fluid preferentially diverts into it. This form of preferential pathway development is consistent with the photographic evidence of Hoefner and Fogler (1988) for the development of a dissolution channeling.

A second analysis strategy is also illustrated in Fig. 28. This vision of preferential pathway development involves the transient growth of an existing continuous pore that traverses the domain at time zero. The modeling of this concept of preferential pathway development will be discussed in Section 4.3.

Both of these analysis strategies are based on simplistic descriptions of the phenomena that actually produce preferential pathways. Both, however, provide an analytical framework that may be tested using laboratory observations. Given that one can solve for the implied domain impacts, the results may be used to identify likely strategies for a more comprehensive analysis.

Similar to the analysis of the impacts of a wall separation, let the total flow $Q_T(t)$ of a constant head permeameter cell be partitioned into flow through the porous domain Q_d and flow through the preferential pathway $Q_p(t)$. For the "pore penetration" model, Q_d remains constant. For the "pore growth" model, $Q_d = Q_d(t)$ because the area of the pore changes with time. At any time (t) the contribution of each fraction to the total cell discharge may be computed as follows.

$$Q_T(t) = Q_d(t) + Q_p(t) = q_d[\pi R^2 - \pi R_p(t)^2] + v_p(t)\pi R_p(t)^2 \tag{60}$$

where q_d is the Darcy fluid velocity through the porous media, $v_p(t)$ is the average fluid velocity through the preferential pathway at time t, R is the permeameter cell radius, and $R_p(t)$ is the preferential pathway radius. For the solutions presented here, it will be assumed that the permeability of the soil is not significantly altered by the aggressive permeant. The apparent bulk hydraulic conductivity of the soil sample is computed from the transient test cell discharge.

$$K(t) = \frac{LQ_T(t)}{\pi R^2(\phi_1 - \phi_2)} = \frac{Lq_d[R^2 - R_p(t)^2] + Lv_p(t)R_p(t)^2}{R^2(\phi_1 - \phi_2)} \tag{61}$$

Clearly, if $Q_p(t)$ is significant relative to $Q_d(t)$, the preferential flow pathway will have a significant impact on the observed response. If this is not the case, then a preferential flow model is probably not required. Experience has indicated that the impact is either very

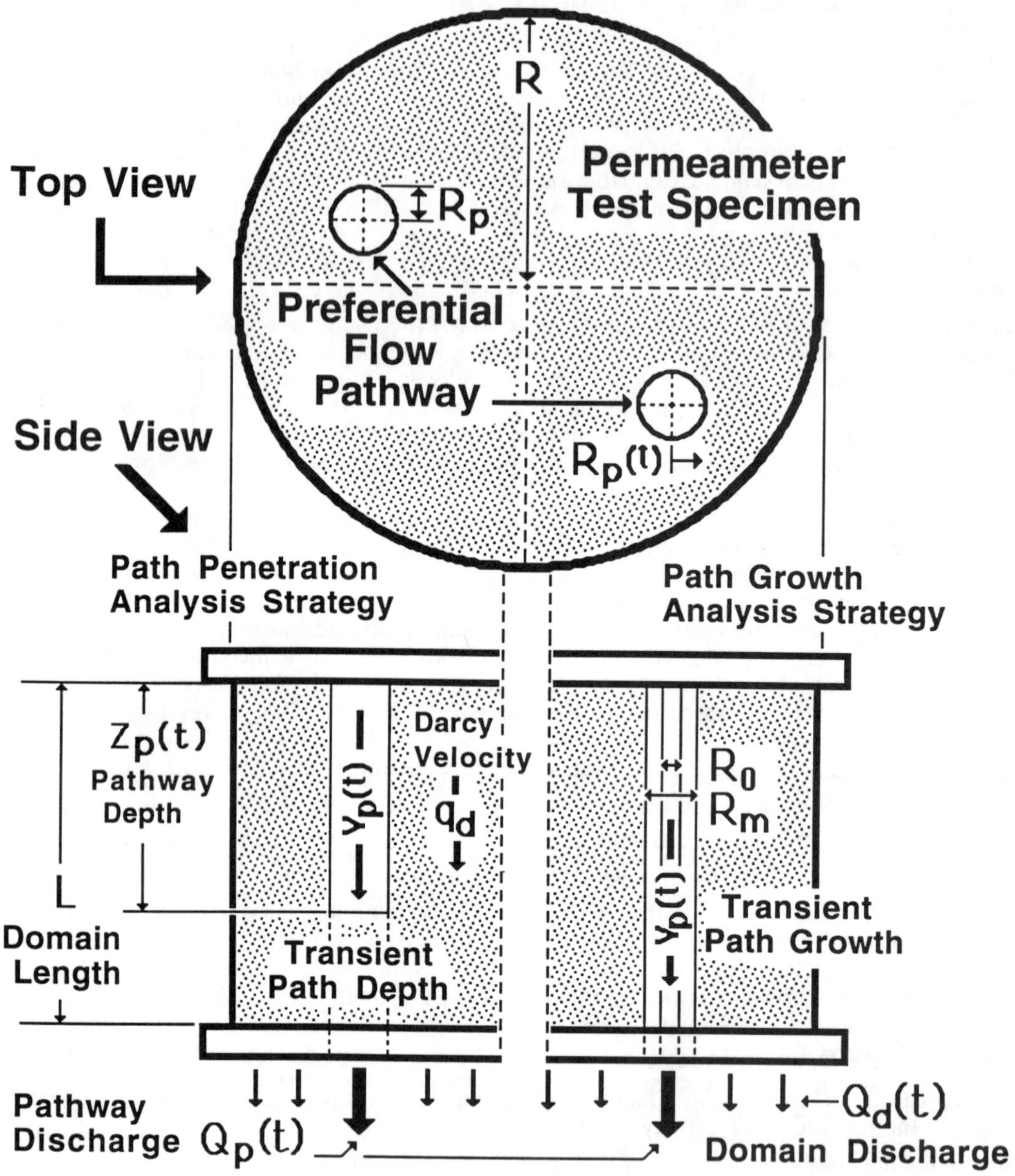

Fig. 28 - Schematic illustration of pore penetration and pore growth preferential flow pathway (Jennings and Shukla, 1992).

important or is negligible.

For the "Path Penetration" model formulation, the fluid velocity within the preferential flow pathway may be evaluated at any time by analogy to flow through a circular conduit. At time t, a momentum balance on the Newtonian permeant within the pathway yields (Bird et al.,1960):

$$\frac{d}{dr}\left\{-r\mu\frac{dv_p(r,t)}{dr}\right\} = \frac{r\rho g(\phi_1 - \phi_p(t))}{Z_p(t)} \tag{62}$$

Subject to:

$$\frac{dv_p(0,t)}{dr} < \infty \tag{63}$$

and

$$v_p(R_p,t) = \frac{k\rho g(\phi_1 - \phi_2)}{\mu L} \tag{64}$$

or

$$v_p(R_p,t) = 0 \tag{65}$$

Here $v_p(r,t)$ is the fluid velocity profile in the pathway of constant radius R_p at time t. The notations "(r,t)" and "(t)" are used to emphasize variables that are functions of space and/or time. Condition (63) requires that the fluid momentum be finite at the center of the pathway. Boundary condition (64) requires that the fluid velocity on the soil-pathway boundary be equal to the Darcy velocity. Alternative boundary condition (65) is the "no slip" condition. Berkowitz (1989) has recently argued that a high pathway velocity would induce a transition zone in the porous media exceeding the Darcy velocity and recommended use of Brinkman's extension of Darcy's law for specifying the boundary condition at radius R_p. Berkowitz (1989) also derived the constraint under which the modified version of (64) would yield significant improvement over the "no slip" condition. Although the results are interesting, they tend to confirm that under most conditions, use of Eq. (65) is justified. Although this is less conceptually satisfying, it is justified because of the inherent magnitude of the preferential flow velocity. The area of the preferential pathway should be small relative to area of the test cell. If the pathway velocity is not much greater than that of the Darcy velocity, the pathway will have no significant impact on the total discharge. If the pathway does have a significant impact, (i.e. has a significantly high velocity) adding the small Darcy velocity contribution to this already large value will not significantly alter its impact. Therefore, if the preferential flow is significant, the boundary condition contribution probably is not.

It should be noted that the governing differential equation of Eq. (62) is similar to, but distinct from the differential equation governing flow through a wall separation (ref. Eq. (38)). Although the left-hand sides of both equations are identical, there is an additional independent variable term (r) on the right-hand side of Eq.(62). Although one might expect this to add complexity to the solution, it does not. The additional term of Eq. (62) adds to the symmetry of the problem and leads to a less complex solution.

Solution of Eq. (62) subject to conditions (63) and (64) yields the following velocity profile.

$$v_p(r,t) = \frac{\rho g}{4\mu}\left(\frac{\phi_1 - \phi_p(t)}{Z_p(t)}\right)(R_p^2 - r^2) \quad + \quad \frac{k\rho g}{\mu}\left(\frac{\phi_1 - \phi_2}{L}\right) \tag{66}$$

The average velocity along the preferential pathway $v_p(t)$ may be computed as before by averaging $v_p(r,t)$ across pathway radius R_p. Substitution into Eq. (43) and integrating yields the following average.

$$v_p(t) = \frac{\rho g R_p^2}{8\mu}\left(\frac{\phi_1 - \phi_p(t)}{Z_p(t)}\right) + \frac{k\rho g}{\mu}\left(\frac{\phi_1 - \phi_2}{L}\right) \tag{67}$$

In this result, the penetration of the preferential flow pathway $Z_p(t)$ and the potential at the leading edge of the pathway $\phi_p(t)$ are unknown. These may be determined by equating flow through the developing pathway to the flow through the unaltered soil below the pathway.

$$v_p(t) = q_p(t) \tag{68}$$

Since flow through the unaltered soil is governed by the following form of Darcy's law,

$$q_p(t) = \frac{k\rho g}{\mu}\left(\frac{\phi_p(t) - \phi_2}{L - Z_p(t)}\right) \tag{69}$$

Equations (67), (68) and (69) may be combined to solve for the potential driving flow through the unaltered soil.

$$\phi_p(t) - \phi_2 = \frac{(L - Z_p(t))\,(\phi_1 - \phi_2)\,(8kZ_p(t) + LR_p^2)}{8kLZ_p(t) + LR_p^2(L - Z_p(t))} \tag{70}$$

However, the depth of the developing pathway $Z_p(t)$ remains unknown. To quantify $Z_p(t)$ a new constitutive relationship must be defined. Here it will be assumed that the rate of pathway penetration is proportional to the average fluid velocity within the pathway. This allows the rate of pathway penetration to increase as it moves deeper into the soil due to the increasing fraction of the driving gradient applied to the shrinking length of unaltered soil. This is equivalent to specifying that the rate of pathway penetration is proportional to the rate at which aggressive permeant mass reaches the leading edge of the pathway.

$$\frac{dZ_p(t)}{dt} = \alpha_p v_p(t) = \alpha_p q_p(t) \tag{71}$$

Here α_p is the pathway propagation rate constant. Equations (69), (70) and (71) form an initial value problem that may be solved subject to the initial condition $Z_p(0) = 0$ for the transient pathway depth.

$$\left(\frac{(R_p^2 - 8k)\mu L}{8\alpha_p k^2 \rho g(\phi_1 - \phi_2)}\right) Z_p(t) - \left(\frac{\mu R_p^2 L^2}{8\alpha_p k^2 \rho g(\phi_1 - \phi_2)} + \frac{\mu(R_p^2 - 8k)R_p^2 L}{64\alpha_p k^3 \rho g(\phi_1 - \phi_2)}\right) \ln\left[\frac{8k}{LR_p^2} Z_p(t) + 1\right] - t = 0 \tag{72}$$

Although this is an implicit function of $Z_p(t)$, it may be solved numerically at any time t for the depth of the pathway. Equations (67), (70) and (72) supply all the information required to evaluate the unknown velocity in Eq. (60) and complete the solution for $Q_T(t)$ and the bulk hydraulic conductivity.

The duration of the pathway development event T_p is the time required for the pathway to reach the bottom of the permeameter cell. This may be evaluated from Eq. (72) by substituting $Z_p(T_p) = L$.

$$T_p = \left(\frac{(R_p^2 - 8k)\mu L^2}{8\alpha_p k^2 \rho g(\phi_1 - \phi_2)} \right) - \left(\frac{\mu R_p^2 L^2}{8\alpha_p k^2 \rho g(\phi_1 - \phi_2)} + \frac{\mu (R_p^2 - 8k) R_p^2 L}{64\alpha_p k^3 \rho g(\phi_1 - \phi_2)} \right) \ln\left[\frac{8k}{R_p^2} + 1 \right] \tag{73}$$

When this whole solution procedure is repeated using the alternative "no-slip" boundary condition of Eq. (65), the resulting solution is as follows.

$$v_p(t) = \frac{\rho g R_p^2}{8\mu} \left(\frac{\phi_1 - \phi_p(t)}{Z_p(t)} \right) \tag{74}$$

$$\phi_p(t) = \frac{R_p^2 \phi_1 L + (8k\phi_2 - R_p^2 \phi_1) Z_p(t)}{R_p^2 L + (8k - R_p^2) Z_p(t)} \tag{75}$$

$$Z_p(t) = \frac{-L + \left\{ L^2 + \left(\frac{8k}{R_p^2} - 1 \right) \frac{2\alpha_p k \rho g}{\mu} (\phi_1 - \phi_2) t \right\}^{1/2}}{\left(\frac{8k}{R_p^2} - 1 \right)} \tag{76}$$

$$T_p = \frac{L^2 \left\{ \left(\frac{8k}{R_p^2} \right)^2 - 1 \right\}}{\left(\frac{8k}{R_p^2} - 1 \right) \frac{2\alpha_p k \rho g}{\mu} (\phi_1 - \phi_2)} \tag{77}$$

The major difference between the two solutions can be seen by comparing Eq. (67) and (74). Note that, as one might expect, the average pore velocities differ only by the magnitude of the Darcy velocity specified on the boundary. If this is a significant fraction of the total velocity, then the first formulation should be used. However, since the area of the preferential flow pathway will generally be quite small, if the impact of the pathway is significant, the velocity will have to be much larger than the Darcy velocity. It would seem that under these conditions the second formulation would be acceptable. There is some motivation to use the second formulation because Eq. (76) is significantly less complicated than Eq. (72).

The kind of hydraulic conductivity response that can be simulated with a pore penetration formulation is illustrated in Fig. 29. The experimental data are for the impact of

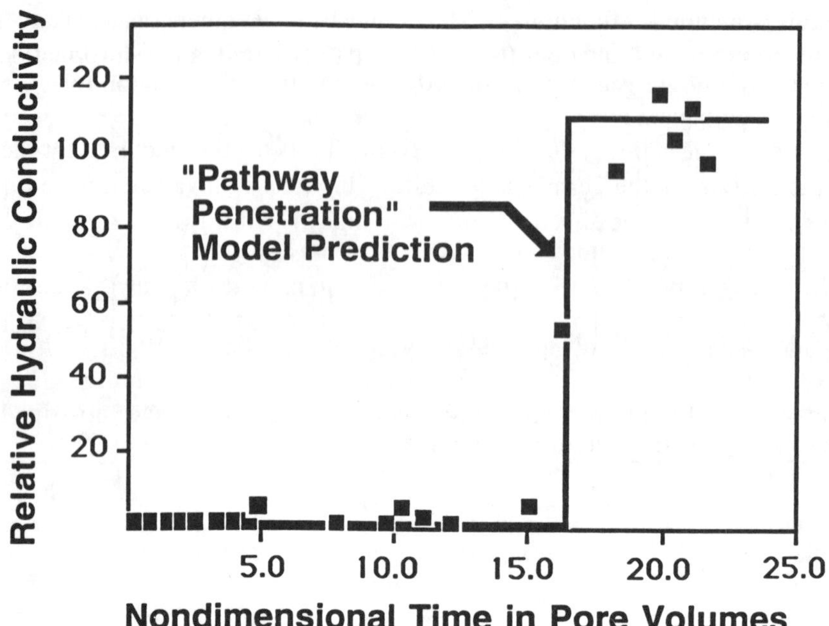

Fig. 29 - Aggressive permeant interaction (calcium bentonite soil - sulfuric acid) characteristic of domain response that may indicate the development of a pore penetration preferential pathway (Nasiatka et al.,1981).

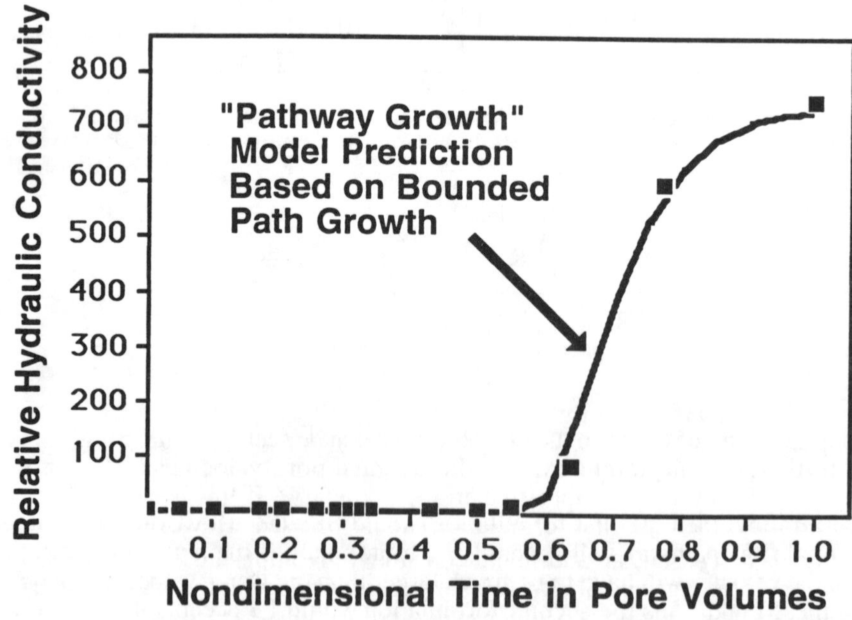

Fig. 30 - Aggressive permeant interaction (noncalcareous smectite - acetone) characteristic of domain response that may indicate the development of a pore growth preferential pathway (Anderson et al., 1985).

a concentrated sulfuric acid solution on a calcium bentonite clay (Nasiatka et al.,1981). The test was conducted in a fixed wall, falling head permeameter at a relatively low hydraulic gradient. The authors did not discuss preferential flow development, but this is common in acid leaching problems. Figure 29 illustrates the ability of a pore penetration formulation to simulate this type of permeameter cell response. Jennings and Ravi (1992) have recently demonstrated that the step-change-like response is an inherent feature of this type of formulation. This occurs because the unaltered soil below the penetrating pathway serves to "plug" the pathway quite effectively up to the time of pathway breakthrough. However, at T_p, the pathway opens completely and can allow an almost unlimited amount of fluid to bypass the soil. Thus, at time T_p, the step change in apparent bulk hydraulic conductivity can be very large. Although it may seem to be counter-intuitive, this type of formulation is apparently unable to capture the behavior if the change is more gradual on the pore volume time scale. Simulating preferential flow impacts that develop more gradually in time requires an alternative formation mechanism. Pathway growth models are one possibility.

4.3 Pore Growth Preferential Flow Pathway Models

The pathway growth modeling strategy (see Fig. 28) is an alternative vision of how preferential pathways develop in soils. In this concept, one assumes that there is an existing continuous pore that traverses the domain at time zero and that is subject of aggressive permeant attack. As the aggressive permeant invades this pore, it begins to increase the size of the cross-section, thus providing more free space for fluid flow. The fluid velocity within the pathway may also be evaluated at any time t by a momentum balance. For this case, the momentum balance yields:

$$\frac{d}{dr}\left\{-r\mu\frac{dv_p(r,t)}{dr}\right\} = \frac{r\rho g(\phi_1 - \phi_2)}{L} \tag{78}$$

which has the symmetry of Eq. (62) and the distinct analytical advantage of time-invariant coefficients. Equation (78) may be solved subject to boundary conditions (63) and (64) or (65) given that the pore radius R_p is now a function of time (i.e. $R_p = R_p(t)$). Using the boundary condition of Eq. (64) yields the following solution,

$$v_p(r,t) = \frac{\rho g}{4\mu}\left(\frac{\phi_1 - \phi_2}{L}\right)(R_p(t)^2 - r^2) + \frac{k\rho g}{\mu}\left(\frac{\phi_1 - \phi_2}{L}\right) \tag{79}$$

which may be integrated over space to yield the average velocity $v_p(t)$.

$$v_p(t) = \frac{\rho g}{2\mu L}(\phi_1 - \phi_2)(8k + R_p(t)^2) \tag{80}$$

Note that the function $R_p(t)$ is as yet unknown. Similar to all of the other preferential pathway models discussed, an additional constitutive relationship is required to complete the formulation. Jennings and Shukla (1992) experimented with four possible forms of the required relationship.

$$\frac{dR_p(t)}{dt} = \beta_1(R_m - R_p(t)) \tag{81}$$

$$\frac{dR_p(t)}{dt} = \beta_2 \tag{82}$$

$$\frac{dR_p(t)}{dt} = \frac{\beta_3}{R_p(t)} \tag{83}$$

$$\frac{dR_p(t)}{dt} = \frac{\beta_4}{v_p(t)} \tag{84}$$

Coefficients β_1 through β_4 are rate constants and R_m is the maximum possible size of the pathway. Equation (81) requires that the rate of pathway growth be proportional to remaining growth potential at any time. Therefore, the pathway will begin to grow at its maximum rate and growth will slow exponentially as the pathway reaches its maximum diameter. Equation (82) is a simple zero order kinetics mode that will yield a linear, unbounded pore growth. Equations (83) and (84) are bounded growth rate formulations that specify a rate inversely proportional to either the pathway radius or pathway velocity.

Each of these rate expressions may be solved subject to an initial condition such as Eq. (85) to yield the required radius function.

$$R_p(t \geq t_p) = R_o \tag{85}$$

Here t_p represents the threshold time required for the aggressive permeant to invade the pore and begin altering its size, and R_o represents the initial size of the invaded pore.
The solutions of Eqs (81) through (84), each subject to Eq. (85), are as follows.

$$\text{Eq. (81),(85)} \Rightarrow \quad R_p(t) = R_m - (R_m - R_0)\exp[-\beta_1(t - t_p)] \tag{86}$$

$$\text{Eq. (82),(85)} \Rightarrow \quad R_p(t) = R_0 + \beta_2(t - t_p) \tag{87}$$

$$\text{Eq. (83),(85)} \Rightarrow \quad R_p(t) = [R_0^2 + 2\beta_3(t - t_p)]^{1/2} \tag{88}$$

$$\text{Eq. (84),(85)} \Rightarrow \quad R_p(t) = [R_0^3 + \frac{24\mu L\beta_4}{\rho g(\phi_1 - \phi_2)}(t - t_p)]^{1/3} \tag{89}$$

Any of these functions may then be substituted into Eq. (80) to complete the solution.

Figure 30 illustrates the type of preferential flow-influenced bulk hydraulic conductivity response that can be captured with a pore growth formulation. Figure 30 illustrates data on the physical consequences of full strength acetone permeating a non-calcareous smectite (Anderson et al., 1985). The test was conducted in a fixed wall permeameter using a hydraulic gradient of 361.6. Note that significant changes in the hydraulic conductivity were observed well before one pore volume of fluid had penetrated the sample. Significant changes in bulk hydraulic conductivity prior to one pore volume are clear indications of preferential flow activity. Domain impact models such as DIP formulations cannot predict responses significantly prior to one pore volume because at least one pore volume of aggressive permeant must enter the domain before it has had an opportunity to interact with all of the soil. It may alter the properties of the soil near the x=o plane early in time, but there will not be a global response in the flow until the soil up to x=L has reacted.

Flow through preferential pathways is not bound by this constraint. This led Anderson et al., (1985) to conclude that the formation and subsequent flow through cracks and micropores was the probable mechanism for the increased permeability. Note that the bounded growth formulation used (i.e. Eq. (80) and (86)) is quite capable of capturing this type of behavior.

Figure 31 presents a second example of a response that can be captured with a preferential flow data. However, unlike the example in Fig. 30, this response occurs later in time. Figure 31 presents data for the interaction between aniline and a sand-bentonite (7 % by weight) mixture in a flexible wall permeameter at a hydraulic gradient of 100 (Evans and Kugleman, 1985). The mechanism responsible for the impact was thought to be volume change of the bentonite which (presumably) decreased the size of the particles plugging the pore space of the sand. The results are for full strength aniline. Aqueous solutions of aniline at concentrations of 15 and 30 g/l produced no significant changes in permeability. Although the authors did not report evidence of preferential flow, this would seem to be another plausible explanation of the response.

Figure 32 presents a final example based on data of Uppot and Stephenson (1988) for the impact of methanol permeating a montmorillonite clay in a flexible wall permeameter at a hydraulic gradient of 831. Note that at a time of approximately 3.5 pore volumes the apparent hydraulic conductivity increased by a factor of 4. Although these results could have been produced by a global domain impact, subsequent inspection of the soil indicated that it was "visibly cracked". Although the confining pressure applied to the wall prevented cracks from opening wide, Uppot and Stephenson (1988) concluded that the cracks were the result of "hydraulic fracture" and that preferential flow through them was responsible for the increase in hydraulic conductivity. Note that the bounded growth preferential pathway model does quite well at simulating the type of behavior observed.

It should be of concern that there are several possible formulations for preferential flow pathway models, all of which may be able to simulate similar impacts. It is rewarding to know that these models can simulate behavior that cannot be simulated by domain impact models such as the DIP models of Section 3. However, the converse is not true. There are hydraulic conductivity responses that occur at times in excess of one pore volume that might be simulated equally well by a domain impact model or a preferential flow model.

It should also be of concern that a substantial number of model coefficients have been introduced. There are up to 5 unique coefficient groups in DIP domain impact models. There can be nearly as many in preferential flow models. All of the coefficients have been defined, but this falls rather short of providing techniques for measuring them. After all, although some of them can be measured independent of the aggressive permeant interactions (e.g. initial preferential pore radius), some of them cannot be measured independently of the phenomenon they are designed to quantify. Interface penetration or pore growth rate coefficients may fall into this class. How then is a potential user to determine which modeling strategy is most appropriate, and how should one go about identifying the appropriate coefficient values?

We do not yet have infallible answers to these very important questions, but are making a great deal of progress toward their resolution. A significant amount of work has been done on the “inverse” problem of parameter estimation and quality of fit analysis for aggressive permeant interactions. This will be explored in the next section.

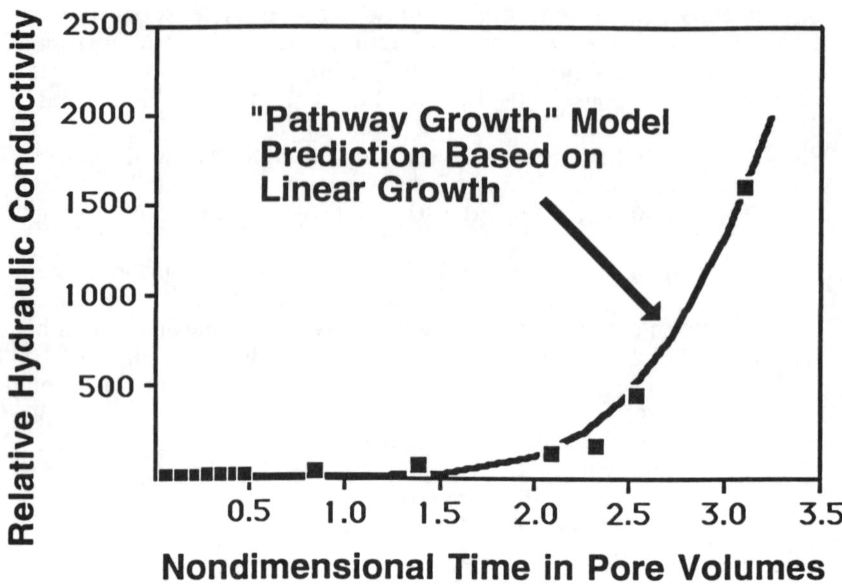

Fig. 31 - Aggressive permeant interaction (sand/bentonite soil - aniline) characteristic of a domain response that may indicate the development of a pore growth preferential pathway (Evans and Kugleman, 1985).

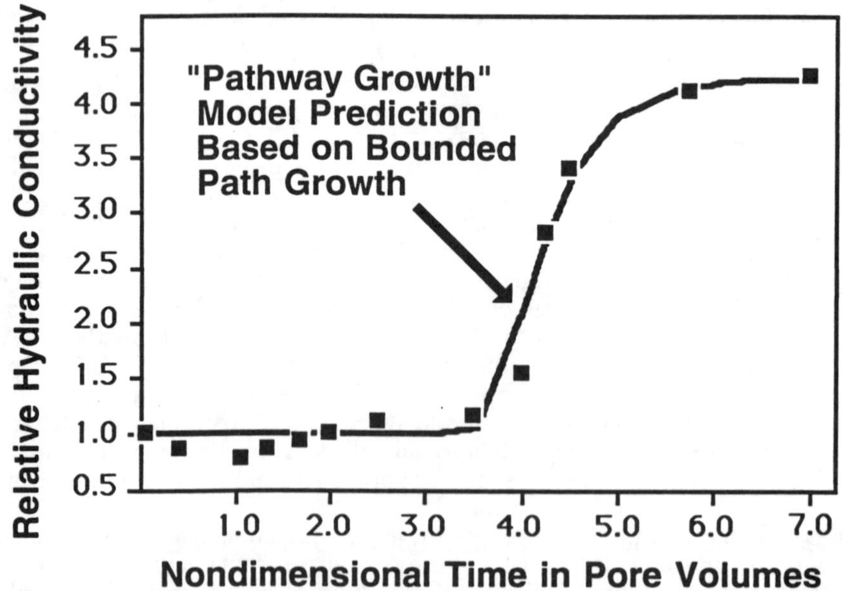

Fig. 32 - Aggressive permeant interaction (montmorillonite - methanol) characteristic of a domain response that may indicate the development of a pore growth preferential pathway (Uppot et al., 1988).

5. Aggressive Permeant Model Parameter Estimation

The mechanisms of aggressive permeant interactions are discussed in Section 2. Strategies to model the domain impacts of these interactions are discussed in Section 3. Strategies to model preferential flow pathway developments are discussed in Section 4. However, all of this material makes use of variables (coefficients, parameters, constants, etc.) that must be measured before the analytical tools can be applied. In this section we will consider techniques for extracting the required information from existing experimental data. It would be desirable to quantify the required parameters from fundamental material constants, but the state-of-the-art in aggressive permeant analysis is not yet this advanced. At the present time, most of the work is concentrating on extracting the required information from the results of permeameter studies that yield data on the bulk hydraulic conductivity response as a function of time.

Ravi and Jennings (1990) presented algorithms for extracting discrete interface penetration model coefficients from transient permeability data. The techniques are based on a Modified Levenberg-Marquardt non-linear least squares analysis. Although these techniques are relatively well-known, the difficulty of their application is compounded by the fact that interface penetration models are a series of analytically distinct functions joined at discrete but implicit time (phase) boundaries that are functions of the unknown coefficients. The problem is also one of constrained optimization since the magnitude of the extracted coefficients must satisfy "physical" parameter constraints such as non-negativity. Finally, the experiments required to generate K(t) data are difficult to control over the long time periods required and the non-dimensional time scale of pore volumes is non-linear in real time. The result is normally a sparse set of K(t) observations that can suffer from a rather high degree of experimental randomness and be poorly distributed in real time. The problems of experimental randomness are also compounded by the fact that the bulk hydraulic conductivity often changes over orders of magnitude during the course of the experiment. Therefore, the randomness about relatively large values of K(t) may dominate over significant but smaller values.

The algorithms of Ravi and Jennings (1990) were based on a penalty function approach (see Bard, 1974) to impose parameter constraints. However the algorithms had difficulty producing an infallible set of optimum coefficients (often they converge to local optima) and some data produced non-unique solutions. In this presentation we will discuss a different modification to the basic Levenberg-Marquardt strategy. The algorithms require more computational effort, but seem to yield better results.

5.1 Levenberg-Marquardt Parameter Estimation

The general "least squares" parameter estimation problem may be stated as follows (Dennis and Schnabel 1983).

$$\text{Minimize:} \qquad \Phi(\Gamma) = \frac{1}{2}\sum_{i=1}^{M} r_i(\Gamma)^2 \tag{90}$$

Here $r_i(\Gamma)$ denotes the ith component of the residual function r (i.e. the difference between some postulated functional relationship and the ith measurement of the function's dependent variable), and Γ denotes a vector of the parameters in the postulated functional relationship. In this application we will assume that there is a known set of bulk hydraulic conductivity data (K_i ; i=1,M), and a known aggressive permeant interaction model $f(\Gamma,t)$ that is a function of N parameters in the parameter set Γ (i.e. $\Gamma = (\gamma_1, \gamma_2, \ldots \gamma_N)$, and time.

However, for this application we will add the consideration that not all data points may deserve to be given equal weight and that the values of Γ must respect constraints imposed by their definition. For example, hydraulic conductivity values may not be negative or zero. Therefore, in this application we will seek to identify the values of the parameters in Γ that satisfy the following.

$$\text{Minimize:} \quad \Phi(\Gamma) = \frac{1}{2}\sum_{i=1}^{M} w_i[f(\Gamma, t_i) - K_i]^2 \tag{91}$$

$$\text{Subject to:} \quad \pi_j(\Gamma) \geq 0\,; \quad \forall j = 1, J \tag{92}$$

Here w_i is the weight assigned to the ith known bulk hydraulic conductivity datum and π_j is a set of J constraints that must be imposed on the values of the parameters in Γ.

The numerical techniques available for minimizing the function of Eq. (91) may be broadly classified into either gradient methods or direct search methods. Gradient methods attempt to use information about the derivative of the residual to aim the search in a desirable search direction. Direct search methods are designed to operate without explicit gradient information. The modification of the Levenberg-Marquardt method (Marquardt, 1963) employed here belongs in the class of gradient search methods. The original method is as follows:

Let **J** denote the Jacobian matrix of $\Phi(\Gamma)$, with elements defined by Eq. (93).

$$J_{ij} = \frac{\partial}{\partial \gamma_j}[f(\Gamma, t_i) - K_i] \tag{93}$$

For the kth iteration of the solution search, the Levenberg-Marquardt step size δ_k is obtained by solving the following linear system of equations.

$$\left(\mathbf{J}_k^T \mathbf{W} \mathbf{J}_k + \lambda_k \mathbf{I}\right)\delta_k = -\mathbf{J}_k^T \mathbf{W} \mathbf{r}_k \tag{94}$$

Here **W** is a diagonal matrix containing the parameter weights and λ_k is a non-negative scalar referred to as the Levenberg-Marquardt parameter, and **I** is the identity matrix.

Marquardt (1963) demonstrated that there always exists some value $\lambda_k > 0$ such that the step size δ_k will be in a direction that will decrease the magnitude of the residual. As the value of λ_k approaches zero, Eq. (94) reduces to the Gauss-Newton algorithm. As λ_k becomes large, the algorithm takes on the properties of the steepest-descent technique (Ortega and Rheinboldt, 1970). For intermediate values of λ_k, the algorithm yields a search direction that combines the rapid convergence properties of Gauss-Newton techniques with a broader range of convergence characteristic of the steepest-descent methods.

The Levenberg-Marquardt algorithm is implemented as follows:

i. Select an initial guess for the parameter set Γ, and the Levenberg-Marquardt parameter λ_k.

ii. Solve Eq. (94) for the step size δ_k.

iii. Update the parameter estimation in the direction of search,

$$\Gamma_{k+1}=\Gamma_k+\delta_k \tag{95}$$

iv. Check for solution convergence. Stop if the solution has converged.

v. Update the Levenberg-Marquardt parameter according to Marquardt (1963) and return to step ii.

5.2 Modified Levenberg-Marquardt Parameter Estimation.

The Levenberg-Marquardt algorithm of the previous section must be modified to respect the parameter constraints of Eq. (92). It has also been observed that, although the algorithm generally converges from an initial guess of the parameter set values, it does not necessarily converge to the global optimum solution. It is altogether possible and generally quite probable that the algorithm will only attain a local optimum from some arbitrary starting point. To overcome the problem of initial guesses that do not yield the global optimum, Ravi and Jennings (1990) have recommended the use of a Monte Carlo multi-start implementation. Ravi and Jennings (1990) also recommend an intuitive and much more computationally efficient method of enforcing constraints than the penalty function method. If these two recommendations are adopted, the Modified Levenberg-Marquardt algorithm should be implemented as follows:

i. Select an initial guess for the parameter set (i.e. let $\Gamma = \Gamma_0$) and the Levenberg-Marquardt parameter λ_o. The initial guess Γ_0 must be feasible (i.e. $\pi_j(\Gamma_0) \geq 0$). In the multi-start random search technique the values of Γ_0 are selected by Monte Carlo sampling of a prescribed parameter search range using a uniform random number generator. Let Γ_{min} and Γ_{max} define the lower and upper bounds for the search range of each parameter, and let ξ be a vector of N uniformly distributed random numbers $0.0 \leq \xi_n \leq 1.0$. The initial guess may then be computed as follows: ,

$$\Gamma_0 = \Gamma_{min} + \xi\,(\Gamma_{max} - \Gamma_{min}) \quad ; \quad \text{S.T. } \pi_j(\Gamma_0) \geq 0 \tag{96}$$

and this may be repeated as often as desired to yield several feasible starting points.

ii. Evaluate the Levenberg-Marquardt correction vector δ_k for any iteration $k \geq 0$ by solving Eq. (94)

iii. Update the parameter estimation by applying Eq. (95)

iv. Verify that all constraints are satisfied (i.e. $\pi_j(\Gamma_0) \geq 0$). If any of the updated parameters violate a constraint, the corresponding step size component should be halved until the parameter satisfies all constraints. When parameters violate non-linear constraints the corresponding component values from the previous iteration should be retained.

v. Check for solution convergence. Stop if convergence criteria have been satisfied, and let

$$\Gamma = \Gamma_{k+1} \tag{97}$$

vi. If convergence criteria have not been satisfied, update the Levenberg-Marquardt parameter according to Marquardt (1963) and return to step ii.

The penalty function approach has been replaced with step iv. Experience has indicated that for some data that yield large δ_k from arbitrary, feasible Γ_k values, the updated parameter set Γ_{k+1} can step beyond the feasible region boundary and avoid all penalties imposed in the neighborhood of the feasible region boundary. In such cases, the algorithm can iterate to a non-feasible local optimum. The approach described in step iv alters the Levenberg-Marquardt direction of search, but forces the search to remain in the feasible region. This, coupled with a multi-start implementation, appears to be an effective (if not computationally efficient) method of extracting reasonable coefficient values.

5.3 Application of Levenberg-Marquardt Parameter Estimation

The modified Levenberg-Marquardt parameter estimation procedure described in Section 5.2 appears to work quite well for most aggressive permeant data. The Monte Carlo implementation scheme is not particularly efficient, but for practical applications it seems to be wise to trade computational efficiency for robustness. Experience has demonstrated that the algorithm will be successful in extracting "optimum" coefficients for nearly all aggressive permeant hydraulic conductivity data sets. It is certainly true that when the model is not appropriate for the data the quality of the fit will be poor. However, this allows the user to distinguish between successful and unsuccessful modeling strategies.

Consider the final data set presented in Fig. 33. These data are for the interaction between acetone and a soil mixture (sodium montmorillonite, Georgia kaolinite and fine sand), (Acar and D'Hollosy, 1987). The test was conducted in a constant-head, flexible-wall permeameter at a confining stress of 70 kPa (10.1 psi).

Figure 33 presents the laboratory observations, as well as the optimum fits of a 2-, 4-, and 5-parameter Discrete Interface Penetration (DIP) representation of the phenomenon. The numerical values for the optimum coefficients are provided on the figure, as well as the nondimensional optimization residual $\Phi^*(\Gamma)$.

$$\phi^*(\Gamma) = \frac{\phi(\Gamma)}{(M - N)} \tag{98}$$

Here M is the number of data points and N is the number of parameters in the model applied. The data of Fig. 33 show a relatively complicated response (i.e. more than one mechanism at work), but the parameter estimation procedure is still able to detect the optimum fit.

Note that in Fig. 33 the fit achieved for the 2-, 4- and 5-parameter DIP model becomes successively better. This is as one would expect. With more parameters, one should be able to fit the data better. However, this is not always the case. When two models yield

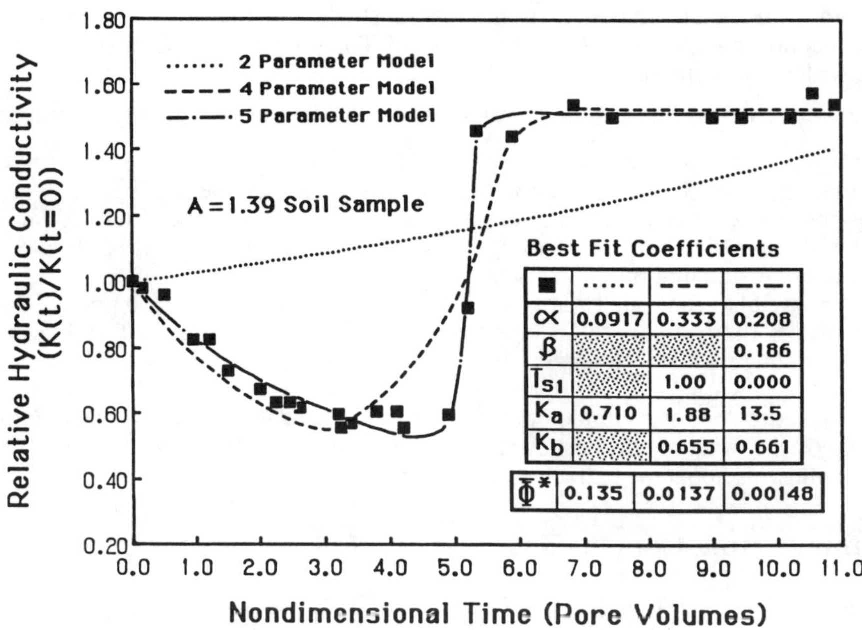

Fig. 33 - Aggressive permeant interaction (soil mixture - acetone) illustrating “fit” improvement for a sequence of DIP models (Acar and D’Hollosy, 1987; Jennings and Rave, 1990).

essentially equal quality, Eq. (97) will favor the model with the least number of parameters.
Numerous other examples have already been presented. Figures 2, 3, 5, 6, 7, 9, 10, 11, 12, 13, 14, 15, 17, 19, 20 and 21 all present optimum model fits that were determined using the modified Levenberg-Marquardt algorithm. Parameter optimization information was omitted from the preferential flow models illustrated on Figures 29, 30, 31, and 32, but work recently completed at The University of Toledo has demonstrated that these too can be fit with the algorithm.

Readers are cautioned that absolute values of the residual $\Phi^*(\Gamma)$ cannot be compared from one figure (i.e. one data set) to another. The magnitude of $\Phi^*(\Gamma)$ is governed by the absolute magnitude of the hydraulic conductivity response and the number of data points. For any one data set it is possible to use the magnitude of the optimum residual to infer which of the possible modeling strategies is most successful, but this should not be compared to the residual of other data sets.

Readers are also cautioned that some data may yield non-unique parameter sets. For some responses, there appear to be more than one set of parameters that yield essentially identical optimum residual. Generally this results from high degrees of randomness in the experimental observations. Ravi and Jennings (1990) have presented extensions of the algorithm that allow one to examine the nature of the optimization space, quantify indices on the linearity of the sampling properties of the Levenberg-Marquardt algorithm, and compute confidence intervals about the extracted parameters.

6. Summary and Conclusions

In this chapter we have presented an introduction to the mechanisms and models currently being applied to describe the phenomena of aggressive permeant interactions with soils. We have demonstrated that a rather substantial body of experimental evidence has been assembled on the physical consequences of these interactions, and that we are beginning to make progress in modeling these observations. Discrete Interface Penetration (DIP) models are successful at describing many of the experimental data. Although the models are phenomenological (as opposed to mechanistic), they are often able to capture the observed behavior quite well. "Calibrated" DIP models may then be used to infer what the consequences would be in real space and time for facilities such as clay barriers used at hazardous waste disposal sites. Here "calibrated" means models for which the essential coefficients have been extracted by parameter optimization but not necessarily measured independently.

Where DIP models are not successful in capturing observed responses, often the reason appears to be that the data were influenced by the formation of preferential flow pathways that "short circuit" the flow field. We are also beginning to make progress in modeling efforts for aggressive-permeant-induced preferential flow pathways. The body of experimental evidence on preferential flow development is less extensive, but now that we know what to look for, we are developing experimental techniques to yield the information necessary to evaluate the significance of these phenomena. It is possible that this is only a laboratory problem (i.e. an unfortunate artifact of some permeameter designs) and will not be a significant field concern. On the other hand, field experience seems to indicate that flow through preferential pathways is extremely important. The question of whether the "field" pathways share the same origins as the laboratory observations has yet to be resolved.

Before concluding, we would like to take this opportunity to speculate on what the future will hold for the analysis of aggressive permeant interactions with soils. There are many possibilities. One might, for example, observe that there is at least one DIP model missing in the obvious continuum of analyses (the 3-parameter model). There are several possible alternative DIP formulations that have not yet been derived or tested, and questions about the constitutive relationships (Eq. (12), etc.) that should be resolved. One might also note that the constitutive models applied in "pore growth" modeling are rather crude. They could easily be improved with a bit more fluid mechanics and chemistry. Whether or not they will then yield analytical solutions is a different question, but they certainly should be explored. One might also note that the DIP modeling strategy presented here is one dimensional. Realistic field applications would probably require multidimensional solutions. These would not be difficult, but would probably be numerical rather than analytical. Finally, the Levenberg-Marquardt parameter estimation algorithm may ultimately be replaced by more efficient, more globally-convergent routines.

Advances will undoubtedly come in many of these areas, but they will not be the most important advances. It is the opinion of these authors that the most important advances in this field, the next generation of models, will be highly mechanistic formulations designed to be solvent and soil specific. DIP models attempt to be universal. They accomplish this, but at the expense of being phenomenological rather than mechanistic. They create a framework for modeling and can be used to infer what might happen under field conditions. Would this be good enough for the design of a hazardous waste disposal facility ? Probably not ! We think that for some particular problems, nothing less than a fully mechanistic model will do. The formulation must account for the detailed chemical and physical interactions between the specific aggressive permeant and the site-specific soil. These will be challenging models to formulate, will demand a great deal more information than we currently have, and will undoubtedly pose serious numerical analysis and computational difficulties. Nevertheless, they are needed if we are to understand these systems, and the systems are too important to ignore. Fully mechanistic models will come. They will not be simple or easy to apply. They will require the development of new laboratory techniques. They will be expensive to apply and will increase the sophistication of the design of systems like hydraulic barriers. They will also perturb the regulatory process in ways that are difficult to anticipate. Nevertheless THEY WILL COME.

7. References

Acar, Y.B. and D'Hollosy, E., (1987), Assessment of pore fluid effects using flexible wall and consolidation permeameters, in *Geotechnical Practice for Waste Disposal 87.* R.W. Woods (ed.), ASCE Geotechnical Special Pub. No. 13.

Acar, Y.B., Hamidon, A., Field, S.D. and Scott, L., (1985), The effect of organic fluids on hydraulic conductivity of kaolinite, in *Hydraulic Barriers in Soil and Rock,* A.I. Johnson et al. (ed.), ASTM STP 874.

Alperovitch, N., Shainberg, I. and Keren, R., (1981), Specific effect of magnesium on the hydraulic conductivity of sodic soils, *Soil Sci.*, 32(4), 543-554.

Anderson, D.C., Brown, K.W. and Thomas, J.C., (1985), Conductivity of compacted clay soils to water and organic liquids, *Waste Mgmt. and Res.,* 3(4), 339-349.

Anderson, D.C., Crawley, W. and Zabcik, J., (1985), Effects of various liquids on clay soil: bentonite slurry mixtures, in *Hydraulic Barriers in Soil and Rock,* A.I. Johnson et al. (ed.). ASTM STP 874.

ASA-ASTM, (1965) *Methods of Soil Analysis. Part 1. Physical and Mineralogical Methods. Agronomy Monograph No. 9*, American Society of Agronomy, WI.

ASTM, (1990), *Annual Book of ASTM Standards. Volume 04.08 - Soil and Rock; Dimension Stone; Geosynthetics*, ASTM, PA.

Bard, Y., (1974), *Nonlinear Parameter Estimation*. Academic Press, Inc. NY.

Berkowitz, B., (1989), Boundary conditions along permeable fracture walls: influence on flow and conductivity, *Water Resour. Res.*, 25(8), 1919-1922.

Bird, R.B., Stewart, W.E. and Lightfoot, E.N., (1960), *Transport Phenomena*, John Wiley & Sons, NY.

Bowders, J.J., (1985), The influence of various concentrations of organic liquids on the hydraulic conductivity of compacted clay, *Dissertation*, University of Texas at Austin, TX.

Bowders, J.J. and Daniel, D.E., (1987), Hydraulic conductivity of compacted clay to dilute organic chemicals, *J. Geotech. Eng.*, 113:1433-1448.

Brown, K.W. and Anderson, D.C., (1981), *Effects of Organic Solvents on the Permeability of Clay Soils*, U.S. EPA, Cincinnati, OH, EPA-600/2-83-016.

Brown, K.W. and Thomas, J.C., (1984), Conductivity of three commercially available clays to petroleum products and organic solvents, *Haz. Waste* , 1(4), 545-553.

Brown, K.W. and Thomas, J.C. and Green, J.W., (1986), Field cell verification of the effects of concentrated organic solvents on the conductivity of compacted soils, *Haz. Waste and Haz. Mat.*, 3(1), 1-19.

Burton, F, Miller, J. and Pound, C. (ed.), (1981), *Land Treatment of Municipal Wastewater, Process Design Manual*, U.S. EPA, Cincinnati, OH, EPA 625/1-81-013.

Cartwright, K., Griffin, R.A. and Gilkeson, R.H., (1977), Migration of landfill leachate through glacial tills, *Groundwater*, 15(4), 294-304.

Chen, S., Low, P.F., Cushman, J.H. and Roth, C.R., (1987), Organic compound effects on swelling and flocculation of Upton montmorillonite, *Soil Sci. Soc. Am.J.* 51, 1444-1450.

Crim, R.G., Shepherd, T.A. and Nelson, J.D., (1979), Stability of natural clay liners in a low pH environment, *Proceedings of the 2nd Symposium on Uranium Mill Tailings Management*, Colorado State University, Ft. Collins, CO.

Daniel, D.E., Anderson D.C. and Boynton, S.S., (1985), Fixed-wall versus flexible-wall permeameters, in *Hydraulic Barriers in Soil and Rock.*, A.I. Johnson et al. (ed.), ASTM STP 874.

Dennis, J.E. Jr. and Schnabel, R.B., (1983), *Numerical Methods for Unconstrained Optimization and Nonlinear Equations*, Prentice Hall, Inc. NJ.

Dudgeon, C.R., (1967), Wall effects in permeameters, *J. Hyd. Div. Proc., ASCE,* HY5: 137-149.

Dunn, R.J., (1983), *Hydraulic Conductivity of Soils in Relation to Subsurface Movement of Hazardous Wastes*, Thesis, University of California at Berkely, Berkeley, CA.

Dunn, R.J. and Mitchell, J.K., (1984), Fluid conductivity testing of fine-grained soils, *J. Geotech. Eng., ASCE*, 110(11), 1648-1665.

Evans, J.C. and Kugelman, I.J., (1985), Organic fluids effects on the strength, deformation and permeability of soil-bentonite slurry walls, in *Proceedings of the 7th Mid-Atlantic Industrial Waste Conference*, I.J. Kugelman (ed.), Technomic Publishing Co. NJ.

Farley, J.T., Miller, B.M. and Schoettle, V., (1970), Design criteria for matrix simulation with hydrochloric-hydrofluoric acid, *J. Pet. Tech.*, 22(4), 433-440.

Fernandez, F. and Quigley, R.M., (1985), Hydraulic conductivity of natural clays permeated with simple liquid hydrocarbons, *Can. Geotech.J.*, 22(2), 205-214.

Fogler, H.S. and McCune, C.C., (1976), On the extension of the model of matric acid stimulation to different sandstones, *AIChE J.,* 22, 799-805.

Foreman, D.E. and Daniel, D.E.,(1986), Permeation of compacted clay with organic chemicals, *J.Geotechn. Eng.,* 112(7), 669-681.

Franzinz, J.B., (1956), Permeameter wall effects. *Trans. Am. Geo. Union.,* 37:735-737.

Frenkel, H., Goertzen, J.O. and Rhoades, J.D., (1978), Effects of clay type and content, exchangeable sodium percentage, and electrolyte concentration on clay dispersion and soil hydraulic conductivity, *Soil Sci. Soc. Amer. J.,* 42(1), 32-39.

Gipson, A.J., (1985), Permeability Testing of Clayey Soils and Silty Sand-Bentonite Mixtures using Acid Liquor, *Hydraulic Barriers in Soil and Rock,* ed. A.I. Johnson et al., ASTM STP 874, 140-154.

Goldenberg, L.C., Magaritz, M., Amil, A.J. and Mamdel, S., (1984), Changes in hydraulic conductivity of laboratory sand-clay mixtures caused by a seawater-freshwater interface, *J. Hydrol*, 70, 329-336.

Goldenberg, L.C., Magaritz, M. and Mandel, S., (1983), Experimental investigation on irreversible changes of hydraulic conductivity on the seawater-freshwater interface in coastal aquifers, *Water Resour. Res.*, 19(1),77-85, 1983.

Green W.J., Lee, G.F. and Jones, R.A., (1981), Clay soils permeability and hazardous waste storage, *J. Water Poll. Control Fed.*, 53(8), 1347-1354.

Green W.J., Lee, G.F. and Jones, R.A. and Palit, T., (1983), Interaction of clay soils with water and organic solvents: implications for the disposal of hazardous waste, *Env. Sci. and Tech.*, 17(5), 278-282.

Griffin, R.A. and Shimp, N.F., (1978), Leachate migration through selected clays, in *Gas and Leachate from Landfills Formation, Collection and Treatment,*(ed.) E.J. Genetellii and J. Cirello, U.S. EPA, Cincinnati, Ohio, EPA-600/9-76-004.

Guin, J.A., Schechter, R.S. and Silberberg, I.H., (1971), Chemically induced changes in porous media, *Ind.and Engi. Chem. Fund.,* 10, 50-54.

Hardcastel, J.H. and Mitchell, J.K., (1974), Electrolyte concentration-permeability relationships in sodium illite-silt mixtures, *Clays and Clay Minerals*, 22(2),143-154.

Head, K.H., (1982), *Manual of Soil Laboratory Testing, Volume 2: Permeability, Shear Strength, and Compressibility Tests*, Pentech Press, London.

Hewitt, C.H., (1963), Analytical techniques for recognizing water-sensitive reservoir rocks, *J. Pet. Tech.*, 15(8), 813-818.

Hill, R.L. and King, L.D., (1982), A permeameter which eliminates boundary flow errors in saturated hydraulic conductivity measurements, *Soil Sci. Soc. Am. J.*, 46:877-880.

Hoefner, M.L. and Fogler, H.S., (1988), Pore evolution and channel formation during flow and reaction in porous media, *AIChE J.*, 34(1), 45-54.

Jennings, A.A. and Ravi, V., (1990), Analytical extensions of the discrete interface penetration model for dynamic permeability analysis, *Soil Sci. Soc.Am. J.,* 54:7-12.

Jennings, A.A. and Ravi, V., (1990), Mechanisms, impacts and modeling of chemically-induced changes in saturated soil hydraulic conductivity, in *Proceedings of the Ohio River Valley Soils Seminar XXI, Environmental Aspects of Geotechnical Engineering*, Cincinatti, OH.

Jennings, A.A. and Ravi,V., (1992), Interface penetration model formulation for analyzing wall-separation-influenced transient hydraulic conductivity data, *Soil Sci. Soc. Am. J.*,56:1023-1032.

Jennings, A.A. and Shukla, A., (1992), Models to detect the influence of preferential flow pathways on dynamic permeability experiments, in *Preferential Flow, Proceedings of the National Symposium*, T.J. Gish and A. Shirmohammadi (ed.), Dec. 16-17, Chicago, IL.

Klute, A., (1965), Laboratory measurement of hydraulic conductivity in saturated soil, in *Methods of Soil Analysis, Part 1, ASA No.9*, 210-233.

Land, C.S. and Baptist, O.C., (1965), Effect of hydration of montmorillonite on the permeability to gas of water-sensitive reservoir rocks, *J. Pet. Tech.*, 17(10), 1213-1218.

Lentz , R.W., Horst, W.D. and Uppot, J.O., (1985), The permeability of clay to acidic and caustic permeants, in *Hydraulic Barriers in Soil and Rock*, A.I. Johnson et al. (ed.), ASTM STP 874, 127-139, 1985.

Lund, K., Fogler, H.S. and McCune, C.C., (1976), Acidization IV: experimental correlations and techniques for the acidization of sandstone cores, *Chem. Eng. Sci.*, 31(5), 373-380.

Lund, K., Fogler, H.S., (1976), The prediction of the movement of acid and permeability fronts in sandstone, *Chem. Eng. Sci.*, 31(5), 381-392.

Marquardt, D.W., (1963), An algorithm for least squares estimation of nonlinear

parameters, *SIAM J.*, 11:431-441.
McNeal, B.L. and Coleman, N.T., (1966), Effect of soil composition on soil hydraulic conductivity, *Soil Sci. Soc. Am. Proc.*, 30:308-311.
McNeal, B.L., Norvell, W.A. and Coleman, N.T., (1966), Effect of solution composition on the swelling of extracted soil clays, *Soil Sci. Soc. Am. Proc.*, 30:313-317.
McNeal, B.L and Reeve, R.C., (1964), Elimination of boundary-flow errors in laboratory hydraulic conductivity measurements, *Soil Sci. Soc. Am. Proc.*, 28:713-714.
Mehnert, E. and Jennings, A.A., (1985), The effect of salinity-dependent hydraulic conductivity on saltwater intrusion episodes, *J. Hydrol.,* 80(3/4), 283-297.
Mitchell, J.K. and Madsen, F.T., (1987), Chemical effects on clay hydraulic conductivity, in *Geotechnical Practice for Waste Disposal '87, Proceedings of the ASCE-GT Specialty Conference*, ed. R.D. Woods, Ann Arbor, Michigan, June 15-17, 87-115.
Mundell, J.F. and Jennings, A.A., (1987), Penetration model for dynamic permeability analysis, *J. Env. Eng.*, 113(4), 881-899.
Nasiatka , D.M., Shepherd. T.A. and Nelson, J.D., (1981), Clay liner permeability in low pH environments, in *Symposium on Uranium Mill Tailings Management*, Colorado State University, CO.
Neasham, J., (1977), *The Morphology of Dispersed Clay is Sandstone Reservoirs and its Effects on Sandstone Shaleness, Pore Space and Fluid flow Properties*, Soc. Pet. Engrs., AIME, Denver, CO.
Ortega, J.M. and Rheinboldt, W.C., (1970), *Iterative Solution of Nonlinear Equations in Several Variables*, Academic Press, NY.
Otte, F.E. and Jennings, A.A., (1984), The physical consequences of acid reactions with soil, p 176-185, in *Proceedings of the 5th Annual Indiana Water Resources Symposium*, J.D. Martin (ed.), Bloomington, IN June 20-22.
Parks, C.S. and O'Connor, G.A., (1980), Salinity effects on hydraulic properties of soils, *Soil Sci.*, 130(3), 167-174.
Peterson, S.R. and Gee, G.W., (1985), Interactions between acidic solutions and clay liners: permeability and neutralization, in *Hydraulic Barriers in Soil and Rock*, A.I. Johnson et al. (ed.), ASTM STP 874, 229-245, CA.
Pierce, J.J., Sallifors, G., Peel, T.A. and Witter, K.A., (1987), Effects of selected inorganic leachates on clay permeability, *J. Geotech. Eng.,* 113(4), 915-919.
Quirk, J.P. and Schofield, R.K., (1955), The effects of electrolyte concentration on soil permeability, *J. Soil Sci.*, 6(2), 163-178.
Ravi, V. and Jennings. A.A., (1990), Penetration model parameter estimation for dynamic permeability measurements, *Soil Sci. Soc. Am. J.*, 54:13-19.
Ravi,V.and Jennings, A.A., (1990), Monte Carlo techniques applied to parameter estimation for permeability-altering soil/permeant interaction models, in *Computer Techniques in Environmental Studies III*, Proceedings of the 3rd International Conference on Development and Application of Computer Techniques to Environmental Studies, (ed.), P. Zannetti, Springer-Verlag, NY.
Schecter, R.S. and Gidley, J.L., (1969), The change in pore size distribution from surface reactions in porous media, *AIChE J.,* 15(3), 339-350.
Schramm, M., Warrick, A.W. and Fuller, W.H., (1986), Permeability of soils to four organic liquids and water, *Haz.Waste and Haz. Mat.,* 3(1), 21-27.
Shainberg, I, Rhoades, J.D. and Prather, R.J., (1981), Effect of low electrolyte concentrations on clay dispersion and hydraulic conductivity of a sodic soil, *Soil Sci. Soc. Am. J.,* 45(2), 273-277.
Shainberg, I, Rhoades, J.D., Suarez, D.L. and Prather, R.J., (1981), Effect of mineral weathering on clay dispersion and hydraulic conductivity of sodic soil, *Soil Sci. Soc. Am. J.,* 45(2), 287-291.
Shaughnessy, C.M. and Kunze, K.R., (1981), Understanding sandstone acidizing leads to improved field practices, *J. Pet. Tech.*, 33:1196-1202.

Smith, C.F. and Hendrickson, A.R., (1965), Hydrofluoric acid stimulation of sandstone reservoirs, *J. Pet. Tech.*, 17(2), 215-222.
Somerton, C.W. and Wood, P., (1988), Effect of walls in modeling flow through porous media, *J. Hyd. Eng.*, 114:1431-1448.
Tokunaga, T.K., (1988), Laboratory permeability errors from annular wall flow, *Soil Sci. Soc. Am. J.*, 52:24-27.
Trick, L.C., Crosser, M.L. and Geiser, R.A., (1985), Wastewater treatment by spray irrigation: quality of effluent and impacts on soils and groundwater, in *Proceedings of the 40th Annual Purdue Industrial Waste Conference*, West LaFayette, IN.
Uppot, J.O.and Stephenson, R.W., (1988), Permeability of clay under organic permeants, *J.Geotech. Eng.*, 115(1), 115-131.
Weber, W.J., (1972), *Physicochemical Processes for Water Quality Control*, Wiley-Interscience, N.Y.
Yaron, B. and Thomas, G.W., (1968), Soil hydraulic conductivity as affected by sodic water, *Water Resour. Res.*, 4(3), 545-552.

Chapter 9

Variational data assimilation: some aspects of theory and application

X. Zou,[a] I.M. Navon[b]

[a] *Supercomputer Computations Research Institute, Florida State University, Tallahassee, Florida 32306-4052, USA*

[b] *Department of Mathematics and Supercomputer Computations Research Institute, Florida State University, Tallahassee, Florida 32306, USA*

Abstract

Variational data assimilation is performed using either a T40 18 layer version of the National Meteorological Center spectral model or a limited-area shallow-water equations model, with operationally analyzed fields as well as simulated datasets. Issues of incomplete observations, control of gravitational oscillations, and inclusion of "on-off" physical processes are addressed in the framework of variational data assimilation.

Key words

Variational data assimilation, incomplete observations, gravitational oscillations, adjoint of "on-off" physical processes.

1 Introduction

Numerical modeling has now become a major tool for the study of atmospheric dynamics. These models are constructed based on the physical laws that govern the temporal evolution of the flow and that express the mass, momentum, and energy of the fluid. In numerical form, they appear as a computer code that uses as independent variables (or input), the initial condition from which the model must be integrated and many parameters that define the physical and numerical conditions of the time integration. The dependent variables (or output), of models consist of the temporal sequence of meteorological fields produced by the integration, and also of the many derived quantities that one may wish to compute from that sequence of meteorological fields.

In many real-life situations, meteorologists are led to consider the following problems:

a) To determine values of part of the input parameters, such as initial conditions and/or model parameters, corresponding to given values of output parameters (i.e. observations). Such problems will normally be solved by optimal control of partial differential equations as unconstrained optimization problems. A prescribed scalar function, the cost function, measuring the misfit between the output parameters and the observed quantities will be defined and the optimal values of the input parameters will be found out using various large-scale unconstrained minimization algorithms which require information of function and gradient with respect to the input parameters. Two typical examples of such problems are variational data assimilation and parameter estimation.

b) To determine to which input parameters are some quantities calculated from the output parameters most sensitive. In some situations, one will be interested in the sensitivities of one forecast aspect (called the response) with respect to a large number of input parameters.

In all these situations one is required to determine the gradient of a function or a response with respect to a large number ($\sim 10^6$) of input parameters. The method of adjoint model equations provides an efficient way of solving such problems, which would of course be totally prohibitive if one were to use the only other method that could be considered for numerically determining the gradient: namely the finite-difference scheme.

In this chapter we will discuss variational 4-dimensional data assimilation. For parameter estimation see Smedstad and O'Brien (1991) and Zou, *et al.* (1992b) and for sensitivity analysis using the adjoint method, see Cacuci (1981a,b), Hall, *et al* (1982), Hall and Cacuci (1983) and Zou, *et al.* (1993d) and the references therein.

The complete and accurate specification of the 4-dimensional structure of the atmospheric state is a basic problem in meteorology. The conventional observational network alone is not capable of providing this complete description of the atmospheric state, since the observational density is not high enough in either space or time. For this reason it is desirable to include additional information (i.e. statistical or dynamical information) into the data assimilation system. There are many techniques used in data assimilation. Detailed reviews of data assimilation are given in Navon (1986), LeDimet and Navon (1989),

Ghil and Malanotte-Rizzoli (1991) and Daley (1991). Recently, considerable attention has been focused on variational data assimilation, which fits the data as well as possible in a least-square sense and satisfies the dynamics of the forecast model by adjusting only the initial fields. It originates in the control theory of partial differential equations and produces high-quality analyses of the atmosphere for the production of reliable forecasts. The basis of variational data assimilation is to use all the available information, observations which are distributed more or less regularly in both time and space and vary greatly in their nature and accuracy as well as numerical prediction model itself which describes physical conservation laws, in order to produce the best possible initial state in a least-squares norm optimal sense. The goal aimed at can be described as optimizing a cost function which measures the distance between model states and observations plus some prior knowledge. For the underlying optimal control theory see Lions (1971) and for a general description of its application to meteorology, see LeDimet and Talagrand (1986). Since then, a sizable amount of research has been carried out in this area in meteorology such as by Derber (1985), Lewis and Derber (1987), Courtier and Talagrand (1987), Talagrand and Courtier (1987), Zou, *et al.* (1992a), Thépaut and Courtier (1992), Navon, *et al.* (1992), Zou, *et al.* (1993a,c) and Zupanski (1993a).

Apart from its conceptual simplicity and its adaptability, variational data assimilation possesses many advantages. First, it can be shown that it is equivalent to the optimal Kalman Filter in the linear context and under the hypothesis of a perfect model (Daley, 1991; Lorenc, 1986; Rabier *et al.*, 1992). In the nonlinear case and if the model is still assumed to be perfect, results from variational data assimilation or the extended Kalman Filter are expected to be very close if the tangent-linear hypothesis is valid for the duration of the assimilation period. Comparison of variational data assimilation with simplified classic sequential assimilation by Rabier, *et al.* (1992) shows a significantly better performance of variational data assimilation scheme, as it yields consistently better analyses than its sequential counterpart and as the error grows in the subsequent forecasts at a lesser rate.

The model forecast error can also be considered in the context of variational data assimilation by introducing an additional term in the model equation (Derber, 1989, Zupanski, 1993a). The initialization problem, aiming at establishing balanced initial conditions without high frequency gravity wave noise, can also be incorporated into variational data assimilation by adding a penalty term into the cost function (Thépaut and Courtier, 1992; Zou, *et al.*, 1992a; Zou, *et al.*, 1993a).

The main limitation of variational data assimilation is that it remains rather computationally expensive and extensive testing has to be carried out before its possible operational implementation.

This chapter presents the main aspects underlying a rigorous derivation of variational data assimilation. As this chapter is intended to be self-contained, the basic formalism of the variational problem including the linkage between the nonlinear model, tangent linear model, adjoint model, cost function and the gradient of the cost function with respect to the initial conditions is briefly reviewed in Section 2. This section is divided into four main parts. Section 2.1 presents variational data assimilation theory for nonlinear systems with a standard cost function. The system of nonlinear operator equations and the associated cost function, gradient calculation, and adjoint are introduced and described in Section 2.1; altogether, they are intended to be sufficiently general to include the mathematical representation of a large number of problems arising in a wide variety of fields. Section 2.2 centers on practical coding of adjoint models and their verification. State-of-the-art large-scale unconstrained minimization schemes are described and summarized in Section 2.3, where different limited-memory quasi-Newton methods and truncated Newton methods as well as their relative performance are briefly reviewed. As an illustrative example, the formalisms in Sections 2.1-2.3 are applied in Section 2.4 to perform a twin experiment of variational data assimilation using a shallow-water equations model. Theory and numerical results concerning the impact of incomplete observations on the uniqueness of the solution and the convergence rate of the minimization processes are presented in Section 3. Methods aimed at the specification of suitably balanced optimal initial fields, i.e. applying different penalty terms in the framework of variational data assimilation to control gravitational oscillations, are provided in Section 4. A sequential penalty method is applied to a shallow-water equations model in Section 4.1 and different simple penalty terms constraining directly the time tendency of model variables are applied to a 3-dimensional adiabatic version of the operational primitive equations model at the National Meteorological Center (NMC) in Section 4.2. The computational aspects and the difference in the retrieved initial fields due to the inclusion of "threshold" processes, i.e. large-scale precipitation and deep cumulus convection, are presented in Section 5. These "threshold" processes represent locally nondifferentiable phenomena. Finally, Section 6 summarizes and highlights the main points underlying variational data assimilation and discusses the potential of using adjoint techniques to further extend the scope of this theory for meteorological applications.

2 Basic Framework of Variational Data Assimilation

2.1 Variational Data Assimilation Formalism

Cost function

The objective of variational 4-D data assimilation is to find the solution to a numerical forecast model which will best fit a series of observational fields distributed over some space and time interval. One possible measure of the lack of fit, the cost function J, consists of a weighted least square fit of the model forecast to the observations:

$$J(\mathbf{x}(t_0)) = \frac{1}{2}\sum_{r=0}^{R}\left(\mathbf{C}\mathbf{x}(t_r) - \mathbf{x}^{obs}(t_r)\right)^T \mathbf{W}(t_r)\left(\mathbf{C}\mathbf{x}(t_r) - \mathbf{x}^{obs}(t_r)\right) \tag{1}$$

where t_r represents the time when an observation occurs in the assimilation interval $[t_0, t_a]$, R is the total number of time levels in the assimilation interval when observations are available, $\mathbf{x}(t_r)$ is the N-component vector in space R_N containing values of model variables at time t_r, $\mathbf{x}^{obs}(t_r)$ is the M-component $(M \leq N)$ vector in space R_M containing values of observations at time t_r, $\mathbf{C}$ is a projection operator from the space R_N to the space R_M, and $\mathbf{W}(t_r)$ is an $M \times M$ weighting matrix. If $\mathbf{W}(t_r)$ are taken as unit matrices, J is defined as an $\mathcal{L}_2$ norm of $\mathbf{Cx} - \mathbf{x}^{obs}$. One can also choose an energy norm to define J, which will provide a natural weight for different model variables.

The values of $\mathbf{x}(t_r)$, $r = 0, 1, \ldots, R$ are obtained by integrating a numerical model which can be written in a general form

$$\frac{d\mathbf{x}}{dt} = \mathbf{F}(t, \mathbf{x}) \tag{2}$$

from the initial state $\mathbf{x}(t_0)$. In time discretized form, (2) will assume the following general form

$$\mathbf{x}(t_0 + \Delta t) = \quad \mathbf{F}_1(\mathbf{x}(t_0)) + \mathbf{L}_1\mathbf{x}(t_0) \tag{3}$$

$$\mathbf{x}(t_r + \Delta t) = \quad \mathbf{F}(\mathbf{x}(t_r)) + \mathbf{L}(t_r)\mathbf{x}(t_r - \Delta t) + \mathbf{m}(t_r)\mathbf{x}(t_r), \quad \text{for} \quad r = 1, 2, \ldots, \tag{4}$$

where $\mathbf{F}_1(\mathbf{x})$ and $\mathbf{F}(\mathbf{x})$ are nonlinear operators and $\mathbf{L}_1$, $\mathbf{L}$ and $\mathbf{m}$ are linear operators.

Weighting

Weights in the cost function serve the following purposes: (i) scale the cost function J to become a non-dimensional quantity; (ii) reflect the confidence we have in the quality of the observed data, and (iii) provide also initial conditioning essential in efficient minimization. Weights are usually chosen as the inverses of the covariance matrices of the observation errors. However, these variances are difficult to specify properly and much more research work is necessary in this area. Thépaut and Courtier (1992) used an energy norm to define the cost function. The choice of an energy form defines a physical measure of atmospheric fields as well as a natural weighting for the different model variables. In the variational data assimilation experiments of Navon *et al.* (1992) with the adiabatic version of the NMC spectral model, a diagonal weighting matrix is used with $\mathbf{W}_\chi, \mathbf{W}_\psi, \mathbf{W}_T, \mathbf{W}_{\ln p_s}$ and $\mathbf{W}_q$ as $\mathbf{W}$'s diagonal submatrices. These submatrices represent the weighting factors for the potential velocity (χ), stream function (ψ), temperature (T), surface pressure (p_s) and moisture (q) fields, respectively. The submatrix $\mathbf{W}_\psi$, for instance, is calculated by the following formula

$$\mathbf{W}_\psi = \frac{1}{\max_{i,j,k} |\psi^{obs}_{i,j,k}(t_0) - \psi^{obs}_{i,j,k}(t_R)|^2} \tag{5}$$

with similar expressions for the velocity potential, temperature, surface pressure, and moisture fields, i.e. the inverse of the maximum difference between the two observational fields at times t_0 and t_R. In the experiments of Zou, *et al.* (1992a) using a shallow-water equations model, a constant diagonal weight is used with values of $10^{-4}m^{-2}s^2$ and $10^{-2}m^{-2}s^2$ as the weighting coefficients for geopotential and wind fields respectively. Courtier and Talagrand (1990) used a temporal weighting in which weights given to individual observations varied linearly with time, the total sum of the weights assuming the same value as the reciprocal of squared estimates of the statistical root-mean-square observational errors. The idea is that the model not being perfect, it cannot adjust uniformly to the whole set of observations at intermediate times. In the case of variational data assimilation, where one performs a forecast starting from the final time of the assimilation period, a better adjustment to later observations is obviously preferable, and larger weights are assigned to more recent observations in the definition of the cost function.

Gradient

To find the minimum of the cost function J, most of large-scale efficient minimization algorithms require the user to supply values of the gradient of the cost function with respect to control variables which in our case are the initial conditions.

To calculate the gradient of the cost function with respect to the initial condition, we will define a quantity J':

$$\begin{aligned} J'(\mathbf{x}(t_0)) &\equiv J(\mathbf{x}(t_0) + \mathbf{x}'(t_0)) - J(\mathbf{x}(t_0)) \\ &= \sum_{r=0}^{R} \left(\mathbf{W}(t_r) \left(\mathbf{C}\mathbf{x}(t_r) - \mathbf{x}^{obs}(t_r) \right) \right)^T \mathbf{C}\mathbf{x}'(t_r) + \sum_{r=0}^{R} O(\mathbf{x}'^2(t_r)). \end{aligned} \tag{6}$$

i.e. the change in the cost function resulting from a small perturbation $\mathbf{x}'(t_0)$ about the initial conditions $\mathbf{x}(t_0)$. In limit as $\|\mathbf{x}'\| \to 0$, J' is the directional derivative in the $\mathbf{x}'(t_0)$ direction and is given by

$$J'(\mathbf{x}(t_0)) = (\nabla J(\mathbf{x}(t_0)))^T \mathbf{x}'(t_0). \tag{7}$$

Equating (6) and (7) results in

$$(\nabla J(\mathbf{x}(t_0)))^T \mathbf{x}'(t_0) = \sum_{r=0}^{R} \left(\mathbf{W}(t_r) \left(\mathbf{C}\mathbf{x}(t_r) - \mathbf{x}^{obs}(t_r) \right) \right)^T \mathbf{x}'(t_r). \tag{8}$$

It is clear that if $\mathbf{x}'(t_r)$ can be expressed as a function of $\mathbf{x}'(t_0)$, then the gradient of the cost function with respect to the initial conditions can be found.

Since $\mathbf{x}'(t_r)$ is the perturbation at time t_r in the forecast resulting from the initial perturbation $\mathbf{x}'(t_0)$, it can be obtained by integrating the tangent linear model, which in turn can be obtained by linearizing the nonlinear model (3)-(4):

$$\mathbf{x}'(t_0 + \Delta t) = \frac{\partial \mathbf{F}_1(\mathbf{x}(t_0))\,\mathbf{x}'(t_0)}{\partial \mathbf{x}} + \mathbf{L}_1\mathbf{x}'(t_0) \tag{9}$$

$$\mathbf{x}'(t_r + \Delta t) = \frac{\partial \mathbf{F}(\mathbf{x}(t_r))\,\mathbf{x}'(t_r)}{\partial \mathbf{x}} + \mathbf{L}\mathbf{x}'(t_r - \Delta t) + \mathbf{m}(t_r)\mathbf{x}'(t_r), \quad \text{for } r = 1, 2, \ldots, \tag{10}$$

which may then be rewritten symbolically as

$$\mathbf{x}'(t_r) = \mathbf{P}_r\mathbf{x}'(t_0) \tag{11}$$

where $\mathbf{P}_r$ represents the result of applying all the operator matrices in the linear model to obtain $\mathbf{x}'(t_r)$ from $\mathbf{x}'(t_0)$.

Using the tangent linear model (11), (8) becomes

$$(\nabla J(\mathbf{x}(t_0)))^T \mathbf{x}'(t_0) = \sum_{r=0}^{R} \left(\mathbf{W}(t_r)\left(\mathbf{C}\mathbf{x}(t_r) - \mathbf{x}^{obs}(t_r)\right)\right)^T \mathbf{C}\mathbf{P}_r\mathbf{x}'(t_0). \tag{12}$$

This implies

$$\nabla J(\mathbf{x}(t_0)) = \sum_{r=0}^{R} \mathbf{P}_r^T \mathbf{C}^T \mathbf{W}(t_r)\left(\mathbf{C}\mathbf{x}(t_r) - \mathbf{x}^{obs}(t_r)\right), \tag{13}$$

where $\mathbf{P}_r^T$, $r = 1, 2, \ldots, R$ are the corresponding adjoint operators of the linear operators $\mathbf{P}_r$, $r = 1, 2, \ldots, R$ in the tangent linear model. Therefore, the gradient of the cost function may be obtained by a single integration of the adjoint model from final time t_a to initial time t_0 of the assimilation window with zero initial conditions for the adjoint variables at time t_a while the weighted differences

$$\mathbf{C}^T \mathbf{W}(t_r)\left(\mathbf{C}\mathbf{x}(t_r) - \mathbf{x}^{obs}(t_r)\right), \quad r = R, R-1, \ldots, 0 \tag{14}$$

are inserted on the right-hand-side of the following adjoint model

$$\hat{\mathbf{x}}(t_0) = \mathbf{P}^T \hat{\mathbf{x}}(t_a), \tag{15}$$

whenever an observational time $t_r (r = R, R-1, \ldots, 0)$ is reached, where $\hat{\mathbf{x}}$ represents the adjoint variables.

2.2 Adjoint models

The discretized atmospheric models are implemented by rather long and complicated codes. The models are in permanent evolution and are constantly subject to modifications. The possibility of routinely performing work involving adjoint model integration requires that, whenever a modification is made on the direct model, a corresponding modification should also be made on the adjoint. Experience shows that even minor mistakes in the adjoint model code can render it totally useless. This requires that the adjoint code must be developed directly from the basic direct code (and not, for instance, from the partial differential equations from which the direct code is built). It also implies that the components of the adjoint code will be in one-to-one correspondence with the components of the direct code: to each subroutine of the direct code, there is a mirror subroutine of the adjoint performing the corresponding adjoint task. Since the adjoint operator $\mathbf{P}^T$ is based on the tangent linear operator $\mathbf{P}$, the above discussion implies that discrete operations in the forward nonlinear model have unique corresponding discrete operations in the tangent linear model with only one exception: all the nonlinear terms in the original model are linearized locally.

For an adiabatic primitive equations forecast model, the only nonlinear terms are those due to the quadratic terms introduced by the advection processes. Linearization of this type of nonlinear terms is quite simple. For example, a term $u * T$ in the nonlinear model becomes two terms $u * T' + T * u'$ in the tangent linear model. However, when the nonlinear model includes physical processes, the linearization of these processes is not at all evident due to the high degree of nonlinearity and the appearance of either table lookups or iterative procedures which were employed in the direct code. For example, in the case of table lookups, the values in the tables are built for the nonlinear calculation. The original analytical formulas used to build the lookup tables are then required in order to derive the tangent linear version of the nonlinear model. When an iterative method is used to solve a highly nonlinear equation in the nonlinear model, the tangent linear version of this part can not be derived directly from the nonlinear code, i.e. by linearizing all the procedure of the iterative process. One is required to derive the analytical tangent linear equation of the relevant equation which is solved iteratively in the direct code and then write the independent linear code of this part using the same variable notations as in the direct code. These tasks are rather tedious and complicated since normally one has at his disposal the direct code but does not always possess the corresponding updated complete documentation of the direct code. A check of the correctness of the tangent linear model is crucial in these situations.

To date, adjoint models have been used, in the context of meteorology, only at the experimental level. Meteorologists write adjoint models with few basic underlying principles in mind. The first and foremost basic rule for writing the adjoint code is the matrix method, i.e. each DO loop in the direct code corresponds to a matrix/vector multiplication: $\mathbf{y}^{\text{output}} = \mathbf{A}\mathbf{x}^{\text{input}}$. The adjoint code can be developed based on the equality: $\hat{x}^{\text{output}} = \mathbf{A}^T \hat{y}^{\text{input}}$, where $\hat{x}$ represents the adjoint variables corresponding to $\mathbf{x}$. A simple example of the construction of the discrete adjoint code from the discrete linear code based on this matrix method is provided in the Appendix A of the paper by Navon, *et al.* (1992).

If we view the linear model as the result of the multiplication of a number of operator matrices:

$$\mathbf{P} = \mathbf{A}_1 \mathbf{A}_2 \cdots \mathbf{A}_N, \tag{16}$$

where each matrix $A_i (i = 1, \cdots, N)$ represents either a subroutine or a single *DO* loop, then the adjoint model can be viewed as being a product of adjoint subproblems

$$\mathbf{P}^T = \mathbf{A}_N^T \mathbf{A}_{N-1}^T \cdots \mathbf{A}_1^T, \tag{17}$$

i.e. the adjoint model has to be written backward from the tangent linear model. The second rule is the rule of locality: whenever a local modification is made in the direct model, a corresponding modification must be made locally in the adjoint code. This requires transparent notations, between the direct and adjoint codes, in as far as subroutine and variable names, instruction labeling, etc., are concerned. Some examples may be found in the paper by Talagrand (1991). The third principle of building the adjoint code is the trajectory management when an IF statement is encountered in the direct code. The adjoint model will have to follow the same route of the direct model but backward in time.

The above formulated rules must of course be applied with some discernment, since there are situations where there is simply no need for them. A typical example is the

Fourier transform, which being unitary, has as adjoint its own inverse. There is therefore no need to develop the adjoint of a Fourier transform and it is sufficient in the adjoint code to invoke the inverse transform at the appropriate location.

In other words, the discrete adjoint model can be obtained directly from the discrete linear model, which in turn is obtained from the original nonlinear model by linearization around a model state. The linearization is made locally in the code and the structure of the linear model is exactly the same as that of the nonlinear model. These rules simplify not only the complexity of constructing the adjoint model but also avoid the inconsistency generally arising from the derivation of the adjoint equations in analytic form followed by the discrete approximation since there is no commutativity between adjoint and discretization.

The correctness of the adjoint can be checked at any level of the code of the development of the discrete adjoint model by applying the following identity

$$(\mathbf{Ax})^T(\mathbf{Ax}) = \mathbf{x}^T\left(\mathbf{A}^T(\mathbf{Ax})\right), \tag{18}$$

where $\mathbf{A}$ represents either a single *DO* loop or one/several subroutines and $\mathbf{x}$ represents the input of $\mathbf{A}$.

2.3 Large-scale minimization algorithms

Limited-memory quasi-Newton and truncated Newton methods

Once the values of the cost function J (obtained by integrating the nonlinear model forward) and its gradient with respect to the control variables (obtained by integrating the adjoint model backward in time) are found, the optimal initial condition $\mathbf{x}^*(t_0)$ of minimizing J can be obtained by any unconstrained minimization algorithms which require the user to supply the values of the cost function and its gradient. Since the model already taxes the capability of the largest available supercomputers and the optimization of the initial conditions must conform to the operational requirements of timeliness, the choice of a robust and efficient minimization algorithm is crucial.

Limited-memory quasi-Newton (Shanno and Phua, 1980; Buckley, 1978; Gill and Murray, 1979; Buckley and Lenir, 1983, 1985; Buckley, 1989; Liu and Nocedal, 1989; Nocedal, 1980) and truncated Newton (Nash, 1984a,b, 1985; Schlick and Fogelson, 1992a,b) methods represent two classes of algorithms which are attractive for large-scale problems because of their modest storage requirements. They use a low and adjustable amount of storage and require the function and gradient values at each iteration. Limited-memory quasi-Newton methods can be viewed as extensions of conjugate-gradient methods in which the addition of some modest storage serves to accelerate the convergence rate. Truncated Newton methods attempt to retain the rapid convergence rate of classical Newton methods while economizing storage and computational requirements so as to become feasible for large-scale applications.

Limited-memory quasi-Newton algorithms have the following basic structure for minimizing $J(\mathbf{x})$, $\mathbf{x} \in \mathcal{R}^N$:

1) Choose an initial guess $\mathbf{x}_0$, and a positive definite initial approximation to the inverse Hessian matrix $\mathbf{H}_0$ (which may be chosen as the identity matrix).

2) Compute

$$\mathbf{g}_0 = \mathbf{g}(\mathbf{x}_0) = \nabla J(\mathbf{x}_0), \tag{19}$$

and set

$$\mathbf{d}_0 = -\mathbf{H}_0\mathbf{g}_0. \tag{20}$$

3) For $k = 0, 1, \ldots$, set

$$\mathbf{x}_{k+1} = \mathbf{x}_k + \alpha_k \mathbf{d}_k, \tag{21}$$

where α_k is the step-size obtained by a safeguarded procedure.

4) Compute

$$\mathbf{g}_{k+1} = \nabla J(\mathbf{x}_{k+1}) \tag{22}$$

5) Generate a new search direction, $\mathbf{d}_{k+1}$, by setting

$$\mathbf{d}_{k+1} = -\mathbf{H}_{k+1}\mathbf{g}_{k+1}, \tag{23}$$

6) Check for convergence: If

$$\|\mathbf{g}_{k+1}\| \leq \epsilon \max\{1, \|\mathbf{x}_{k+1}\|\}, \tag{24}$$

stop, where $\epsilon = 10^{-5}$. Otherwise continue from step 3.

In practice, limited-memory quasi-Newton methods update formula (Liu and Nocedal, 1989) forms an approximate inverse Hessian from $\mathbf{H}_0$ and k pairs of vectors $(\mathbf{q}_i, \mathbf{p}_i)$, where $\mathbf{q}_i = \mathbf{g}_{i+1} - \mathbf{g}_i$ and $\mathbf{p}_i = \mathbf{x}_{i+1} - \mathbf{x}_i$ for $i \geq 0$. Since $\mathbf{H}_0$ is generally taken to be the identity matrix or some other diagonal matrix, the pairs $(\mathbf{q}_i, \mathbf{p}_i)$ are stored instead of $\mathbf{H}_k$ (which requires a large memory), and $\mathbf{H}_k\mathbf{g}_k$ is computed by a recursive algorithm. All the limited-memory quasi-Newton methods presented below fit into this conceptual framework. They differ only in the selection of the vector couples $(\mathbf{q}_i, \mathbf{p}_i)$, the choice of $\mathbf{H}_0$, the method for computing $\mathbf{H}_k\mathbf{g}_k$, the line search implementation, and handling of restarts.

Just as limited-memory quasi-Newton methods attempt to combine modest storage and computational requirements of conjugate-gradient methods with the convergence properties of the standard quasi-Newton methods, truncated Newton methods attempt to retain the rapid (quadratic) convergence rate of classic Newton methods while making storage and computational requirements feasible for large-scale applications. Recall that Newton methods for minimizing a multivariate function $J(\mathbf{x}_k)$ are iterative techniques based on minimizing a local quadratic approximation to J at every step. The quadratic model of J at a point $\mathbf{x}_k$ along the direction of a vector $\mathbf{d}_k$ can be written as

$$J(\mathbf{x}_k + \mathbf{d}_k) \approx J(\mathbf{x}_k) + \mathbf{g}_k^T\mathbf{d}_k + \frac{1}{2}\mathbf{d}_k^T\mathbf{H}_k\mathbf{d}_k \tag{25}$$

where $\mathbf{g}_k$ and $\mathbf{H}_k$ denote the gradient and Hessian of J, respectively, at point $\mathbf{x}_k$. Minimization of this quadratic approximation results in a linear system of equations for the search vector $\mathbf{d}_k$ known as the Newton equations:

$$\mathbf{H}_k \mathbf{d}_k = -\mathbf{g}_k. \tag{26}$$

In the modified Newton framework, a sequence of iterates is generated from $\mathbf{x}_0$ by the rule $\mathbf{x}_{k+1} = \mathbf{x}_k + \alpha_k \mathbf{d}_k$. The vector $\mathbf{d}_k$ is obtained as the solution (or approximate solution) of the system (26) or, possibly, a modified version of it, where some positive definite approximation to $\mathbf{H}_k$, $\tilde{\mathbf{H}}_k$, replaces $\mathbf{H}_k$.

When an approximate solution is used, the method is referred to as a "truncated" Newton method because the solution process of (26) is not carried out to completion. In this case, $\mathbf{d}_k$ may be considered satisfactory when the residual vector $\mathbf{r}_k = \mathbf{H}_k \mathbf{d}_k + \mathbf{g}_k$ is sufficiently small. Truncation may be justified since accurate search directions are not essential in regions far away from local minima. For such regions, any descent direction suffices, so the effort expended in solving the system accurately is often unwarranted. However, as a solution of the optimization problem is approached, the quadratic approximation of (25) is likely to become more accurate and a smaller residual may be more important. Thus, the truncation criterion should be chosen to enforce a smaller residual systematically as minimization proceeds. One such effective strategy requires

$$\|\mathbf{r}_k\| \leq \eta_k \|\mathbf{g}_k\| \tag{27}$$

where

$$\eta_k = \min\{\frac{c}{k}, \|\mathbf{g}_k\|\}, \quad c \leq 1. \tag{28}$$

Other truncation criteria have also been discussed (Nash, 1984a,b; Schlick and Fogelson, 1992a,b).

The quadratic subproblem of computing an approximate search direction at each step is accomplished through some iterative scheme. This produces a nested iteration structure: an "outer" loop for updating $\mathbf{x}_k$, and an "inner" loop for computing $\mathbf{d}_k$. The *linear* conjugate-gradient method is attractive for large-scale problems because of its modest computational requirements and theoretical convergence in at most N iterations (Golub and van Loan, 1989). However, since conjugate-gradient methods were developed for positive definite systems, adaptations must be made in the present context where the Hessian may be indefinite. Typically, this is handled by terminating the inner loop (at iteration q) when a direction of negative curvature is detected ($\mathbf{d}_q^T \mathbf{H}_k \mathbf{d}_q < \xi$ where ξ is a small positive tolerance such as 10^{-10}); an exit direction that is guaranteed to be a descent direction is then chosen (Dembo and Steihaug, 1983; Schlick and Fogelson, 1992a,b). An alternative procedure to linear conjugate-gradient for the inner loop is based on the Lanczos factorization (Golub and van Loan, 1989), which works for symmetric but not necessarily positive definite systems. It is important to note that different procedures for the inner loop can lead to a very different overall performance in the minimization.

Several comparisons have been made on different large-scale unconstrained minimization algorithms. Navon and Legler (1987) compared a number of different conjugate-gradient and limited-memory quasi-Newton methods for problems in meteorology and concluded that the Shanno-Phua (1980) limited-memory quasi-Newton algorithm was the most adequate for their test problems. The studies of Gilbert and Lemaréchal (1989) and Liu and Nocedal (1989) indicate that the L-BFGS method is among the best limited-memory quasi-Newton methods available to date. Nash and Nocedal (1989) compared the L-BFGS method with the truncated Newton method of Nash (1984a,b, 1985) on 53 problems of dimensions ranging between 10^2 - 10^4. Their results suggest that performance is correlated with the degree of nonlinearity of the objective function: for quadratic and approximately quadratic problems, the truncated Newton algorithm outperformed L-BFGS, while for most of the highly nonlinear problems L-BFGS performed better.

Zou, *et al.* (1993b) compared four of the state-of-the-art limited memory quasi-Newton methods and two truncated Newton methods on several test problems including problems in meteorology and oceanography. Their results confirm that the L-BFGS algorithm of Liu and Nocedal (1989) seems to be the most efficient, particularly robust and user-friendly. It deals with the critical issue of storage in large-scale problems. The L-BFGS update formula generates matrices using information from the last m Quasi-Newton iterations, where m is the number of quasi-Newton updates determined by the user (generally $3 \leq m \leq 7$). After having used the m vector storage locations for m quasi-Newton updates, the quasi-Newton approximation of the Hessian matrix is updated by dropping the oldest information and replacing it by the newest information. A new search direction, which is an estimate of the relative change to the current variables vector that produces the maximum reduction in the cost function, is then computed. It employs a cubic line search required to satisfy a Wolfe (1969) condition and a unit stepsize is always tried first. This algorithm uses a limited amount of storage and the quasi-Newton approximation of the Hessian matrix is updated continuously. The general algorithm for L-BFGS and other limited-memory quasi-Newton algorithms and truncated Newton methods can be found in the paper of Zou, *et al.* (1993b).

Scaling

The term "scaling" is invariably used in many meteorological applications in a vague sense to discuss numerical difficulties whose existence is universally acknowledged, but cannot be described precisely in general terms. Therefore, it is not surprising that much confusion exists about scaling. One normally considers the effect on problem scaling of replacing the original variables $\mathbf{x}$ of the problem by a set of linearly transformed variables $\mathbf{y}$: $\mathbf{x} = \mathbf{L}\mathbf{y}$, where $\mathbf{L}$ is a fixed non-single matrix. The choice of the value of $\mathbf{L}$ depends upon what does one wishes to scale. There are four scaling methods: by variable, by constraint, by gradient, and by Hessian (Gill, *et al.*, 1981). In theory, such a transformation does not affect the optimal solution; however, the scaling of variables may have an enormous effect on the behaviour of finite-precision calculations. One example of poor scaling is an imbalance between the values of the function and changes in $\mathbf{x}$ in the sense that the function values may change very little even though $\mathbf{x}$ changes significantly, or vice versa. The effect of poor scaling is to invalidate the criteria of the line search used to accept the step length. This will lead to the failure of a step-length algorithm to find an acceptable point along the search direction. Let $F(\mathbf{y})$ denote the function of the transformed variables:

$$F(\mathbf{y}) = J(\mathbf{Ly}). \tag{29}$$

The derivatives of the function F with respect to the transformed variables $\mathbf{y}$ are

$$\nabla_y F(\mathbf{y}) = \mathbf{L}\nabla_x J(\mathbf{x}), \qquad \nabla_y^2 F(\mathbf{y}) = \mathbf{L}^T \nabla_x^2 J(\mathbf{x})\mathbf{L}. \tag{30}$$

One instance of bad scaling can occur when the variables $\mathbf{x}$ have widely differing orders of magnitude. A basic rule for scaling is that the variables of the scaled problem should be of similar magnitude and of order unity because within optimization routines convergence tolerances and other criteria are necessarily based upon an implicit definition of "small" and "large" and thus variables with widely varying orders of magnitude may cause serious difficulties for some minimization algorithms (Gill, *et al.*, 1981). One simple direct way to determine the scaling factor is to use the typical values for different fields (for instance 10^{-5} can be used as the scaling factor of vorticity), i.e.

$$\mathbf{x} = \mathbf{Dy} \tag{31}$$

where d_j is set to a typical value of the jth variable, causing each variable to be of similar "weight" during the optimization. Another form of bad scaling occurs when the partial derivatives of a function with respect to a particular variable are not "balanced". A scaling based on the first derivative can be used where $d_j = (1 + |J(\mathbf{x})|)/(2|\nabla_{x_j} J|)$ (see Gill, *et al.*, 1981).

Therefore, scaling is a crucial issue in the success of nonlinear unconstrained optimization problems and some research has been carried out on scaling nonlinear programming problems. It is well-known that a badly scaled nonlinear programming problem can be almost impossible to solve (see also Navon and de Villiers, 1983; Courtier and Talagrand, 1990). An effective automatic scaling procedure would ease these difficulties, and could also render problems which are well scaled easier to solve by improving the condition number of their Hessian matrix which controls the rate of convergence of the conjugate-gradient and quasi-Newton algorithms (Thacker, 1989).

In meteorological problems, the variables in the control vector have also enormously different magnitudes varying over a range of 8 orders of magnitude. Scaling by variable transformation converts the variables from units that reflect the physical nature of the problem to units that display desirable properties for the unconstrained minimization process. In our experiment with the NMC spectral model, the scaling constants for the different fields are calculated by

$$\mathbf{L}_{\psi_{ii}} = \max_{i,j,k} |\psi_{i,j,k}^{obs}(t_0) - \psi_{i,j,k}^{obs}(t_R)|/2, \tag{32}$$

where ψ represents the stream function field, and similarly for χ (velocity potential), T (temperature), $\ln ps$ (log of the surface pressure), and q (specific humidity).

For complicated functions, difficulties may be encountered in choosing suitable scaling factors. There is no general rule to determine the best scaling factors for all minimization problems, and a good scaling is problem dependent. Further improvement of the condition number can be obtained by a more sophisticated scaling (Gill, *et al.*, 1981; Bertsekas, 1982; Luenberger, 1984).

Preconditioning

Preconditioning is a way to speed-up the convergence and to "relax" ill-conditioning of the Hessian matrix by altering both the condition number as well as the distribution of eigenvalues "clustering" within the spectrum (Axelsson and Barker, 1984). The conditioning of the Hessian matrix at the solution of an unconstrained problem determines the accuracy with which the solution can be computed in different directions. In particular, when the Hessian is ill-conditioned, the cost function will vary much more rapidly along some directions in R_N than along others. An ill-conditioned Hessian at the solution is thus a symptom of bad scaling, in the sense that similar changes in the control variables $\|\mathbf{x}\|$ do not lead to similar changes in the cost function J.

We can briefly describe preconditioning as follows:

The solution of

$$\mathbf{Hx} = -\mathbf{b} \tag{33}$$

can be found by minimizing

$$f(\mathbf{x}) = \frac{1}{2}\mathbf{x}^T\mathbf{Hx} - \mathbf{b}^T\mathbf{x}. \tag{34}$$

The transformation

$$\mathbf{y} = \mathbf{E}^T\mathbf{x} \tag{35}$$

leads to the problem of minimizing

$$\tilde{f}(\mathbf{y}) = f(\mathbf{E}^{-T}\mathbf{y}) = \frac{1}{2}\mathbf{y}^T\tilde{\mathbf{H}}\mathbf{y} - \tilde{\mathbf{b}}^T\mathbf{y} \tag{36}$$

where

$$\tilde{\mathbf{H}} = \mathbf{E}^{-1}\mathbf{H}\mathbf{E}^{-T}, \tilde{\mathbf{b}} = \mathbf{E}^{-1}\mathbf{b}. \tag{37}$$

If the spectral condition number of the preconditioned matrix $\tilde{\mathbf{H}}$ is such that

$$\kappa(\tilde{\mathbf{H}}) < \kappa(\mathbf{H}) \equiv \frac{\lambda_{\max}}{\lambda_{\min}}, \tag{38}$$

the rate of convergence is improved (see Axelsson and Baker, 1984).

The matrix $\tilde{\mathbf{H}}$ has the same eigenvalues as the matrix $\left(\mathbf{EE}^T\right)^{-1}\mathbf{H}$. The matrix $\mathbf{Q} = \mathbf{EE}^T$ is called the preconditioning matrix.

One can prove that

(i) $\mathbf{Q}$ is a good approximation to $\mathbf{H}$ if and only if

$$\mathbf{Q} = \alpha\mathbf{H}, (\alpha \text{ is a scalar}), \tag{39}$$

since in this case, $\kappa(\tilde{\mathbf{H}}) = 1$.

(ii) $\mathbf{Q}$ needs not to be a particular good approximation to $\mathbf{H}$ since if $\mathbf{Q} = \mathbf{I}$, $\kappa(\tilde{\mathbf{H}}) = \kappa(\mathbf{H})$.

For the preconditioning of the conjugate-gradient method, the preconditioning matrix $\mathbf{Q}$ can be taken as $\mathbf{Q} = \mathbf{M}^{-1}$, where $\mathbf{M}$ is obtained by performing r steps of a limited-memory quasi-Newton method from the one-parameter family of quasi-Newton updates. The matrix $\mathbf{MH}$ has r unit eigenvalues. This is roughly equivalent to choosing $\mathbf{Q}$ so that the condition number of $\tilde{\mathbf{H}}$ is as small as possible. In nonlinear problems, the matrix $\mathbf{Q}$ will vary from iteration to iteration. For example a limited-memory approximate Hessian matrix may be used as preconditioning matrix. A preconditioning based on a diagonal scaling is often used in practice, which is a less sophisticated technique that has been applied to general problems. Recently, a new preconditioning method was proposed by Zupanski (1993b), which can be used for any minimization algorithm if the preconditioned steepest descent is used in the first iteration. His preconditioning is based on the requirements that the control variable adjustment is balanced by the calculated gradient norm and that a second order Taylor series expansion approximation is valid in the vicinity of the first guess. A significant improvement was found in his experiments. For recent research on preconditioning methods applied to discretized shallow-water equations model see Navon and Cai (1993) as well as the code NSPCG (Holter, *et al.*, 1991).

2.4 A simple example of variational data assimilation

A simple 2-D limited-area shallow water equations model is used to evaluate a quadratic objective function. The equations in Cartesian coordinates may be written as

$$\frac{\partial u}{\partial t} + u\frac{\partial u}{\partial x} + v\frac{\partial u}{\partial y} - fv + \frac{\partial \phi}{\partial x} = 0 \tag{40}$$

$$\frac{\partial v}{\partial t} + u\frac{\partial v}{\partial x} + v\frac{\partial v}{\partial y} + fu + \frac{\partial \phi}{\partial y} = 0 \tag{41}$$

$$\frac{\partial \phi}{\partial t} + u\frac{\partial \phi}{\partial x} + v\frac{\partial \phi}{\partial y} + \phi(\frac{\partial u}{\partial x} + \frac{\partial v}{\partial y}) = 0, \tag{42}$$

where f is the Coriolis parameter, $u, v,$and ϕ are the two components of the velocity field and the geopotential field, respectively; both are spatially discretized with a centered finite-difference scheme in space and an explicit leapfrog integration scheme in time. A rectangular domain of size $L = 6000km, D = 4400km$ is used along with space and time mesh increments of $\Delta x = 300km, \Delta y = 220km$ and $\Delta t = 600s$, respectively.

This model is widely used in meteorology and oceanography since it contains most of the physical degrees of freedom (including gravity waves) present in the more sophisticated 3-D primitive equation models. It is computationally less expensive to implement and results using this model can be expected to be similar to those obtained from a more complicated primitive equation model.

The objective function is defined as a simple weighted sum of squared differences between the observations and the corresponding prediction model values:

$$J = W_\phi \sum_{n=1}^{N_\phi}(\phi_n - \phi_n^{obs})^2 + W_V \sum_{n=1}^{N_V}[(u_n - u_n^{obs})^2 + (v_n - v_n^{obs})^2], \tag{43}$$

where N_ϕ is the total number of geopotential observations available over the 10 hour assimilation window (t_0, t_R), and N_V is the total number of wind vector observations. The quantities u_n^{obs}, v_n^{obs}, and ϕ_n^{obs} are the observed values for the northward wind component, the eastward wind component and the geopotential field respectively, while the quantities u_n, v_n, and ϕ_n are the corresponding computed model values. W_ϕ and W_V are weighting factors, taken to be the inverse of estimates of the statistical root-mean-square observational errors on geopotential and wind components respectively. Values of $W_\phi = 10^{-4}m^{-4}s^4$ and $W_V = 10^{-2}m^{-2}s^2$ are used. The objective function J is viewed as a function of $\mathbf{x}_0 = (u(t_0), v(t_0), \phi(t_0))^T$.

For the numerical experiments, the observational data consist of the model-integrated values for wind and geopotential at each time step starting from the Grammeltvedt initial conditions (Figure 1, Grammeltvedt (1969)). Random perturbations of these fields, performed using a standard library randomizer RANF on the CRAY-YMP (shown in Figure 2) are then used as the initial guess for the solution. A grid of a resolution 21×21 points in space results in a control variables vector of dimension 1323.

Two different scaling procedures were considered: "gradient" and "consistent". The first scales the gradient of the objective function. The second makes the shallow water equations model nondimensional.

The limited-memory quasi-Newton method of Liu and Nocedal (1989) —L-BFGS— was successful only with gradient scaling. The limited-memory quasi-Newton method of Gill and Murray (1979) —E04DGF — (see NAG, 1990) worked only with the nondimensional shallow water equations model. It appears that additional scaling is crucial for the success of the minimization algorithms.

Due to the different scaling procedures employed in the different minimization methods the minimization is stopped when the following convergence criterion

$$\|\mathbf{g}_k\| \le 10^{-4} \times \|\mathbf{g}_0\| \tag{44}$$

is satisfied.

The performance of the truncated Newton often depends on the specified maximum number of permitted inner iterations per outer iteration (MXITCG). Our experience suggests that different settings for MXITCG have small impact on the performance of the truncated Newton method of Nash but a rather large impact on that of the truncated Newton method of Schlick and Fogelson (see Table 1). This results from our current unpreconditioned implementation for the truncated Newton method of Schlick and Fogelson, since the inner conjugate gradient linear equation solver will require more iterations to find a search direction.

To clarify this idea and to see which differences in performance between the two truncated Newton methods were due to the different truncation criteria, conjugate gradient versus Lanczos, and preconditioning, we performed minimization for the truncated Newton method of Nash without diagonal preconditioning. The results are presented in Table 1. Similar trends are identified for both the truncated Newton method of Nash and the truncated Newton method of Schlick and Fogelson in this case: the cost for large MXITCG is much lower than that for small MXITCG. However, the truncated Newton method of Schlick and Fogelson with MXITCG=50 performs much better than

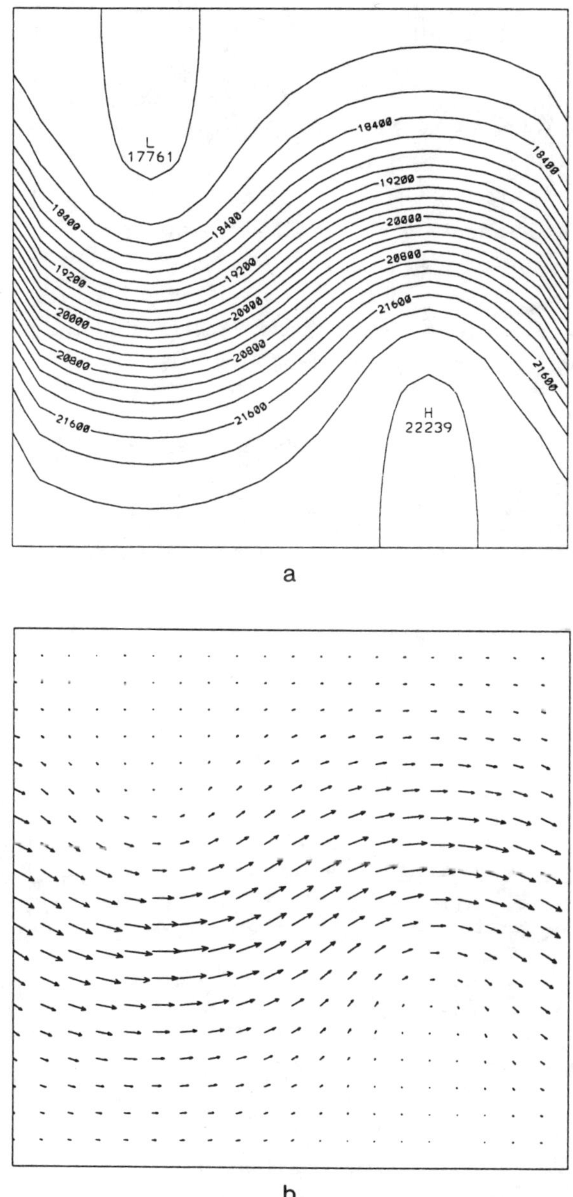

Figure 1: (a) Geopotential fields based on the Grammeltvedt initial condition and (b) the wind field calculated from the geopotential fields in Fig. (a) by the geostrophic approximation. Contour interval is 200 m^2s^{-2} and the value of maximum vector is 29.9 ms^{-1}.

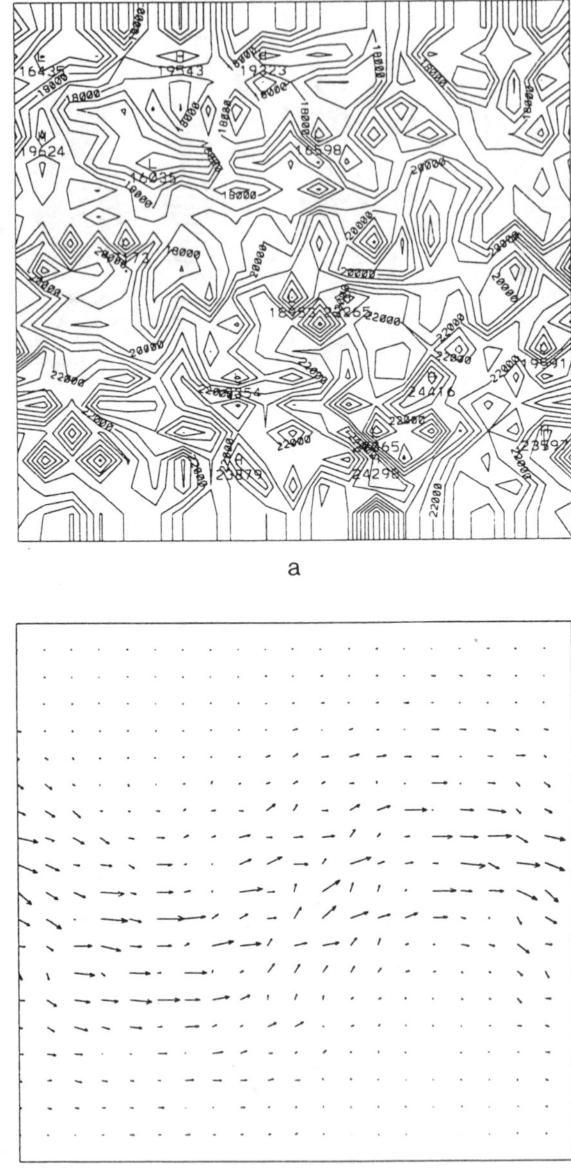

Figure 2: Random perturbation of (a) the geopotential and (b) the wind fields in Fig. 1. Contour interval is 500 m^2s^{-2} and the value of maximum vector is 54.4 ms^{-1}.

Table 1: Initial condition control problem in meteorology

algorithm	MXTITCG	Iter	Nfun	NCG	MTM (total CPU time)	FTM (function calls' CPU time)
TN1	3	19	20	50	12.20	11.31
TN1	50	20	26	54	13.79	12.89
TN1	3	63	64	170	38.82	37.15
(no prec.)	50	39	40	165	32.78	31.53
TN2	3	81	82	242	68.68	67.21
	50	4	5	91	16.41	16.30

Table 2: Meteorological problem with the limited memory quasi-Newton and truncated Newton methods

Algorithm	Iter	Nfun	MTM (total CPU time)	FTM (Function calls' CPU time)
E04DGF	72	203	36.89	33.56
L-BFGS	66	89	15.53	14.76
TN1	19	70	12.20	11.31
TN2	4	96	16.41	16.30

the truncated Newton method of Nash with MXITCG=50 in terms of Newton iterations, conjugate-gradient iterations, function evaluations, and CPU time. This strongly suggests that with a suitable preconditioner for the problem in meteorology, the truncated Newton method of Schlick and Fogelson might perform best. Thus, the use of preconditioning in the truncated Newton method of Nash accelerates performance as expected.

Table 2 presents the performance of the E04DGF, L-BFGS and the two truncated Newton methods. We observe from Table 2 that most of the CPU time is spent on function calls rather than in the minimization iteration. By comparing the number of function calls and CPU time, we find that the computational cost of L-BFGS is much lower than that of E04DGF. In 66 iterations with 89 function calls, L-BFGS converged. In contrast, E04DGF required 72 iterations and 203 function calls to reach the same convergence criterion. This produces rather large differences in the CPU time spent in minimization. L-BFGS uses less than half of the total CPU time required for E04DGF. The truncated Newton methods are competitive with L-BFGS and the truncated Newton method of Nash is slightly better than L-BFGS.

The differences between the figures showing the retrieved initial wind and geopotential and the one in Figure 1 are imperceptible (figures omitted). Table 3 gives the maximum differences between the retrieval and the unperturbed initial conditions. An accuracy of at least 10^{-3} is reached for both the wind and geopotential fields using any of the aforementioned four minimization codes.This clearly shows the capability of the unconstrained

Table 3: Maximum absolute differences between the retrieval and the unperturbaed initial wind and geopotential fields using the limited memory quasi-Newton and truncated Newton methods

Algorithm	$(u'^2 + v'^2)^{1/2}$	$\lvert\phi'\rvert$
E04DGF	7.5×10^{-3}	1.2×10^{1}
L–BFGS	3.8×10^{-2}	9.0×10^{-1}
TN1	8.9×10^{-3}	5.4×10^{1}
TN2	5.8×10^{-3}	4.1×10^{1}

limited-memory quasi-Newton and the truncated Newton methods to adjust a numerical weather prediction model to a set of "observations" distributed in both time and space.

Recently, Wang, *et al.* (1993) indicated that the performance of the truncated Newton method of Nash can be significantly improved by using a more accurate Hessian/vector product which is required when solving the Newton equation (26) to obtain the search direction. A second order adjoint model was integrated backwards in time instead of using a finite-difference approximation to obtain the value of the Hessian/vector product.

3 Incomplete observations

3.1 Uniqueness of the solution

In real situations, meteorological observations are temporally and spatially distributed, and are inhomogeneous both in quality as well as in quantity. The question arises as to what will happen if we have only a limited number of the observations in space and time. This is mathematically equivalent to determining the necessary condition for a unique solution of the problem of optimal control of distributed parameters.

The solution U^* for the problem (1) will be unique if the functional J is strictly convex., i.e. if

$$J(\lambda U + (1-\lambda)V) < \lambda J(U) + (1-\lambda)J(V), \quad \lambda \in [0,1], \; U, V \in R_N. \tag{45}$$

A geometric interpretation of (45) is that the graph of J is under any chord between U and V.

If J is only convex, i.e. the inequality in (45) is not strict, and then the minimum is not necessarily unique and the functional J may have no minima, one or several minima or a continuum of minima. J may also have stationary points which are not minima, but rather maxima or saddle points.

If J has a second derivative then J will be convex if its Hessian matrix H is positive semi-definite and J will be strictly convex if H is positive definite.

A difficulty encountered with non-convex functionals is that there is no global characterization of the absolute minimum and only a local analysis can be carried out.

Therefore, the uniqueness of the solution of minimizing problem (1) depends on the structure of the Hessian matrix H. (1) has a unique solution U^* if the Hessian matrix H is positive definite.

By definition, the functional J depends upon the observations, therefore, computing the Hessian of J, where J is defined by (1), permits one to estimate the link between the observations and the retrieved fields.

In the linear case, the forecast model can be written as

$$\frac{d\mathbf{x}}{dt} = \mathbf{A}\mathbf{x}, \tag{46}$$

where $\mathbf{A}$ is an $N \times N$ constant matrix independent of $\mathbf{x}$.

An explicit expression for the gradient $\nabla J(U)$ and the Hessian matrix $\mathbf{H}$ can be derived by

$$\nabla J(U) = P(0) = \sum_{r=0}^{R} e^{\mathbf{A}^T t_r} \mathbf{C}^T (\mathbf{C} e^{\mathbf{A} t_r} \cdot U - \mathbf{x}^{obs}), \tag{47}$$

$$\mathbf{H} = \sum_{r=0}^{R} e^{\mathbf{A}^T t_r} \mathbf{C}^T \mathbf{C} e^{\mathbf{A} t_r}, \tag{48}$$

where $\mathbf{H}$ is a symmetric matrix ($\mathbf{H}^T = \mathbf{H}$).

It was proved (Zou, *et al.*, 1992a) that $\mathbf{H}$ is positive definite if and only if the rank of

$$\mathcal{A} = \begin{bmatrix} C \\ CA \\ \cdot \\ \cdot \\ \cdot \\ CA^{n-1} \end{bmatrix} \tag{49}$$

is equal to N.

We see from (48) or (49) that the Hessian $\mathbf{H}$ is independent of the observation itself, depending only on the model (represented by the operator $\mathbf{A}$) and on the operator $\mathbf{C}$ mapping the meteorological variables into the space of observations. Therefore, the uniqueness of the solution is dependent on the model and the mapping operator.

In the nonlinear case, however, J is no longer quadratic with respect to the control variable U, where U represents model initial state. Due to the nonlinearity of the forecast model, no global results may be obtained. However since J is bounded from below we know that there exists at least one local minimum U^* which is characterized by $\nabla J(U^*) = 0$. A sufficient condition for U^* to be a unique local minimum is that $H(U^*)$ be positive definite. Since U^* is not known in practice, one can only estimate uniqueness of the solution by calculating the value of the Hessian at the initial guess point assuming that the initial guess is not far from the solution. Due to the large dimension of the control variable (10^4 to 10^6) in variational data assimilation, the computation of a full Hessian may be prohibitive for real applications. However, one can estimate the minimum eigenvalues $\lambda_{\min}(\mathbf{H})$ of the Hessian of the cost function by using an iterative power method or a Rayleigh quotient method (Golub and van Loan, 1989), both of which require only the value of the Hessian/vector product, which can be obtained by integrating the so-called second order adjoint model (Zou, *et al.*, 1992a, Wang, *et al.*, 1992). If $\lambda_{\min}(\mathbf{H}) > 0$, $\mathbf{H}$ is positive definite and the solution of J is unique.

3.2 Results with incomplete observations

In this subsection, some numerical experiments, similar to the one in Section 2.4, are carried out to estimate the influence of the distribution of available observations on the uniqueness of the solution and on the rate of convergence to the local minimum.

First we decreased the number of the observational fields. Suppose that only the wind fields (u and v) are observed and there are no observations of the geopotential field (ϕ). The numerical results show that the minimization of the cost function defined in (43) without the first summation was able to retrieve a unique minimum, i.e. not only perfect initial fields for the u and v components but also a smooth balanced ϕ field. If only the geopotential field is observed, the minimization of the cost function was also able to retrieve a unique minimum for u and v which is consistent (i.e. in geostrophic balance) with the Grammeltvedt (1969) initial ϕ field. Using a linearized shallow-water equations, one can derive that the rank of $\mathcal{A}$ (the linearized model operator) is equal to N for both cases. This shows that our numerical results are consistent with the theoretical conclusion of (49), which determines the sufficient conditions for the existence of a unique minimum of the cost function.

Next we decreased the number of grid points where observations are available. If we reduce the observational data in grid space (i.e. in the horizontal space dimension) by alternating the available observational data in both the x and y directions, for instance by assuming that data was available only every two grid points in each direction (i.e. the number of observations was decreased from 21×21 to 11×11), the minimization process fails. Namely after 71 iterations, the minimization stops due to rounding errors before a smooth solution was obtained. No reasonable retrieval is obtained and the objective function and the norm of gradient decrease by only about 2-3 orders of magnitude, which is roughly half of the orders of magnitude known to be necessary from our experiment to reach a smooth solution. If we continue to further decrease the number of observations in both space directions to every 4 grid points for instance, no significant difference from the case of every 2 grid points is observed and also no satisfactory convergence and retrieval are obtained. It seems that the performance of the minimization depends on the number of available observations at the grid points. At least, this is the case for the limited-area shallow-water equations random perturbation test, where the control variables are only the initial conditions.

One could wonder whether the results from the experiment in which data is given every two grid points could be partially improved if instead of a random perturbed field, a first guess field lacking small scale structure, e.g. a complete flat field, was used to start the minimization. Minimization starting from a flat initial guess yields similar results. This shows that the failure of the minimization with observations available only every two grid points is independent of the fact that we used a randomly perturbed noisy initial guess.

Finally observational data are reduced in time dimension. Suppose that observations are available, say for instance, every 2, 10, 30 or 60 time steps instead of at each time step. We observe that the performance of the minimization is very similar to the case where observations were available at every time step, with the convergence rate of the former being slightly slower than that of the later. A satisfactory retrieval of initial wind

Table 4: Values of the maximum and minimum eigenvalues (λ_{max}, λ_{min}) and the condition number ($\kappa(\mathbf{H})$) of the Hessian of the cost function at the initial guess point using the power method and the shifted power method (with the convergence criteria: $|\lambda_{k+1} - \lambda_k| < 10^{-8}$)

Problems	λ_{max}	λ_{max}	$\kappa(\mathbf{H})$
complete observations	2.20×10^{-4}	4.89×10^{-7}	449.2
less observations in time space (30 steps)	1.84×10^{-4}	1.08×10^{-7}	1695.5
less observations in grid space (2 grids)	2.90×10^{-4}	-1.66×10^{-5}	17.53

and the geopotential height fields is obtained even for all the cases. The difference fields between the retrieved and unperturbed initial wind and geopotential are several orders of magnitude smaller than the initial fields.

In Table 4 we calculated the maximum and minimum eigenvalues and the condition number of the Hessian at the initial guess of the minimization and we find that the reduction of observations in the grid space produces a negative minimum eigenvalue of the Hessian. Thus, we obtain an indefinite Hessian, pointing to the existence of either multiple minima or saddle points. The condition number in the case that observations are available at each grid point but only every 30 time steps (case 2) is larger that in the case of complete observation, which explains that the convergence rate in the case of less observations in time space is slightly slower than that in the case of complete observations in time space.

4 Control of gravitational oscillations

Gravitational oscillations with unrealistically large amplitudes may be generated in numerical models due to the insertion of erroneous data or due to errors in models. Methods aimed at the specification of suitably balanced initial fields, which will not give rise to spurious gravity oscillations, are known as initialization methods. Variational data assimilation, by its nature, can be used to assimilate data which was not previously initialized. Problems arise as to how to efficiently control high frequency gravity wave oscillations in variational data assimilation when assimilating noisy data.

Courtier and Talagrand (1990) indicated that gravity wave noise can be efficiently eliminated by adding a penalty term, defined by the normal modes of the model, to the cost function and carrying out the variational data assimilation. The penalty method follows the logic of tending to reduce the time tendency of the gravity wave component of the flow to a very small value. However this approach involves a sizable computational effort, namely, it requires the availability of both the normal modes of the model and that of the adjoint of the normal mode initialization (NMI) procedure. Other penalty terms controlling the total time tendency of the flow were proposed by Zou *et al.* (1992a, 1993a).

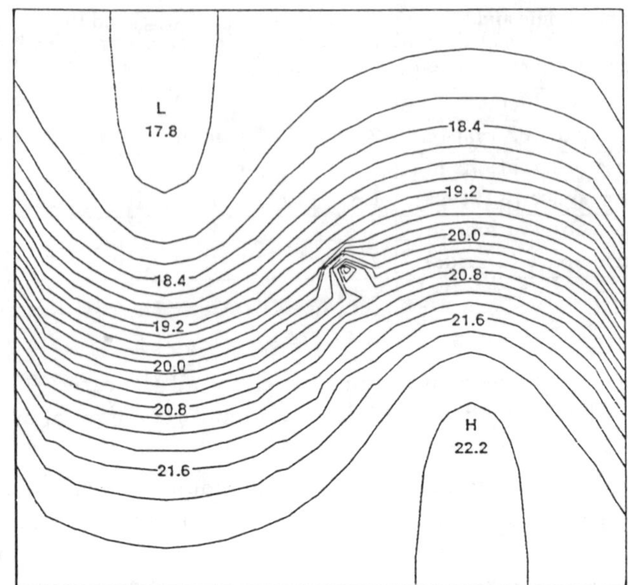

Figure 3: Case no. 1 initial geopotential field produced by one grid point local perturbation on the Grammeltvedt initial geopotential field condition no. 1. Contour interval is 200 m^2s^{-2} and values on the isolines are scaled one grid point local perturbation on the Grammeltvedt initial geopotential field condition no. 1. Contour interval is 200 m^2s^{-2} and values on the isolines are scaled by 10^3.

The proposed method requires less memory and computational effort than the method of combined penalty and NMI proposed by Courtier and Talagrand (1990).

In the following, both the shallow-water equations model and the primitive equations model are used to test the performance of the penalty method on controlling gravitational oscillations.

4.1 A 2-dimensional problem

In this subsection experiments similar to the one described in Section 2.4 are carried out using the shallow-water equations model. In order to assess and compare the performance of the quadratic penalty method in damping the small scale gravity waves present in the initial fields, a small perturbation was added to the initial height field defined in Section 2.4 in the following way

$$\phi'(i,j) = \phi(i,j) + \alpha_{i,j} \times \phi(i,j), \quad i,j = 8,12 \tag{50}$$

where $\alpha_{10,10} = 0.9\%$, $\alpha_{i,j} = 0.45\%$ when i or j equal 9 or 11, and $\alpha_{i,j} = 0.3\%$ when i or j equal 8 or 12, where i and j relate to grid locations in the x and y directions, respectively.

The isoline graphs of the perturbed height fields ϕ' are plotted in Figure 3. We see that the height field thus obtained contains one local grid point perturbation. This perturbed initial condition was used to produce model-generated observations. The variational data assimilation with or without the penalty term is implemented by starting from a model atmosphere at rest and minimizing the cost function defined in (43). When the prescribed convergence criterion

$$\|\vec{g}_k\| \leq \epsilon \max\{1, \|\mathbf{x}_k\|\}, \tag{51}$$

is met, the minimization process was stopped, where $\vec{g}_k$ is the gradient of the cost function at the k-th iteration and $\epsilon = 10^{-14}$, a number close to the machine accuracy of the CRAY-YMP, was chosen based on expecting to obtain a perfect solution, i.e. one in which no visible difference with the initial condition used to produce the observations to be assimilated is observed. However the weights used to define the cost function may also be interpreted as variances of the data. Errors in the retrieved fields with this convergence criteria are, however, much smaller than the error implied by these weights. The cost function is reduced more than the value of the sum of mean square data errors and it corresponds to an overfitting. This is quite acceptable in an "identical twin" approach with an abundance of data available which corresponds to our case. In a real data experiment where model and data might be inconsistent (biased), this would lead to the appearance of spurious large-amplitude high-frequency gravity-wave noise in the solution.

Now, instead of minimizing the cost function defined in (43), a penalized cost function J_r is minimized, where J_r is defined as

$$J_r = J + r \left\| \frac{\partial \phi}{\partial t} \right\|^2 . \tag{52}$$

The gradient of the penalized cost functions can be easily calculated by integrating the same adjoint model with different forcing terms added to the adjoint model, that is, besides adding the usual forcing terms $2W(t_r)(\mathbf{x}(t_r) - \mathbf{x}^{obs}(t_r))$ in the adjoint model whenever an analysis time t_r $(r = 0, 1, \cdots, R)$ is reached, we also add at every time step the following additional terms

$$2r\mathbf{A}^* \frac{\partial \phi}{\partial t}, \tag{53}$$

to the right-hand side of the adjoint model.

For comparison, a minimization of the cost function J_0, which does not include a penalty term, was carried out to serve as a benchmark. The minimization terminated successfully after 104 iterations and 153 function calls. The cost function decreased by 10 orders of magnitude while the norm of gradient decreased by 6 orders of magnitude. The variational assimilation was thus able to perfectly retrieve the original initial conditions (not shown).

Then a number of minimizations are carried out which minimize J_r, $(r \neq 0)$ to damp the perturbation shown in Figure 3. One might have thought at the beginning that the gravity waves may be damped to any desired accuracy by simply choosing a very large value of the penalty parameter r, and then carrying out a single unconstrained minimization. However from our experiments, we found out that for smaller values of the penalty parameter r (say for instance $r < 10^5$) the retrieved geopotential field was not smooth enough, while for larger values of r $(> 10^5)$ the minimization failed. For a value of $r = 10^5$, a single minimization cycle performed very well at first but after

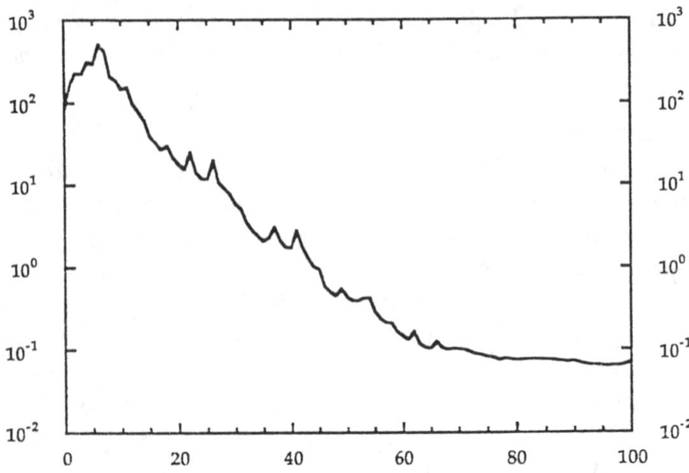

Figure 4: Variation of the value of $||\partial\phi/\partial t||^2$ with the number of iterations during the minimization of the penalized cost function with a penalty parameter $r = 10^5$.

105 iterations with 141 function calls, the minimization failed again before the same convergence criteria (51) was attained. However, upon examination of the geopotential field after 105 iterations, we found out that a smooth initial geopotential field has already been obtained. A measure of the efficiency of the penalty method is given by the following corresponding variation of the quantity $||\partial\phi/\partial t||^2$ which decreased from $0.77 \times 10^2 m^4 s^{-6}$ to $0.70 \times 10^{-1} m^4 s^{-6}$ in the course of the minimization (see Figure 4), i.e. a decrease of three orders of magnitude in the squared time tendency of the geopotential field, pointing out to a successful filtering of high-frequency gravity-waves.

The sensitivity of the performance of the minimization for different values of penalty parameter r is not surprising since it is well-known that the condition number of the Hessian matrix evaluated at the minimum increases as r becomes larger (the so-called ill-conditioning effect of the penalty method) (Gill, *et al.*, 1981, see also Appendix A of the paper by Zou *et al.*, 1992a). The conditioning of the Hessian matrix has a special significance in this case. If the initial value of r is "too large", even a robust unconstrained minimization algorithm (as the one used in our case) will typically experience great difficulty in its attempts to compute the minima due to the slow convergence induced by the increasingly larger condition number of the Hessian of the penalized cost function. Therefore, in order to solve the problem by a penalty function method, a sequence of unconstrained minimization problems has to be solved, with moderately increasing values of the penalty parameter (a method called the exterior quadratic penalty method, see Fiacco and McCormick, 1968 and Rao, 1983). In the exterior penalty function method, each successive $\vec{x}^*(r_l)$ is used as the starting point for solving a minimization problem with the next increased value of the penalty parameter, until an acceptable convergence criterion (i.e. a satisfactory solution) is attained for the first time (where $\vec{x}^*(r_l)$ is the minimum point obtained using the preceding penalty parameter r_l). Another elegant method avoiding the ill-conditioning of penalty methods is the augmented Lagrangian method (Navon

Table 5: Numerical results of the variational data assimilation with application of the exterior penalty function method to assimilate observations with one local perturbation (4 cycle)

k	r_k	Iter	Nfun	J^*/J_0	$\lvert\nabla J^*/\nabla J_0\rvert$	ϵ
1	10^2	48	74	4.46×10^{-6}	1.41×10^{-4}	10^{-12}
2	10^3	16	26	7.85×10^{-1}	1.76×10^{-1}	10^{-13}
3	10^4	19	28	4.20×10^{-1}	1.02×10^{-1}	?5×10^{-13}
4	10^5	26	37	3.96×10^{-1}	3.13×10^{-2}	10^{-14}

and de Villiers, 1983). An exterior penalty function numerical experiment with $r_1 = 10^2$, $c = 10$, $l = 1,4$ was carried out, in which each subsequent minimization cycle uses an increasingly stringent convergence criteria as shown in Table 5 (also see Navon and de Villiers, 1983). Again most of the large scale characteristic features are retrieved after the first minimization cycle (Figs 5a-b). The gravity wave oscillations were damped out gradually till the fourth minimization cycle ended with $r_4 = 10^5$. The solution for the geopotential field is practically the same as that in previous exterior experiment. The total number of iterations and function evaluations are 102 and 165, respectively. The computational cost of this procedure is comparable with the computational cost of the base minimization.

If we examine the performance of the above experiments, one can easily find out that the minimization with a small penalty term or without a penalty term at all but with a looser convergence criteria can retrieve most of the large-scale features of the initial condition fields, while a minimization with larger penalty terms can damp out almost all the small scale gravity wave features present in the meteorological fields. This implies that only two successive minimizations are required to obtain a smooth solution devoid of gravity wave oscillations.

We conducted therefore, another experiment in which the first unconstrained minimization was carried out without a penalty term being included in the cost function and which was terminated when a convergence criteria of 10^{-12} (instead of the stringent one of 10^{-14}) was reached. The minimization solution of the case where no penalty term was included is then taken to serve as the first guess for the next minimization cycle with a large penalty term being included in the cost function ($r = 10^5$). The minimization of the penalized cost function was then terminated when a convergence criterion of $\epsilon = 10^{-14}$ was reached. We call this two-stage minimization the "short-cut" penalty method. We found that the distribution of the retrieved initial height fields after the two minimizations is the same as in Figure 5(d). As far as the computational cost is concerned (see Table 6), this experiment required a total of only 86 iterations and 120 function calls for the whole procedure, i.e. it was computationally cheaper than the base experiment.

To illustrate the effectiveness of this two stage minimization procedure on damping the gravitational oscillations, a number of 24 h forecasts (14 h real forecast due to a length of 10 h assimilation window) starting from the variationally data assimilated initial state with and without the inclusion of a quadratic penalty term in the cost function were carried out. In Figure 6 we plot the time variation of the geopotential field at a given

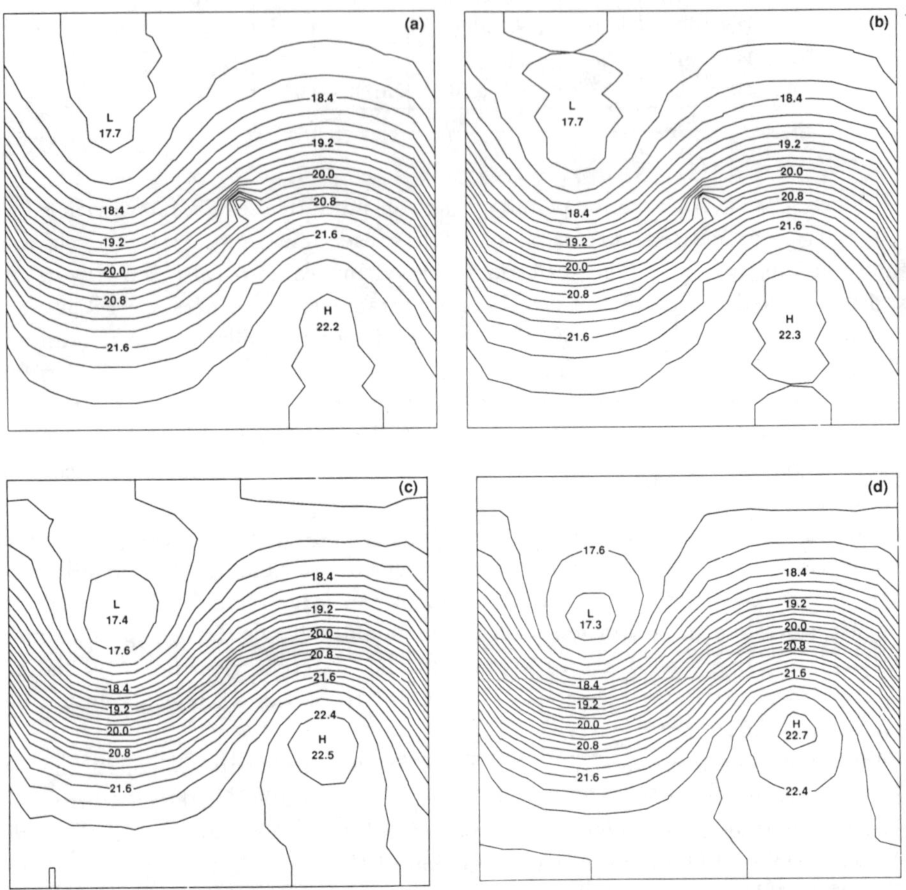

Figure 5: Retrieved initial geopotential height after each cycle of the exterior penalty function method with a penalty parameter set to (a) $r_1 = 10^2$, (b) $r_2 = 10^3$,(c) $r_3 = 10^4$ and (d) $r_4 = 10^5$, respectively. The Case No. 1 observations were assimilated, where the initial guess for the first minimization cycle is set to zero, the solution of the ith penalized minimization was used as the initial guess for the $(i+1)^{th}$ minimization, $i = 1, \ldots, 3$. Contour intervals and scaling factor for the values on isolines are the same as in Figure 3.

Table 6: Numerical results of the variational data assimilation with combination of the non-penalized assimilation and penalized assimilation to assimilate observations with one local perturbation

k	r_k	Iter	Nfun	J^*/J_0	$\|\nabla J^*/\nabla J_0\|$	ϵ
1	0	52	75	4.46×10^{-6}	1.41×10^{-4}	10^{-12}
2	10^5	34	45	7.85×10^{-1}	1.76×10^{-1}	10^{-14}

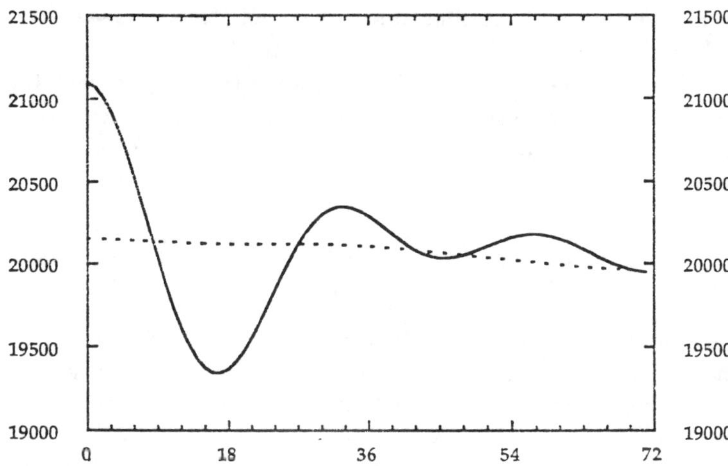

Figure 6: Time variation of the geopotential fields for 24 h at grid point (10,10) after variational data assimilation with (dotted line) and without (solid line) a penalty term being included in the cost function when the Case No. 1 observations are assimilated.

grid point for a variationally assimilated integration without the inclusion of a penalty term as well as the results from the integration with the penalty term being included in the cost function. From the the hourly evolution of the geopotential field for the short-range forecast (24 h) with the penalized variational data assimilation and a run without the inclusion of a penalty term, we observe the occurrence of a major impact for the first 6-10 hours of the numerical integration. A perfect damping of the short gravity waves is observed, matching in every respect similar results obtained by the use of nonlinear normal mode initialization. However the present method has several advantages, the first being the fact that no explicit knowledge of the normal modes of the model is required, the second being that by using the penalty function approach it is possible to take into account data distribution in time. However, the variational data assimilation procedure itself (without the inclusion of any penalty term) remains computationally expensive. The initialization here results as a by-product of the adjoint model integration and involves little additional computational cost beyond the computational cost required by the variational data assimilation.

The results obtained compare favorably with those obtained by other researchers using implicit normal mode initialization of the shallow-water equations model over a limited area domain (See Juvanon du Vachat, 1986; Semazzi and Navon, 1986; and Temperton, 1988).

4.2 A 3-dimensional problem

Variational 4-D data assimilation combined with a penalty method constraining time derivatives of the surface pressure and divergence is implemented on an adiabatic version of the NMC 18 level primitive equations spectral model (NMC, 1988) with surface drag and horizontal diffusion.

High-frequency fluctuations in the surface pressure tendency reflect the presence of external gravity waves. A way to suppress these external gravity waves consists in imposing as a constraint the vanishing of the time tendency of the surface pressure: $\frac{\partial \ln p_s}{\partial t}$. A quadratic penalty term of the form

$$J_{r_1} = r_1 \sum_{i=0}^{R-1} \left(\frac{\partial \ln \vec{p}_s(t_i)}{\partial t} \right)^T \left(\frac{\partial \ln \vec{p}_s(t_i)}{\partial t} \right), \tag{54}$$

was thus augmented the cost function J, where $\ln \vec{p}_s$ is an M dimensional vector of the logarithm of the surface pressure, M is the total number of Gaussian grid points used in the definition of the cost function, and R is the number of time steps in the time interval spanning the window of assimilation. Variational data assimilation with a penalty term J_{p1} included in the cost function means that while minimizing the distance between the model solution and data, the high frequency oscillations presented in the surface pressure field are also controlled.

We may also impose a constraint on the time tendency of the divergence to damp high-frequency oscillations of internal gravity waves in a manner similar to (54). We define a penalty term

$$J_{r_2} = r_2 \sum_{i=0}^{R-1} \left(\frac{\partial \vec{D}(t_i)}{\partial t} \right)^T \left(\frac{\partial \vec{D}(t_i)}{\partial t} \right), \tag{55}$$

where $\vec{D}$ is the $M \times K$ dimensional vector of divergence and K is the total number of vertical levels.

The penalized cost function thus assumes the following form

$$J_{r_1,r_2} = J + J_{r_1} + J_{r_2}. \tag{56}$$

The gradient of the penalized cost function can be obtained by integrating the same adjoint model as the one used in obtaining the gradient of the non-penalized cost function. The only difference is that more forcing terms are added to the right-hand-side of the adjoint equations model besides the weighted differences between the model solution and data (Zou, *et al.*, 1993a).

Numerical experiments are devised as follows: initial conditions for the model and analyzed data were obtained from the Global Data Assimilation scheme at NMC. The variational data assimilation was carried out adiabatically, as well as with a surface drag and horizontal diffusion terms included in the model. Two uninitialized state vectors, separated by a 6 hour time interval, were taken as analyzed data to be assimilated. The cost function was defined as

$$J(\mathbf{x}(t_0)) = \frac{1}{2}\left(\mathbf{x}(t_0) - \mathbf{x}^{obs}(t_0)\right)^T \mathbf{W}\left(\mathbf{x}(t_0) - \mathbf{x}^{obs}(t_0)\right) + \frac{1}{2}\left(\vec{X}(t_R) - \vec{X}^{obs}(t_R)\right)^T \mathbf{W}\left(\vec{X}(t_R) - \vec{X}^{obs}(t_R)\right) \quad (57)$$

where $\mathbf{x}$ is the model state vector, $\mathbf{W}$ is a diagonal weighting matrix whose values are calculated as the inverse of the maximum squared difference between the two analyzed data fields at times t_0 and t_R. The time difference between the two analyzed times is 6 hours. A limited-memory quasi-Newton method (Navon and Legler, 1987) due to Liu and Nocedal (1989) was employed throughout the minimization experiments.

In order to illustrate the performance of the penalty method we calculated the amount of residual gravity wave component in the retrieved initial state, which was 0.076. The amount of residual gravity wave component before and after the application of nonlinear NMI are 2.56 and 0.15 respectively, which provides a benchmark here for assessing the efficiency of filtering gravitational oscillations by the simple penalty method developed within the framework of variational data assimilation. We see that the solution of the minimization with both penalty terms J_{r_1} and J_{r_2} being added to the cost function contains a lesser amount of residual gravity component than the one contained in the nonlinear NMI initialized data.

The minimization of J_{r_1,r_2} with $r_1 = 10^{13}$ and $r_2 = 10^{19}$ performed very well with both J_{r_1}/r_1 and J_{r_2}/r_2 decreasing by about 2 orders of magnitudes. To show that the gravitational oscillations were damped out after minimizing the penalized cost function J_{r_1,r_2} and to confirm that after the high-frequency gravity wave oscillations have been completely eliminated by the penalty procedure while the Rossby modes are not degraded, and furthermore that once these oscillations have been eliminated, they do not seem to re-occur, a closer inspection of the integration results was carried out.

A 3 days forecast was carried out from the nonlinear NMI data and the retrieved initial states without a penalty term and with the two penalty terms, using the adiabatic version of the NMC spectral model with surface drag and horizontal diffusion. Figure 7 shows the time variation of the surface pressure at Point 1 (a point in the Indian Ocean) and Point 2 (a point near the Rocky Mountains) during this period. It is seen that the value from the retrieved initial state without a penalty term oscillates around a slowly varying value from the retrieved initial state with penalty terms at both points (solid and dotted lines in Figure 7). This fact indicates that the gravity oscillations present in the integration with the non-penalized retrieval neither grow nor influence significantly the slowly varying part of the fields. This fact is confirmed by a comparison between the penalization result and the nonlinear NMI result (dashed line in Figure 7). The time variation of the surface pressure is nearly the same at Point 2 for the results obtained from the retrieved initial state with penalty terms and from that obtained from the nonlinear normal mode initialized state, with the former being even smoother than the latter. At Point 1 the time variation of the penalty result is smooth and exhibits a 12 hours oscillation. However, after nonlinear NMI both the large amplitude 12 hours oscillations

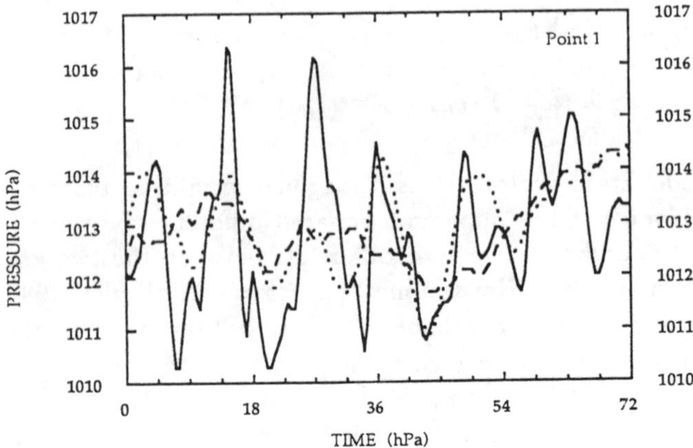

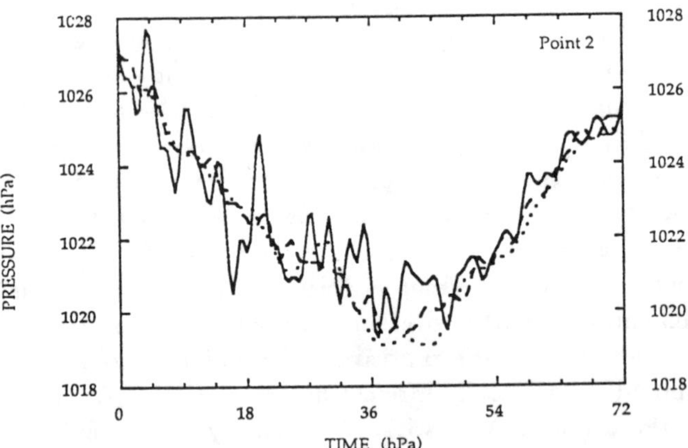

Figure 7: Time variation of 3 days forecasts at Point 1 and Point 2. Solid line: no penalty term. Dotted line: with penalty terms $J_{r_1} + J_{r_2}$. Dashed line: nonlinear NMI.

and the low amplitude high-frequency oscillations are reduced. After two and half days the time variation is nearly same.

In view of these experimental results, we are led to the conclusion that adequate penalization of the cost function is able to suppress undesirable high frequency gravity wave oscillations. The constraints of requiring the vanishing of the surface pressure and divergence at each time step were implemented weakly by the penalty method. The Rossby waves are not affected by the penalty procedure. It is also worth noting that very little additional computational cost is required to calculate the values of the penalized cost function and its gradient, and that the number of iterations and function evaluations is about the same as that required by the minimization without a penalty term. The advantage of using of simple penalty terms over penalty terms including the model normal modes results in a simplification of the procedure, allowing a more direct control over the model variables, and the possibility of using weak constraints to eliminate high frequency gravity wave oscillations. This approach does not require direct information about the model normal modes. One of the encouraging results obtained is that the introduction of the penalty terms does not slow down the convergence rate of the minimization process.

5 Inclusion of "on-off" processes in variational data assimilation

In the last several years most attention on variational methods for 4-D data assimilation has been focused on 2-D models (LeDimet and Talagrand, 1986; Derber, 1985; Lewis and Derber, 1985; Hoffman, 1986; Talagrand and Courtier, 1987; Courtier and Talagrand, 1987; Zou *et al*, 1992a) or 3-D adiabatic models (Thépaut and Courtier, 1992, Navon *et al.*, 1992 , and Chao and Chang, 1992). The issue of the impact of "on-off" threshold processes on variational data assimilation was addressed only recently (Douady and Talagrand, (1990); Zou, *et al.*, 1993c; Vukićecić (1993); Zupanski, 1993), when the inclusion of the adjoint of diabatic processes is considered.

The specific problem related to the presence of "threshold", or "on-off" processes, triggered when a given parameter attains a prescribed critical value, is how to code the adjoint of these processes and how will the minimization process behaves for locally nondifferentiable terms. The mathematical modeling of these physical processes defies a convenient quantification at the level of the resolution presently used in numerical forecast models. Different parameterization schemes are used and some involves special computational treatment due to the complicated nonlinear relationship existing between variables and the high cost of the computer time. Examples of these are, in the deep cumulus convection, the calculation of the dew point temperature from the vapor pressure, the calculation of the equivalent potential temperature (θ) from temperature and pressure fields at the lifting condensation level (LCL), and the derivation of the temperature and specific humidity of an ascending parcel from its potential temperature and pressure. In order to alleviate the computational load, the operational model employs several pre-calculated tables along with linear or bilinear interpolations. For the tangent linear and adjoint models, one can no longer use the original tables which were employed in the direct nonlinear model and linearize the nonlinear model technically and locally in the code as one does in developing the tangent linear model of the adiabatic model. The values in the tables are built for the nonlinear calculation. The original (analytical) formulas used to build the lookup tables are required to compute the first derivatives. This is quite different from the development of the adjoint of the adiabatic model, where the only nonlinear terms present are those due to the quadratic terms. The "on-off" processes appearing in the code as IF statements depend on the model variables' value which produce a finite number of first-order discontinuous points. These points represent the "on-off" switches in time and space domains. The adjoint and gradient calculations assumes that the initial small perturbation does not change when and where the "on-off" switch occurs. Therefore, the adjoint model integration follows the same path backward as the nonlinear model's integration, i.e. the switches are determined by the basic state.

Once these problems of building the adjoint and calculating the value of the gradient of the cost function defined by a model constraint involving "on-off" physical processes are solved, the following questions arise: what will the computational load due to these processes be, and whether the minimization will still work for first order discontinuous points and moreover what will be the influence on the solution of the optimized initial conditions?

As an initial estimation to these problems, some preliminary experiments of variational data assimilation were carried out using a version of 18 vertical levels, triangularly truncated at wavenumber T40, of the NMC operational spectral model. In the experiments, the "observations" consisted of two complete states of vorticity, divergence, temperature, surface pressure field and moisture from the NMC global operational data assimilation system, 6-hour apart, which are denoted by $x^{\text{obs}}(t_0)$ and $x^{\text{obs}}(t_R)$. The adjustment was performed on the 6-hour interval $[t_0, t_R]$ preceding t_R. The problem is to find an optimal initial state $x^*(t_0)$ which minimizes the cost function J given by (57).

The computational cost of calculating the gradient of the cost function using the adjoint technique is independent of the dimension of the problem, meaning

$$\frac{\mathcal{T}(J, \nabla J)}{\mathcal{T}(J)} \leq c \tag{58}$$

where $\mathcal{T}(J)$ and $\mathcal{T}(J, \nabla J)$ measure the CPU time to compute the cost function J and the pair of $(J, \nabla J)$, respectively, and c is a constant independent of n. This constant is small, between 2 and 4, depending on the number of operations performed in the direct model. Table 7 displays the CPU time used by the NMC T40, 18 layer model, its tangent linear model and its adjoint model for a six hour time integration, with or without the large-scale precipitation and deep cumulus convection, respectively. The time step is 1800s. All the "on-off" routes and the nonlinear model solutions at each time step were stored in memory during the direct integration to avoid major recomputation costs in the adjoint. We observe that the required CPU time increased from 16.65s for an adiabatic model to 29.15s for the partially diabatic model due to the "on-off" physical processes. The tangent linear model integration is slightly more expensive than that of the nonlinear model integration, while the integration of the adjoint model is most expensive but still less than twice the CPU time required for the integration of the nonlinear model.

Table 7: CPU time used by the NMC T40, 18 layer primitive equation model, its tangent linear model and its adjoint model for a six hour time integration with a time step of 1800 seconds

CPU time (second)	nonlinear model	linear model	adjoint model
adiabatic	16.65	21.79	25.14
with "on-off" processes	29.15	31.10	34.49

Table 8: CPU time used for calculating the cost function J and its gradient with respect to the initial state ∇J

CPU time (second)	$\mathcal{T}(J)$	$\mathcal{T}(\nabla J)$	$\mathcal{T}(\nabla J)/\mathcal{T}(J)$
adiabatic	17.31	27.35	2.6
with "on-off" processes	30.01	37.73	2.3

Table 8 displays the values of c for the gradient calculation. For the adiabatic version of the model, c is equal to 2.6. When the "on-off" physical processes are included in the direct model and the corresponding adjoint model, $c = 2.3$. In addition, we note that the cost function calculation adds 1 second of CPU time, in both cases, to the model's integration. An additional 3 seconds are required for including the forcing term in the adjoint model. Therefore, the direct access of data in the nonlinear model and adjoint model, for both the model solution and the "on-off" route, is computationally very efficient.

The first test consisted of two parallel assimilations with and without the large-scale precipitation and cumulus convection, both including the horizontal diffusion and a simplified surface drag (see Navon et al., 1992). The minimization was found to be similar in both cases with the value of the cost function decreasing to about 15% of its original value. The norm of gradient decreased about two orders of magnitude. The total CPU times required for 60 iterations of the minimization with and without the physical processes were 76 and 44 minutes, respectively. The maximum memory used was 7.4MW and 6.4MW respectively.

From this experiment, we conclude that "on-off" threshold processes when included in the model do not hinder the minimization process. Theoretically, if the gradient of the cost function calculated by integrating the adjoint model is always a one-sided gradient, i.e. the model integration follows the same "on-off" route at each iteration, the large-scale unconstrained minimization will still work. We note however that this is not the case in our minimization experiment, and despite this fact, the minimization process performed successfully. Figures 8(a–c) show three examples of the large-scale saturation, evaporation, and the cumulus convection during the first 3 iterations of minimization, respectively. In

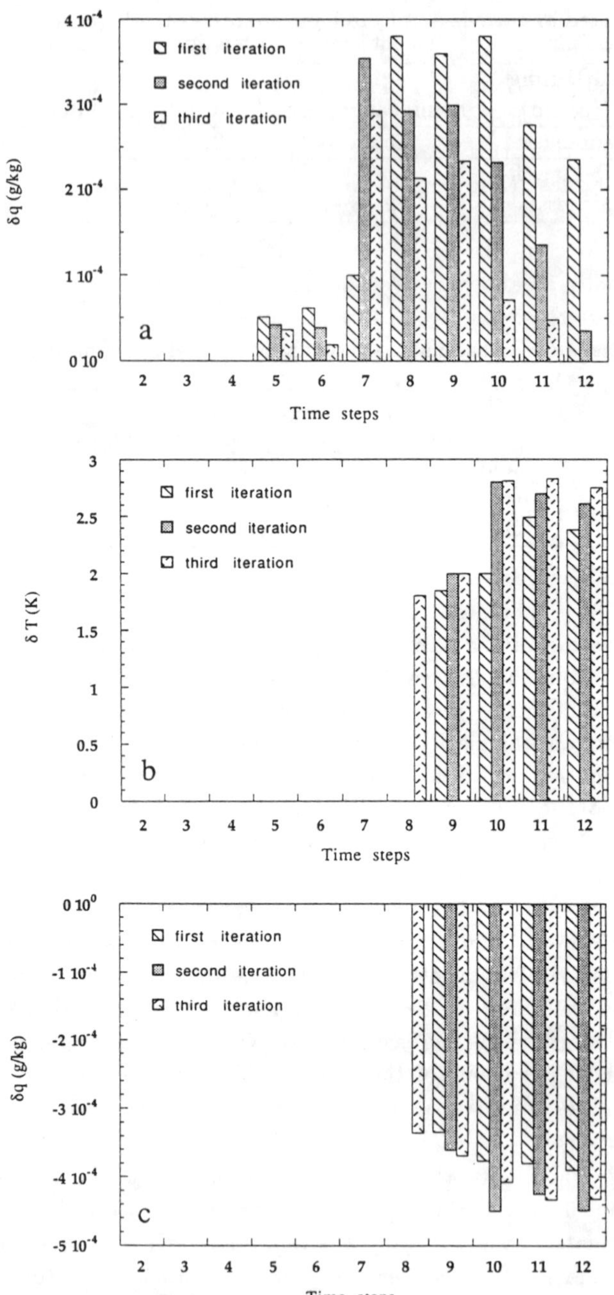

Figure 8: (a) Large scale supersaturation amounts at the point $(90^oE, 35^oN, 225mb)$ as a function of time. (b) Cumulus convective heating of the environment at the point $(25^oW, 38^oN, 920mb)$ as a function of time. (c) Cumulus convective moistening of the environment at the point $(25^oW, 38^oN, 920mb)$ as a function of time.

Figure 8(a) we plotted the amount of δq_c (the increment of specific humidity due to the large-scale precipitation process) when large-scale saturation occurs ($q > q_s$) at a model grid point located at $(90^o E, 35^o N, 225mb)$. We observe that the saturation process was turned on from the fourth leap-frog time step during the first and second iterations. During the third iteration, this process was also turned on at the fourth time step but it was turned off at the last time step of the assimilation window. We can draw a similar conclusion from Figures 8(b,c) for the cumulus convection process.

In the following, we compare the differences between the retrieved initial states from the adiabatic and partially adiabatic runs. The root-mean-square of the retrieved divergence field was found to increase slightly when the large-scale precipitation and the cumulus convection processes were included in the variational data assimilation (Figure omitted).

In order to observe the differences in the retrieved initial conditions with or without the inclusion of the precipitation processes, we now consider the horizontal distribution of the model's variables displayed over an area encompassing North America and plotted in Figures 9(a,b) the difference of the divergence, vorticity, temperature and the moisture fields at high and low levels respectively. In the area where the divergence in the higher layers and the convergence in lower layers increases due to the inclusion of the "on-off" processes in the minimization, (we shall call it the active region), the temperature and the vorticity were also found to have increased. At the lower levels, as the convergence increased, the temperature decreased and the moisture increased. In addition, a northwest-southeast wave train is observed in the vorticity field which propagates toward the equator.

Figure 10 displays the retrieved initial stream functions near 500mb (level 9). We observe that there is a trough upstream of the active region, and that the differences in the minimization solution with or without the physical processes are strongly correlated with the weather system.

We also carried out 6 h forecasts of the precipitation from the two initial conditions (Figure 11). We find that the initial condition obtained by the variational data assimilation including the "on-off" processes produces larger amounts of precipitation and greater area coverage over most of the map. However, the amount of precipitation near Lake Erie is smaller than the amount forecasted from the initial condition obtained without the inclusion of the physical processes. Since we do not have verifying data, we cannot conclude which result is better. We have however demonstrated that differences near meteorologically active areas do result from the two respective minimizations.

D at 856 mb

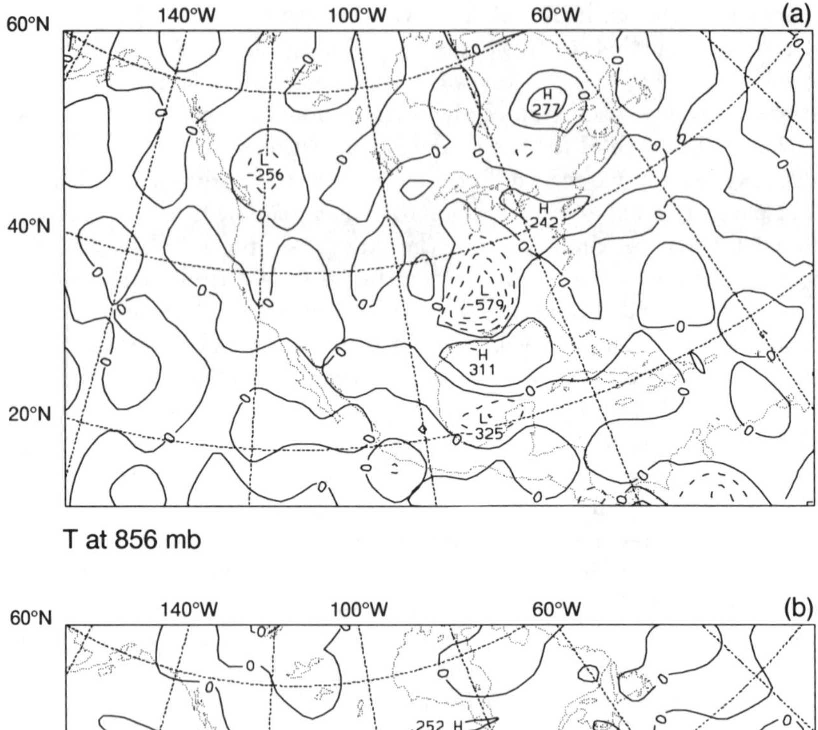

T at 856 mb

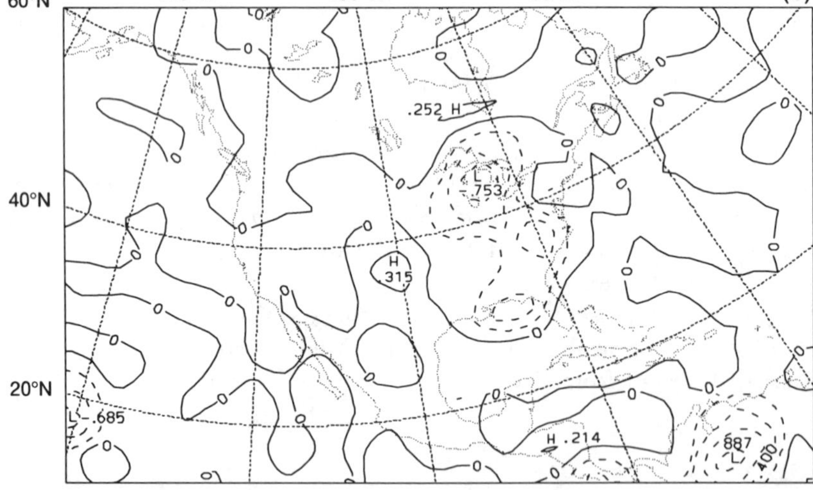

Figure 9: The differences of divergence(D), temperature(T), vorticity(ζ) and the moisture(q) fields at (a–d) the 856mb level and (e-h) the 225mb level between the two retrieved initial states. Contour intervals for divergence, vorticity, temperature and moisture are $8.0 \times ^{-7} s^{-1}$, $2.0 \times 10^{-6} s^{-1}$, $0.07K$ and $8.0 \times 10^{-5} gkg^{-1}$ in the upper layer and $1.5^{-6} s^{-1}$, $5.0 \times 10^{-6} s^{-1}$, $0.2K$ and $8.0 \times 10^{-5} gkg^{-1}$ in the lower layer respectively. Values on the isolines of divergence and moisture fields are scaled by 10^8 and 10^6 respectively. Values on the isolines of vorticity in the upper and lower layers are scaled by 10^8 and 10^7 respectively.

ζ at 856 mb

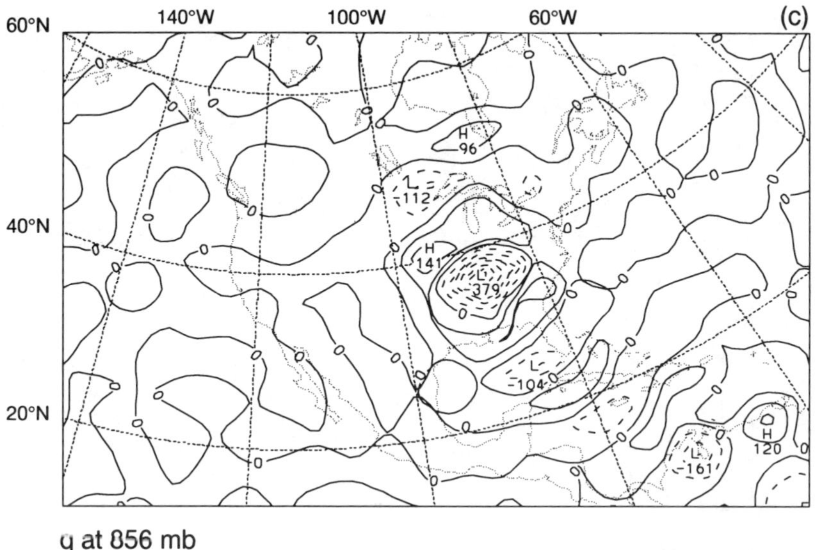

q at 856 mb

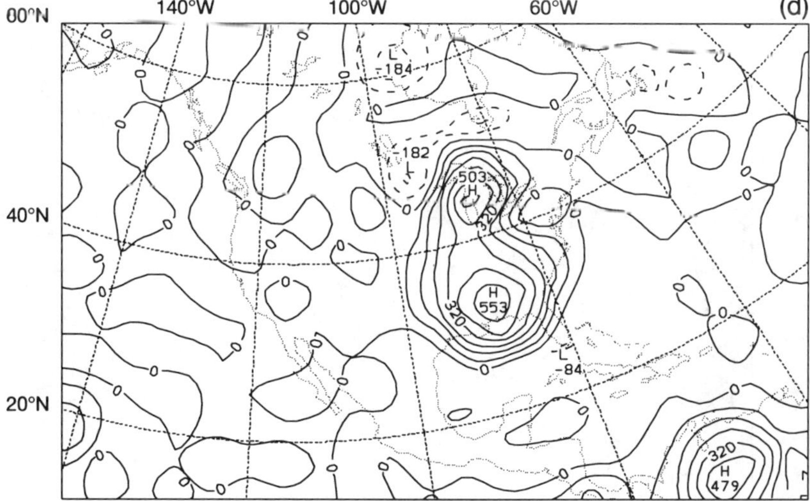

Figure 9: (Continued)

D at 225 mb

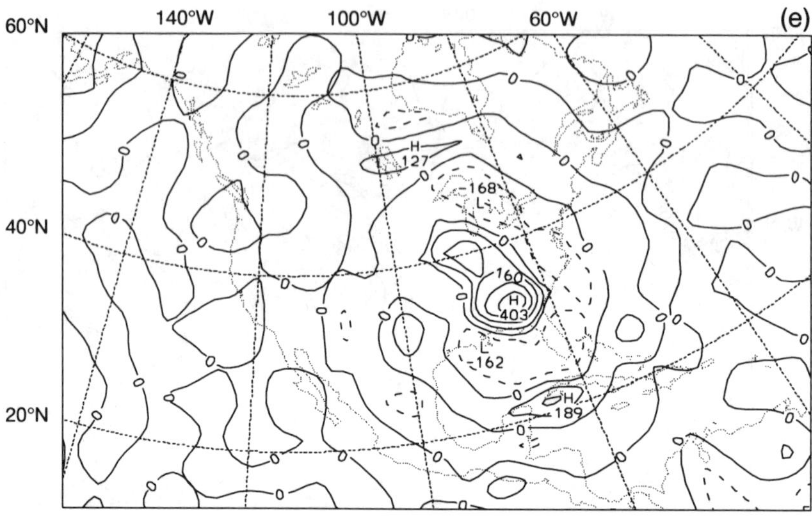

T at 225 mb

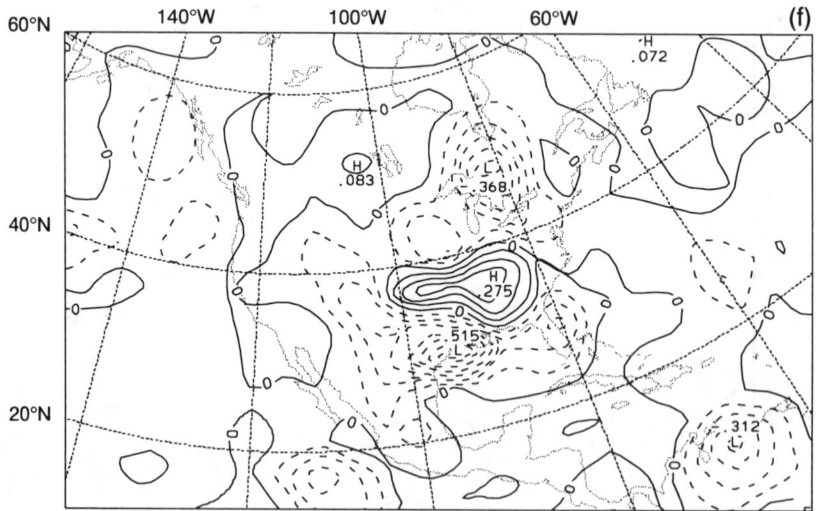

Figure 9: (Continued)

ζ at 225 mb

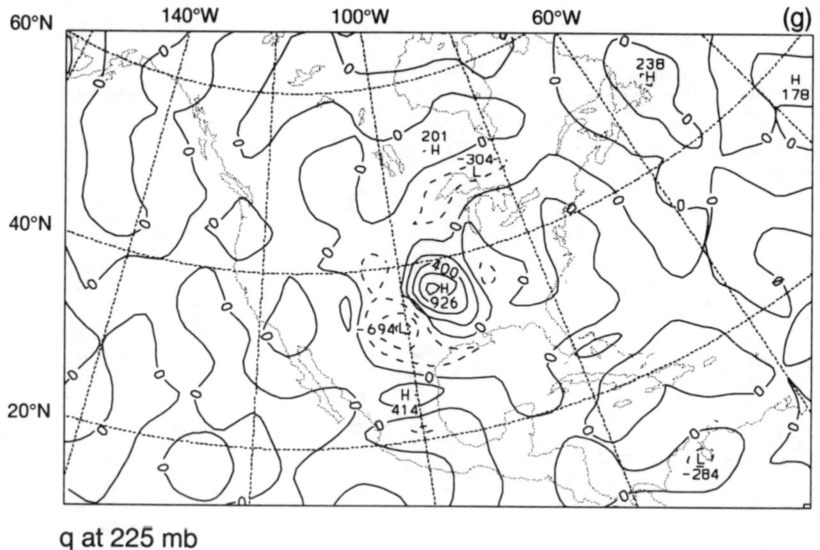

q at 225 mb

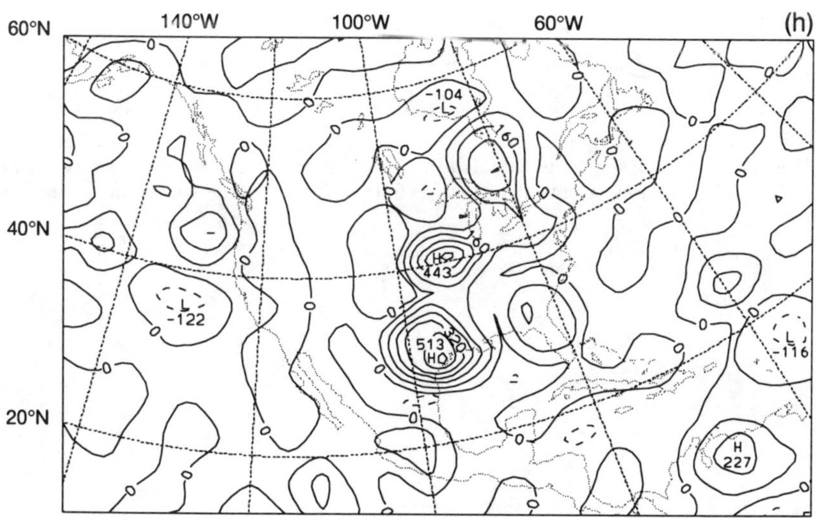

Figure 9: (Continued)

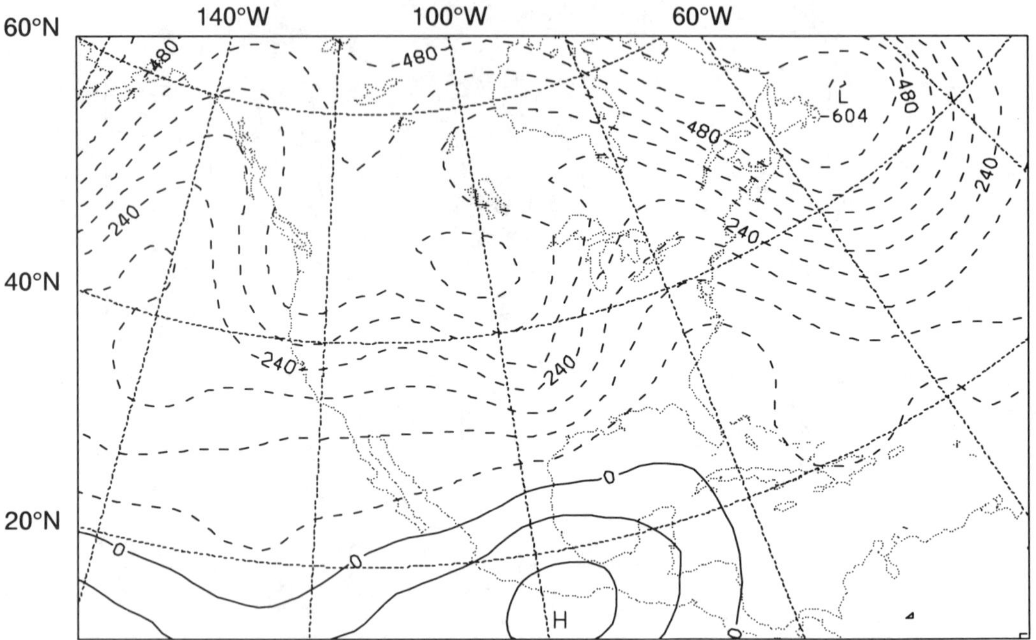

Figure 10: The distribution of the retrieved initial stream function at 500mb obtained by the minimization including the "on-off" processes. Contour interval is $6.0 \times 10^6\ m^2 s^{-1}$ and values on the isolines are scaled by 10^{-5}.

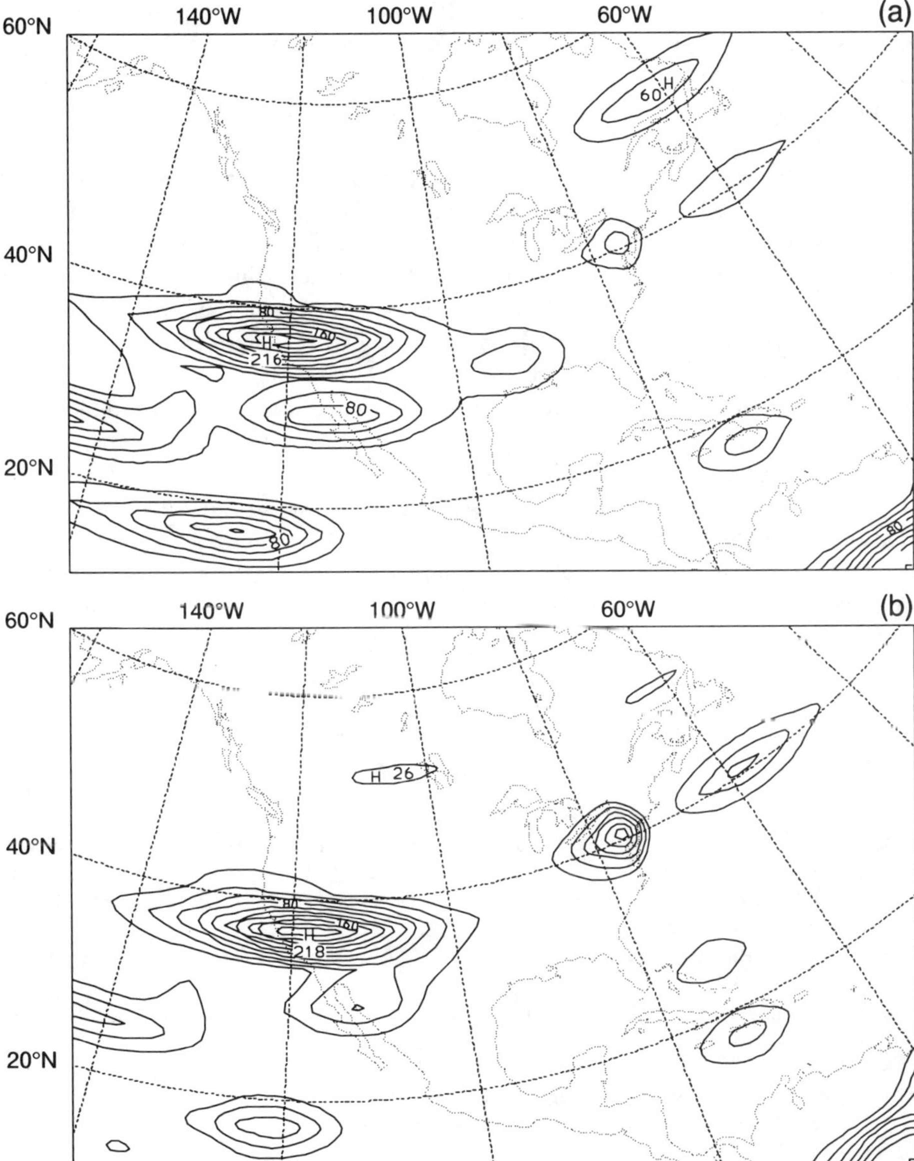

Figure 11: The total 6 hour precipitation forecasted from the initial state with(a) or without (b) the "on-off" processes. Contour interval is 2.0×10^{-2} m and values on the isolines are scaled by 10^3.

6 Summary and Conclusions

The methods and concepts of optimal control theory have been employed to formulate a variational 4-dimensional data assimilation for meteorological problems described by systems of coupled nonlinear equations and the minimization of nonlinear cost functions. The adjoint method is the most economical method to use, if the physical problem involves a large data base.

It is of practical interest to mention that, in particular applications, additional constraints may be required to be imposed on the cost function. Inclusion of the penalty term in the cost function while performing the minimization on uninitialized fields leads to better numerical conditioning and ensures that the minimizing solution will not contain an unacceptable amount of high frequency gravity waves. Other constraints such as energy and enstrophy conservation, model bias, and observational errors can be included in the cost function in different situations.

Inclusion of the "on-off" physical processes in the nonlinear and adjoint models does not hinder the minimization processes. Along this line, further research is needed related to the inclusion of additional physical processes in the adjoint model, such as the planetary boundary layer processes, vertical diffusion, topographic variations, shallow convection and radiation. Once the full adjoint of the operational model including the physics package is obtained, it can be used to tune the forecast model by carrying out an adjoint sensitivity study or a parameter estimation. As such the adjoint model constitutes a useful diagnostic tool and helps explore the physical details of the numerical model.

References

Axelsson, O. & Barker, V. A. (1984) *Finite Element Solution of Boundary Value Problems: Theory and Computation*, Computer Science and Applied Mathematics, Academic Press, 111 Fifth Avenue, New York, New York 10003, 437pp.

Bertsekas, D. P. (1982) *Constrained Optimization and Lagrange Multiplier Methods*, Computer Science and Applied Mathematics, Edited by Werner Rheinboldt, Academic Press, 111 Fifth Avenue, New York, New York 10003, 395pp.

Buckley, A. G. (1978) A combined conjugate-gradient quasi-Newton minimization algorithm, *Mathematical Programming*, **15**, 200-210.

Buckley, A. G. (1989) Remark on algorithm 630. *ACM Transactions of Mathematical software*, **15**, 262-274.

Buckley, A. G. & Lenir, A. (1983) QN-like variable storage conjugate gradients. *Mathematical Programming*, **27**, 155-175.

Buckley, A. G. & Lenir, A. (1985) Algorithm 630-BBVSCG. A variable storage algorithm for function minimization. *ACM. Trans. Math. Software*, **11**, 103-119.

Cacuci, D. G. (1981a) Sensitivity theory for nonlinear systems: I. Nonlinear functional analysis approach. *J. Math. Phys.*, **22**, 2794-2802.

Cacuci, D. G. (1981b) Sensitivity theory for nonlinear systems: II. Extensions to additional classes of responses. *J. Math. Phys.*, **22**, 2803-2812.

Chao, W. C. & Chang, L. P. (1992) Development of a 4-dimensional variational analysis system using the adjoint method at GLA. Part I: Dynamics. *Mon. Wea. Rev.*, **120**, 1661-1673.

Courtier, P. & Talagrand, O. (1987) Variational assimilation of meteorological observations with the adjoint equation - Part I. Numerical results. *Q. J. R. Meteorol. Soc.*, **113**, 1329-1347.

Courtier, P. & Talagrand, O. (1990) Variational assimilation of meteorological observations with the direct and adjoint shallow-water equations. *Tellus*, **42A**, 531-549.

Daley, R. (1991) Atmospheric Data Analysis. *Cambridge Atmospheric and Space Science Series, eds. Cambridge University Press*, 457pp.

Dembo, R. S. & Steihaug, T. (1983) Truncated-Newton algorithms for large-scale unconstrained optimization. *Mathematical Programming*, **26**, 190-212.

Derber, J. C. (1985) The variational 4-D assimilation of analysis using filtered models as constraints. *Ph. D. Thesis, University of Wisconsin-Madison, Madison, Wisconsin 53706*, 142pp.

Derber, J. C. (1989) A variational continuous assimilation technique. *Mon. Wea. Rev.*, **117**, 2437-2446.

Douady, D. & Talagrand, O. (1990) The impact of threshold processes on variational assimilation. World Meteorological Organisation, Proceedings from International Symposium on Assimilation of Observations in Meteorology and Oceanography, Clermont-Ferrand, France, July 9-13, 486-487.

Fiacco, A. V. & McCormick, G. P. (1968) *Nonlinear Programming: Sequential unconstrained minimization techniques*, John Wiley and Sons, New York and Toronto, 210pp.

Ghil, M. & Malanotte-Rizzoli, P. (1990) Data assimilation in meteorology and oceanography. *Adv. Geophys.*, **33**, 141-266.

Gilbert, J. C. & Lemaréchal, C. (1989) Some numerical experiments with variable-storage quasi-Newton algorithms. *Mathematical Programming*, **45**, 407-435.

Gill, P. E. & Murray, W. (1979) Conjugate-gradient methods for large-scale nonlinear optimization. *Technical Report SOL 79-15*, Systems Optimization Laboratory, Department of Operations Research, Stanford University, 61pp.

Gill, P. E., Murray, W. & Wright, M. H. (1981) *Practical Optimization*, United States Edition published by Academic Press, INC., Orlando, Florida 32887, 401pp.

Golub, H. G. & van Loan, C. (1989) Matrix Computation. Second Edition, Johns Hopkins Series in the Mathematical Sciences, The Johns Hopkins University Press, 642pp.

Grammeltvedt, A. (1969) A survey of finite-difference schemes for the primitive equations for a barotropic fluid. *Mon. Wea. Rev.*, **97**, 387-404.

Hall, M. C. G., Cacuci, D. G. (1983) Physical interpretation of the adjoint functions for sensitivity analysis of atmospheric models. *J. Atmos. Sci.* , **40**, 2537-2546.

Hall, M. C. G., Cacuci, D. G. & Schlesinger, M. E. (1982) Sensitivity analysis of a radiative-convective model by the adjoint method. *J. Atmos. Sci.*, **39**, 2038-2050.

Hoffmann, R.N. (1986) A four-dimensional analysis exactly satisfying equations of motion. *Mon. Wea. Rev.*. **114**, 388-397.

Holter, W. H., Navon, I. M. & Oppe, T. C. (1993) Parallelizable preconditioned conjugate-gradient methods for the Cray Y-MP and the TMC CM-2, *submitted to International J. of High Speed Computing.* Tech. Rep., FSU-SCRI-92-128, available at SCRI, Florida State University, Tallahassee, FL 32306-4052, 49pp.

Juvanon du Vachat, R. (1986) A general formulation of normal modes for limited-area models: Applications to initialization. *Mon. Wea. Rev. 114*, 2478-2487.

LeDimet, F. X. & Navon, I. M. (1989) Variational and optimization methods in meteorology - A review, Research Report, Supercomputer Computations Research Institute, Florida State University, Tallahassee.

LeDimet, F. X. & Talagrand, O. (1986) Variational algorithms for analysis and assimilation of Meteorological Observations: Theoretical Aspects. *Tellus*, **38A**, 97-110.

Lewis, J. M. & Derber, J. C. (1985) The use of adjoint equations to solve a variational adjustment problem with advective constraints. *Tellus* **37A**, Vol.4, pp. 309-322.

Lions, J. L. (1971) Optimal control of systems governed by partial differential equations. *Translated by S. K. Mitter, Springer-Verlag, Berlin-Heidelberg*, pp 396.

Liu, D. C. and Nocedal, J. (1989) On the limited memory BFGS method for large scale optimization, *Mathematical Programming*, **45**, 503-528.

Lorenc, A. C. (1986) Analysis methods for numerical weather prediction, *Q. J. R. Meteorol. Soc.*, **112**, 1177-1194.

Luenberger, D. G. (1984) *Linear and nonlinear programming*, Addison-Wesley Publishing Company, Second Edition, 491pp.

NAG (1990) Fortran library reference manual (Mark 14), Numerical Algorithms Group. **3**. [Available from NAG Limited, Oxford OX2-8DR, United Kingdom, or NAG Inc., Downers Grove, IL 60515-5702.]

Nash, S. G. (1984a) Newton-type minimization via the Lanczos method, *SIAM J. Numer. Anal.*, **21**, 770-788.

Nash, S. G. (1984b) Solving nonlinear programming problems using truncated-Newton techniques. *Numerical Optimization, Proceedings of the SIAM Conference on Numerical Optimization*, Boulder, Colorado, June 12-14, 1984. 119-136.

Nash, S. G. (1985) Preconditioning of truncated-Newton methods. *SIAM J. Sci. Stat. Comput*, **6**, 599-616.

Nash, S. G. & Nocedal, J. (1989) A numerical study of the limited memory BFGS method and the truncated-Newton method for large scale optimization. *Tech. Rep.*, Northwestern University, **NAM 02**, 1-19.

Navon, I. M. (1986) A review of variational and optimization methods in meteorology, in: Y. Sasaki, ed., Variational Methods in Geosciences (Elsevier, New York), 29-34.

Navon, I. M. & Cai, Y. (1993) Domain decomposition and parallel processing of a finite element model of the shallow water equations, *Computer Methods in Applied Mechanics and Engineering*, **Vol 106,No 1–2**,179- 212.

Navon, I. M. & De Villiers, R. (1983) Combined penalty multiplier optimization methods to enforce integral invariant conservation. *Mon. Wea. Rev. 111*, 1228-1243.

Navon, I. M. & Legler, D. M. (1987) Conjugate-gradient methods for large-scale minimization in meteorology. *Mon. Wea. Rev.*, **115**, 1479-1502.

Navon, I. M., Zou, X., Derber J. & Sela, J. (1992) Variational data assimilation with an adiabatic version of the NMC spectral model. Mon. Wea. Rev., **120**, 1433-1446.

NMC (1988) Documentation of the Research Version of the NMC medium range forecasting model. *National Meteorological Center, W/NMC2, Room 204, Development Division, World Weather Building, 5200 Auth Road, Camp Springs, Maryland 20746, U.S.A.*

Nocedal, J. (1980) Updating quasi-Newton matrices with limited storage. *Mathematics of Computation*, **35**, 773-782.

Rabier, F. & Courtier, P. (1992) Four-dimensional assimilation in the presence of baroclinic instability. *Quart. J. Roy. Meteor. Soc.*, **118**, 649-672.

Rao, S. S. (1983) *Optimization: Theory and Applications*, Published in the Western Hemisphere by Halsted Press, a Division of John Wiley & Sons, INC., New York, 747pp.

Schlick, T. & Fogelson, A. (1992a) TNPACK — A truncated Newton minimization Package for large-scale problems. I. Algorithm and usage, *ACM Trans. Math. Softw.*, **18**, 46-70.

Schlick, T. & Fogelson, A. (1992b) TNPACK — A truncated Newton minimization Package for large-scale problems. II. Implementation Examples. *ACM Trans. Math. Softw.*, **18**, 71-111.

Semazzi, F. & Navon,I. M. (1986) A comparison of the bounded derivative and the normal mode initialization methods using real data. *Mon. Wea. Rev. 114*, 2106-2121.

Shanno, D. F. & Phua, K. H. (1980) Remark on algorithm 500 - a variable method subroutine for unconstrained nonlinear minimization. *ACM Trans. on Mathematical Software*, **6**, 618-622.

Smedstad, O. M. & O'Brien, J. J. (1991) Variational data assimilation and parameter estimation in an equatorial Pacific Ocean model. *Prog. Oceanog.*, **26**, 179-241

Talagrand, O. (1992) The use of adjoint equations in numerical modeling of the atmospheric circulation, in: A. Griewank and G. F. Corliss, ed., Automatic Differentiation of Algorithms: Theory, Implementation, and Application (Society for Industrial and Applied Mathematics), 169-180.

Talagrand, O. & Courtier, P. (1987) Variational assimilation of meteorological observations with the adjoint vorticity equation - Part I. Theory. *Q. J. R. Meteorol. Soc.*, **113**, 1311-1328.

Temperton, C. (1988) Implicit normal mode initialization. *Mon. Wea. Rev. 116*, 1013-1031.

Thépaut J. N. & Courtier, P. (1992) Four-dimensional variational data assimilation using the adjoint of a multilevel primitive equation model. Q. J. R. M. S., **117**, 1225-1254.

Thacker, W. C. (1989) The role of the Hessian matrix in fitting models to measurements. *J. Geophy. Res.*, **94**, 6177-6196.

Vukićecić, T. (1993) Linearization and adjoint of parameterized moist diabatic processes, *Tellus*,**45A**, 493-510.

Wang, Z., Navon, I. M., LeDimet, F. X. & Zou, X. (1992) The Second Order Adjoint Analysis: Theory and Application. *Meteorol. Atmos. Phys.*, **50**, 3-20.

Wang, Z., Navon, I. M. & Zou, X. (1993) The adjoint truncated Newton algorithm for large-scale unconstrained optimization, *submitted to Optimization Methods and Software*

Wolfe, P. (1969) Convergence conditions for ascent methods.*SIAM Review* , **11**, 226-235.

Zou, X., Navon, I. M. & LeDimet, F. X. (1992a) Incomplete observations and control of gravity waves in variational data assimilation. *Tellus*, bf 44A, 273-296.

Zou, X., Navon, I. M. & LeDimet, F. X. (1992b) An optimal nudging data assimilation scheme using parameter estimation. *Q. J. of Roy. Met. Soc.*, **118**, 1163-1186.

Zou, X., Navon, I. M. & Sela, J. (1993a) Control of gravitational oscillations in variational data assimilation, *Mon. Wea. Rev.*, **121**, 272-289.

Zou, X., Navon, I. M., Berger, M., Phua, Paul K. H., Schlick, T. & LeDimet, F. X. (1993b) Numerical experience with limited-memory quasi-Newton and truncated Newton methods. *SIAM J. on Optimization*, **Vol 3**, 582-608. Also Tech. Rep., FSU-SCRI-90-167, available at SCRI, Florida State University, Tallahassee, FL 32306-4052, 44pp.

Zou, X., Navon, I. M. & Sela, J. G. (1993c) Variational data assimilation with moist threshold processes using the NMC spectral model. *Tellus*, **45A**, 370–387.

Zou, X., Barcilon, A. , Navon, I. M., Whitaker, J. & Cacuci, D. G. (1993d) An adjoint sensitivity study of blocking in a two-layer isentropic model, *Mon. Wea. Rev.*, **Vol 121**, 2833-2857.

Zupanski, Dusanka, (1993) The Effects of Discontinuities in the Betts-Miller Cumulus Convection Scheme on Four-Dimensional Variational Data Assimilation, *Tellus* ,**45A**, 511-524.

Zupanski, Milija, (1993a) Regional four dimensional variational data assimilation in a quasi-operational forecasting environment.*Mon. Wea. Rev.*,**121**, 2396-2408.

Zupanski, Milija, (1993b) A Preconditioning Algorithm for Large Scale Minimization Problems,*Tellus*, **45A**, 478-492.

Chapter 10

Expert system support for environmental decisions

J.R. Davis,[a] G. Guariso[b]

[a] *CSIRO Division of Water Resources, Canberra, Australia*
[b] *Center for Environmental Informatics (CIRITA), Politecnico di Milano, Milan, Italy*

Abstract

Expert systems differ from traditional mathematical models in their ability to handle qualitative information that is very common in environmental decision problems. Their architecture and most important features are briefly revised in this chapter to show how they can support decision makers. Their possible applications are presented through a set of actual examples, which span from the possibility of prototyping models in the early phases of the decision process, to supporting model choice and calibration or completely replacing the analytical model(s) when the process has already reached the design and implementation phases. The lesson learnt from these experiences shows that these systems have already reached a development that may considerably help environmental managers, but research is still necessary to fully exploit their capabilities.

Key words

Decision support systems, Artificial Intelligence, rule-based systems, knowledge engineering, natural resources management.

1 Introduction

According to Simon (1960) any decision process is composed of four basic phases: intelligence, design, choice, and implementation.

Intelligence here means to undertake an initial search within the decision context. This involves looking up necessary data, isolating the raw material of the decision process (including the parties to the decision) and, most importantly, identifying the problem domain as precisely as possible. In the design phase, different but feasible solutions to the problem are created and analyzed to determine the extent to which each solution solves the problem. From these alternative solutions, a plan is selected in the choice phase and, finally it is executed and its performance is monitored in the last phase. Carrying out any one of these phases may clarify aspects of the preceding phase and thus several cycles may be necessary to reach a viable decision.

Even accepting this simple characterization of decision making, one peculiarity of environmental problems is immediately obvious, and that is *complexity*.

The intelligence phase is complicated by the amount of data and information, arriving from different sources, relevant to the problem to be solved. The presence of both a spatial dimension (because of the inhomogeneous distribution of problems over space) and of a time dimension (because present environmental conditions are usually related to past inputs) characterize the large majority of environmental problems. Indeed, the problem to be solved is, in many instances, not clear at all. One does not need to enter the philosophical debate of whether a problem is an independent objective reality or depends on the proponent's perception, to understand that the definition of an environmental problem may be extremely difficult in practice. Consequently, environmental problems are often referred to as being "poorly structured" or "completely lacking a structure".

If a decision task can be structured in an algorithmic way (i.e. as a recipe), a computer program can solve it and so assist the decision maker. These types of problems (called procedural problems because the recipe provides a procedure to follow in order to get a solution) are common in the automation of manufacturing systems or in many other industrial sectors. On the other hand, when the problem formulation is more vague, or the decision has to be made in a new context or the problem is too complex for existing algorithms, a computer program cannot solve the problem by itself. A human needs to be involved in the solution. For example, it is computationally impossible to describe the development of a fish population at the level of individuals, but one can approximate it with a simpler demographic model at a higher level. The structuring of the problem, i.e. the choice of the level of description, must be performed by a human using a general solution strategy such as: analogy to find similar, already well-known problems; redefinition of the problem into a form using different, but known, terms; deduction of a particular strategy from an existing one; or approximation of the problem using another level that is easier to describe. The computer, being a deterministic machine, can solve only well-structured and defined problems.

The complexity of the design phase arise from our lack of basic knowledge of environmental processes. The exact mechanism for, and extent of, ozone depletion by CFC are still matters of debate. The same is true for the relationship between increasing concentrations of carbon dioxide and atmospheric temperature. Even evapotranspiration from a crop, a topic long studied, cannot be described by a general and well-founded formula. Simple models proposed in the past have proved unable to interpret the complex behaviours of living species, while we are still unable to properly understand the behaviour of the more complex models developed in recent last years. Consequently, predicting the effects of alternative solutions to environmental problems is inevitably an unreliable activity.

The multifaceted nature of environmental problems always emerges in the choice phase. They invariably involve different groups of people, with different perceptions, goals, and powers. Multi-objective and multi-attribute methods (e.g. Goicoechea et al., 1982; Hwang and Yoon, 1982; Keeney and Raiffa, 1976) have been proposed to deal with multiple parties to environmental decisions, but these techniques have not been very successful in practice. In essence, the choice phase is a political process, where the criteria and choice strategies are difficult to enunciate let alone quantify. For example, humans are more adept than computers in deciding whether the benefits to one societal group arising from more development offset the costs to another group from increased pollution. This political process often requires negotiation between many parties and may last for long periods.

Computer science has developed tools capable of dealing with both the quantitative and with the qualitative aspects of these phases of decision making. Mathematical models have a long tradition of being applied to the quantitative aspects of environmental problems as demonstrated by the books in this series. They are the most advanced tools available for both predicting and choosing between alternative solutions in those cases where a problem can be completely formalized in quantitative terms. Unfortunately this is very rare.

In practice the qualitative aspects of environmental problems usually dominate the quantitative aspects. It is important to understand that "qualitative" is not a synonym for "approximate". There are qualitative considerations in all decisions that are precise and systematic. For example, the result of the choice phase "this alternative is preferable to the others" is a qualitative but exact result that does not leave any room for discussion.

As noted by Fishwick (1989) the term "qualitative" is normally used with two different meanings in engineering. One occurs in the intelligence phase and refers to the definition of the problem, the choice of the region or parties involved, and the selection of the decision variables and constraints. These choices are necessarily qualitative even when developing a traditional quantitative model and are based on the subjective judgement of the analyst or on the result of a complex political discussion. The second meaning refers to the type of variables used in the model. Variables that take categorical values (e.g. red, green, yellow) are called qualitative variables, and they cannot be easily incorporated in quantitative models. This chapter discusses how these qualitative, but objective, aspects of environmental problems can be included in computer systems.

The most widely known computer tools for treating qualitative information are expert systems. Expert systems are available commercially and have already been employed in environmental applications. However, Artificial Intelligence (AI), the branch of computer science dealing with processing qualitative as opposed to quantitative information, has also developed other ways of treating qualitative information. They will not be dealt with in this chapter, given their minimal impact to date, but the reader can find descriptions of techniques such as influence diagrams and belief networks in Klir (1991), bond graphs in Karnopp et al. (1990), and qualitative modelling and simulation in Rajagopalan (1986).

2 Expert System Architecture

Computers were designed from their earliest days to manipulate numbers. However, some of the early developers of computers also saw their potential for mimicking human cognitive behaviour (McCorduck, 1979). AI research was characterised by the manipulation of qualitative (or symbolic) information using the rules of logic rather than (as in mainstream computer science) the manipulation of quantitative information using the rules of mathematics.

Much of the early AI research focused on the development of general algorithms for solving problems but this approach was not successful. Real progress was not made until the AI community came to accept that humans were good at problem solving, not because they used general problem solving approaches, but simply because they possessed extensive knowledge about a problem developed from experience. Thus the AI task switched to the development of programs that could store and access problem-specific knowledge. Ironically, simple, everyday problems turned out to be too difficult to be solved by these programs because of the diverse and extensive knowledge we use to do such simple things as cross the street and decide what to buy at the supermarket. But these new AI programs were successful with highly specialised problems where a limited, but rare, knowledge was needed. And so these knowledge intensive programs were called expert systems.

It is now accepted within the AI community that the term expert systems, was an unfortunate choice because it focused attention upon the supposed expert level of performance of these programs. In fact, the only expert systems that have reached truly expert performance have been those operating in very tightly restricted, well codified problem areas. In spite of this limited success, the distinctive software architecture of expert systems offers advantages to environmental modellers irrespective of the depth of expertise they draw upon.

The earliest practical expert systems were developed for analysing experimental data to infer the structures of chemical compounds (Buchanan and Feigenbaum, 1978), mineral prospecting (Duda et al, 1979) and medical diagnosis (Buchanan and Shortliffe, 1984). Of these, the architecture of the EMYCIN expert system (derived from the MYCIN medical expert system) has become the dominant paradigm. Its architectural features are shown in Table 1.

Table 1: List of EMYCIN features

Use of rule structures to capture expert knowledge
Separation of the knowledge base from the inference engine
Use of declarative rather than procedural knowledge
Use of an uncertainty handling mechanism
Use of a trace and explanation feature
Design as a shell program

Variations have been developed over the last 15 years on each of these canonical features but it is remarkable how they have persisted in the design of modern expert systems. We will now briefly describe each of these features.

2.1 Knowledge representation

Much of the knowledge that we need for problem solving is qualitative in nature. The early developers of expert systems devised a rule-like IF-THEN syntax for representing such knowledge. Some current systems use a three part IF-THEN-ELSE syntax. Each part consists, in turn, of one or more (parameter operator value) triplets that are joined conjuctively. For example, a chunk of knowledge about land management might be encoded as

IF (land-use is horticulture)
  AND (slope gt 10%)
  AND (soil-type is-one-of [GA, GB, FA])
THEN (recomm-mgmt-prac is cover-crops)

In this rule, 'land-use' is the parameter, "horticulture" is the value and "is" is the operator in the first triplet.

Rules are a suitable representation when the knowledge is modular and well-codified. Each rule represents an independent chunk of knowledge. This is both a strength and a weakness. It is a strength because the total knowledge (the knowledge base) can be developed incrementally but a weakness because it forces a restricted structure onto what may be complex interconnections between chunks of knowledge.

Frames are an alternative form of knowledge representation where the chunks of knowledge are represented as objects, each with attributes (called slots). Each slot has an associated method for filling it; for example, a rule could fire to fill a slot. Frames can be linked hierarchically and values of slots (even slots themselves in some systems) can be inherited. Thus, frame-based systems offer a richer and, often, more natural knowledge representation than do rule-based systems at the cost of interdependence between chunks of knowledge and complexity of description.

2.2 Separation between knowledge and inference

The expert knowledge in an expert system is held in a separate file to the program, the inference engine, that applies the knowledge at run time (Figure 1). The knowledge base is often an ASCII file although it can be a coded version of the near-English statement of the knowledge (such as the above rule). Thus, unlike a conventional mathematical model where the specialised knowledge is encoded in equations which are then compiled into the program, the knowledge in an expert system is maintained in a source code form.

The knowledge base can be likened to a data base where the data items are replaced with chunks of knowledge. Just as the data in the data base can be modified without affecting the program, the knowledge in the knowledge base can be modified (e.g. new information added) without having to recompile the expert system or, indeed, without having to even know a computer language. This is of particular advantage if a modular form of representation, such as rules, is used. The disadvantage is that the source code needs to be interpreted at run-time by the inference engine and this can be slow, particularly as the knowledge base needs to be searched for relevant chunks of knowledge.

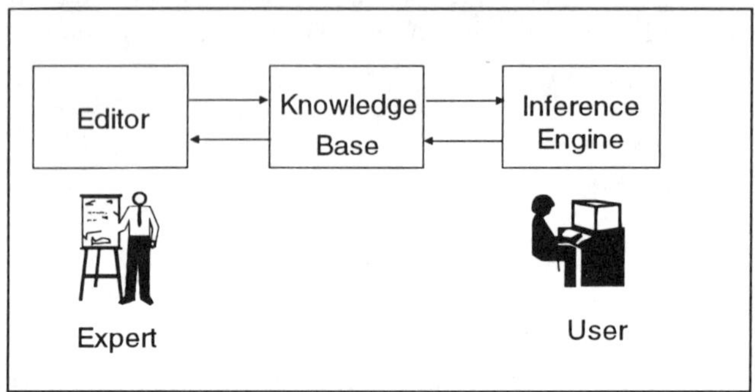

Figure 1: Modern expert systems preserve the distinction between the repository of the expert knowledge, the knowledge base, and the mechanism for using the knowledge, the inference engine. Most expert systems also provide an editor to help the expert build up the knowledge base.

The inference engine can be written either to take known facts and apply the rules to see if further valid facts can be inferred or to see whether a hypothesised fact is true, given the known facts and the information contained in the knowledge base. The former is called forward chaining whilst the latter is backward chaining. Backward chaining is the appropriate control strategy where there are a relatively small number of possible answers to be checked (it is also easier to control), but forward chaining is more appropriate for exploratory problems.

2.3 Declarative vs procedural knowledge

Knowledge can be divided into two broad classes --- procedural and declarative (Winograd, 1975). Procedural knowledge tells us how to solve a problem; declarative knowledge describes a problem. Conventional programs are written procedurally; i.e. by starting at the entry point and following the flow of the program one will arrive at an answer to the problem. In contrast, the knowledge in an expert system's knowledge base is often declarative; that is, it consists of statements about the problem. Thus, if a rule representation is being used, each rule is a statement about the problem (although each rule can also be seen as a small IF-THEN procedure). Solving problems where the information is held declaratively reduces to locating relevant chunks of knowledge. The program to solve such problems becomes, essentially, a search program.

The advantage of expressing knowledge declaratively is that chunks of knowledge do not need to be stored in any order, nor does the knowledge need to be complete or unambiguous. The search procedure can incorporate mechanisms for handling missing information and resolving conflicts between different chunks of knowledge. The main disadvantage is the slowness since every step on the way to a solution involves a search of the knowledge base. Hence efficient search procedures are central to effective expert systems.

2.4 Uncertainty handling

Few mathematical models calculate the uncertainty associated with their predictions (unless the model is based on statistical relationships such as regressions) because the calculations, using probability theory, quickly become very complex. Instead, statistical simulation or sensitivity testing are used to assess the reliability of predictions. The absence of uncertainty estimates raises the implication that mathematical models are accurate.

Expert systems, following the EMYCIN design, incorporate a mechanism for calculating the uncertainty associated with solutions. EMYCIN used confirmation theory (Shortliffe and Buchanan, 1984) for uncertainty handling; modern experts systems also use fuzzy logic, Dempster-Shafer and Bayesian network (Giarrantano and Riley, 1989) theories of uncertainty.

2.5 Trace and explanation

Expert systems normally keep track of their search through the knowledge base for relevant chunks of knowledge. From this "trace" they can reconstruct an explanation which is essentially a history of the way the solution was built up from separate chunks of knowledge. Explanations will differ for different problems since dissimilar chunks of knowledge will be used for each problem. Although explanation is potentially a very powerful feature of expert systems, it has been found to be of limited practical use because it does not provide the depth of explanation that humans require. Current research into explanation employs a separate knowledge base for generating explanations to the knowledge base used to predict solutions to problems (Wick and Thompson, 1992).

It is difficult for procedural, mathematical programs to explain their answers for two reasons. Firstly, the knowledge is encoded in mathematics and so an explanation based on equations would be hollow to most users ("The answer is 43 because I divided 88 by 2 and subtracted 1"). Secondly, the same procedure (i.e. the same equations) is followed to solve all problems and so the explanation would have the same form for all answers (although the values of the variables would change).

2.6 Shell programs

Because the knowledge base is maintained separately to the inference engine it is possible to develop "shell" programs that are empty of knowledge but which contain all the mechanisms needed to acquire, maintain and apply knowledge. EMYCIN (Empty MYCIN) was the first such expert system shell. Many such shells are now available commercially and it is now only worth developing one's own expert system from scratch for specialised problems.

A shell program contains many component parts and each component can be implemented in different ways. The more expensive shell programs provide choices for each component. For example, the uncertainty handler may allow a choice of fuzzy logic or confirmation theory whilst the conflict resolver may allow precedence or rule generality to be used (Davis and King, 1977).

3 Historical Summary

Interest in environmental applications of expert systems arose in the early 1980s. The first influential paper was probably that of Starfield and Bleloch (1983). In a review of agricultural applications Moninger and Dyer (1988) list 70 environmental applications of expert systems. By 1989, Davis and Clark (1989) had identified 200 papers dealing with applications of expert systems to a range of environmental problems, and Carrascul and Pau (1992) described 110 applications of expert systems in agriculture and food processing alone. There have been regular specialised conferences on the application of AI techniques in natural resource management (Moninger and Dyer, 1988) and specialised streams on AI applications to environmental problems at general AI conferences (e.g. IJCAI-93). A journal devoted to environmental application of AI techniques, *AI Applications: Natural Resource, Agriculture and Environmental Science*, has been produced since 1986.

In spite of this academic interest, there is little evidence of widespread practical use of expert systems in environmental areas although it is difficult to tell whether the adoption rate is lower than for environmental software in general. No large scale study has been undertaken of the reasons for this lack of practical interest in environmental expert systems but possible reasons are:

- Poorly managed applications of a new technology by enthusiasts
- Unrealistic expectations by potential users because of the newness of the technology
- Poor integration of expert systems with conventional software
- Lack of understanding of the best ways to employ expert systems in environmental problem solving

Environmental scientists have probably been going through the enthusiastic amateur stage. Many systems reported in the literature were developed, often with minimal input from users, mainly for the developers to try out the technology. Similarly, potential users believed that the systems would achieve a higher level of performance that is possible in amateur systems. Whilst it is relatively easy to develop a simple knowledge base, it takes many months, even years, to develop, test and document a knowledge base that is of practical use (Bachant and McDermott, 1984). Such systems need to be carefully planned and implemented using conventional methods from software engineering (Stock, 1988).

For many applications expert systems need to be integrated with conventional mathematical models. As we have seen most problems involve a mixture of quantitative and qualitative knowledge. Not only is mathematics the appropriate language for manipulating quantitative, continuous knowledge but there has been a massive investment in developing and applying mathematical models to environmental issues. Thus, expert systems are more likely to be adopted if they extend the usefulness of mathematical models rather than attempt to replace them. Many shells allow external models to be called from within the expert system but a seamless integration of expert systems and mathematical models whereby qualitative and quantitative knowledge can be employed in a consistent fashion has yet to be achieved in commercial shells.

Finally, it is not yet widely appreciated how the qualitative knowledge in expert systems can be best used for environmental problem solving. Are they best used in the intelligence, design, choice, or implementation phases of problem solving? Should specialistic knowledge be used to supply input data to models, to set up some of the more complex models, in the models to develop solutions, or to interpret the output of models? Below we describe different ways in which expert systems have been used to help environmental managers solve problems.

4 Selected Examples of Applications

4.1 Model Prototypes

It is common for decision makers to want to pose new types of questions to a model as a result of answers to previous questions. This is intrinsic to environmental decision making because the lack of problem structure (Section 1.0) implies that the pertinent questions are not clear at the beginning of the process, but arise interactively during use of a model.

It is apparent that decision support systems based on traditional mathematical models (e.g. De Simoni et al., 1986; Loucks and Da Costa, 1991) cannot provide the flexibility in the intelligence phase necessary to cope with the unstructured nature of environmental problems. Mathematical models, being procedural, are intrinsically constrained in the type of questions that they can answer. For example, a differential equation assigns a value to the derivative of a state variable, say, the level of a reservoir or the density of an animal population. However, the decision maker may want to know the variance of the state variable, its mean value in a certain period, a combination of state variables or only the values that exceed a certain threshold. While there is sufficient knowledge embedded in the model to answer all these questions, the way the knowledge is encoded strongly limits its possible uses. This has been recognized by many modellers who have proposed alternative modelling styles (Kuipers, 1986; Mattson, 1988; Milanese et al.,1991). The way in which the knowledge recorded in expert systems is formalised may also be used to improve the expressivity and flexibility of environmental models.

4.1.1 Eco

Long standing research has been carried out in this area in the ECO project at the University of Edinburgh. Muetzelfeldt et al (1989) suggest that a model should be formalized into rules and facts so that the values of states, components, variables in the model can be easily obtained. The user can also pose inverse questions such as asking for the necessary values of input parameters in order to obtain a certain output. In this work, the authors used PROLOG, a computer language which allows both a declarative and a procedural programming style, to develop a "model blueprint" and "value predicates". The model blueprint is a knowledge base containing a description of the model structure. It can be searched to answer questions related to the model's purpose. The value predicates answer questions about the values of variable in the model using either recursion or invoking external programs to perform numerical evaluations. The rules used to perform these tasks are of the type

IF variable V appears in formula F
THEN value of V = expand(F w.r.t. V)

where the predicate "expand" allows the resolution of the formula with respect to a certain variable. In the same way, a differential equation may be translated into a rule of the type

IF variable V at time t has value v
AND the variation of V at time t is kV
THEN the value of V at t+k is V+kV

where the predicate "variation" computes the derivative of the variable. Obviously, writing a model in this form is very inefficient, although the inference capabilities of the PROLOG compiler can be exploited by a suitable definition of predicates (Muetzelfeldt et al., 1989).

This approach provides an environment where queries about model structure as well as questions about the values of variables can both be answered by the same inference process. The cost is the low speed of computation and the difficulty of developing a friendly user interface, but these are usually not significant issues in the prototyping phase, when both the model of the system under study and the possible needs of the decision maker are still in an exploratory phase.

4.2 Support for the use of mathematical models

Models are used in decision making to predict the consequences of alternative solutions and so enter into the design phase of the decision process. The choice of which model(s) to use for prediction can be seen as either part of the intelligence phase or an early part of the design phase.

The use of expert systems as support to mathematical modelling dates back to 1981 when a prototype expert system was developed to help users select the parameters needed to run the HSPF hydrologic model (Gashing et al., 1981). Many of the environmental models developed within public research institutions and universities had grown to a dimension and complexity outside the reach of a normal user. Thus it was natural, with the advent of commercial expert system shells, that expert systems were seen as a powerful way of bringing to the user, knowledge used by experienced modelers when choosing a mathematical formulation or determining values for parameters.

This first system was followed by several others such as EXSRM (Engman et al., 1986), Flood Advisor (Fayegh and Russell, 1986), RHEX (French and Bhaskar, 1987) and Expert SWMM (Baffaut and Delleur, 1990) which were all attempts to improve the usability and effectiveness of well-known hydrological models, usually written in FORTRAN.

With the continuing progress in environmental models and improvements in user interfaces for implementing these models, expert systems have been increasingly used for selecting between different models rather than as an interface to a single model. For example, there are more than one hundred air pollution models (see, for instance: Zannetti, 1990) and several hundred groundwater models currently available (International Ground Water Modeling Center, 1992) and most users need assistance to choose an appropriate model. An expert system can play two roles in this case: the system can try to match the characteristics of the problem to the characteristics of the models and suggest which models are suitable, or the system (knowing the characteristics of all the models) can help the user formulate proper queries to the selected model. These uses are exemplified in the following sections.

4.2.1 Qual2e Advisor

The most widely known example of this approach is probably the advisor developed by US EPA (Barnwell et al., 1986) to assist users with the complex water quality model QUAL2E. Development of this model started in 1970 and it has been refined and revised several times (Roesner et al., 1981). It has been widely distributed and used within and outside the USA. Working with this model is a complex task; it requires more than 100 different types of input, ranging from provision of field data, to choosing options for modelling certain subsystems (for instance, the relation between light intensity and algal growth), to the partitioning of the river into reaches. QUAL2E Advisor helps the user in defining these inputs, by suggesting parameter values and options on the basis of information provided by the user.

The expert system has been developed on a personal computer using the M.1[1] shell and is intended for environmental engineers in local EPA offices or other public institutions with limited experience in using mathematical models. The interaction thus assumes knowledge of precise technical terms when referring to water quality, but the user can employ more general statements to choose between alternative mathematical formulations.

```
Are any lake or ponds or other section of stream with ill-defined cross-
sections involved in the simulation ? (y or n)
[user]>y
Choose one of the following descriptions as best fitting the shape of
the channel cross section for the flow of interest:
1. Rectangular with nearly vertical sides such that width does
   notincrease markedly with increasing streamflow
2. Natural channels in cohesive soils, tending to large width-to-depth
   ratios and  stable, well defined banks,typical of the eastern or
   northwestern United States, where width increase moderately with
   increasing streamflow
3. Wide dish-shaped channels in semi-arid regions, having banks with
   very shallow  slopes, where width increase dramatically with
   increasing streamflow.
Enter 1, 2 or 3
[user]>3
```

Figure 2: Sample of the QUAL2E Advisor interface.

As an example, let us suppose that the user wants to define the hydraulics of a reach. There are two options: either a trapezoidal section to relate depth, velocity and width of the stream or some empirical coefficients whose values are related to the flow-rate. The first option is chosen by the Advisor if the user can enter the slope of the two banks, the bottom width and the channel slope. If these data are not available, the system uses the empirical approach and suggests values for the unknown coefficients. In fact, since the coefficients are already available for a number of different river morphologies and since these morphologies strongly depend upon the type of soil in which the river flows (e.g. steep banks are typical in cohesive soils), it is sufficient to ask the user for characteristics of the local soils to estimate an appropriate coefficient. Thus, the determination of a coefficient which may be of little meaning to the user becomes a simple question on the soil type. Two rules which translate part of this procedure are:

---

[1]M.1 is a product of Teknowledge Inc., Palo Alto, CA.

IF there are lakes or reservoir in the stretch
THEN empirical approach to compute flow velocity is chosen

IF empirical approach to compute flow velocity is chosen
AND soil type = non-cohesive
THEN velocity coefficient = 0.34
AND depth coefficient = 0.36
AND width coefficient = 0.29

These rules are activated by the inference engine, when the user answers questions posed by the Advisor (Figure 2).

In practice, the result of the consultation is the compilation by the Advisor of a large file defining all the data inputs and options for QUAL2E. It is interesting to note that the user would probably find difficulties in interpreting the long and complex output file from QUAL2E; a system that helps users interpret complex output is described in Section 4.5.1.

4.2.2 Gmt

In order to help a user choose the most appropriate simulation model for assessing groundwater resources, it is necessary to formalize all the information related to the aquifer under consideration and the domain of applicability of the models. Information on aquifer models can be easily structured into three basic groups.

The first group are general aquifer characteristics that can be either inferred from data in a database (or a GIS) or provided by the user. For instance, if the dispersivity parameters in the database are equal in all directions, the aquifer is classified as isotropic. The user may want to further characterize it as mono- or multi-layered, leaky or not-leaky, etc. All the information of this first group (the aquifer conditions, the nature of the possible pollutants to be analyzed, the decay rates of chemical reactions, etc.) may be represented by keywords, chosen from a limited set. The second group are pollution source(s). The structure and the position of each source are necessary pieces of information. The structure of the source may be classified by keywords such as point, linear, square. Position can be given by coordinates for point sources, plus additional shape information (orientation, length, thickness, etc.) in case of more complex structures. The third group are the boundary conditions of the aquifer under consideration. These reflect relevant geological features (such as impervious boundaries and open air channels) which can either be defined by the user or directly extracted from a database/GIS.

This information defines the problem and is sufficient for choosing a suitable model. A set of rules, which process all three groups of information has been implemented within the Groundwater Managers Toolkit (GMT) (De Leo et al., 1990). GMT is connected to a database containing hydrological, geophysical and biochemical data about the groundwater resources of a region and can access ten different models each applicable to a specific aquifer condition. The package is implemented under DOS using C, PASCAL and FORTRAN languages. All necessary models are invoked by GMT as external procedures and data are exchanged through files.

Typical rules used in GMT are:

(1) IF hydraulic conductivity is constant over space
    AND storage coefficient is constant over space
  THEN the aquifer is homogeneous

(2) IF no boundary condition has been defined
  THEN the aquifer is infinite

(3) IF the aquifer is homogeneous
    AND the aquifer is infinite
    AND the pollution source is point wise
  THEN apply "analytical_model_B".

The premise of the first rule can be determined by looking at information of the first type in the database and checking if the values of the hydraulic conductivity and the storage coefficient are constant. The premise of the second rule establishes boundary conditions from information from the user. The third rule determines a model based on the conclusions of the first two rules together with additional information provided by the user about the pollutant source type.

External procedures may be invoked to check whether the conditions of some rules are true. For instance, to check if "permeability is constant over space", the rule can test whether the variation coefficient (i.e. the ratio between standard deviation and mean of the permeability measure) is smaller than a fixed percentage (such as 5%). Averages and standard deviations are computed by external procedures called automatically by the rule base.

GMT tries to avoid the common but frustrating result that there is no model that exactly fits the characteristics of the aquifer. In these circumstances, most expert systems would be unable to draw any conclusion about the most suitable model. It is easy to modify the rule set to allow a more extensive treatment of missing knowledge. Matching of the aquifer characteristics and model characteristics can be relaxed to mean, for instance, all but one attribute. In practice, this approach is implemented through a number of rules each referring to all attributes except one, and the introduction of a dummy value ("indifferent") for model attributes which matches any value of any attribute in the aquifer description.

Thus, a typical conclusion of a search to apply to a model for a finite aquifer and a linear source would be:

```
No model found for the given aquifer
If source is point wise, use ModelY
If aquifer is infinite, use ModelZ.
```

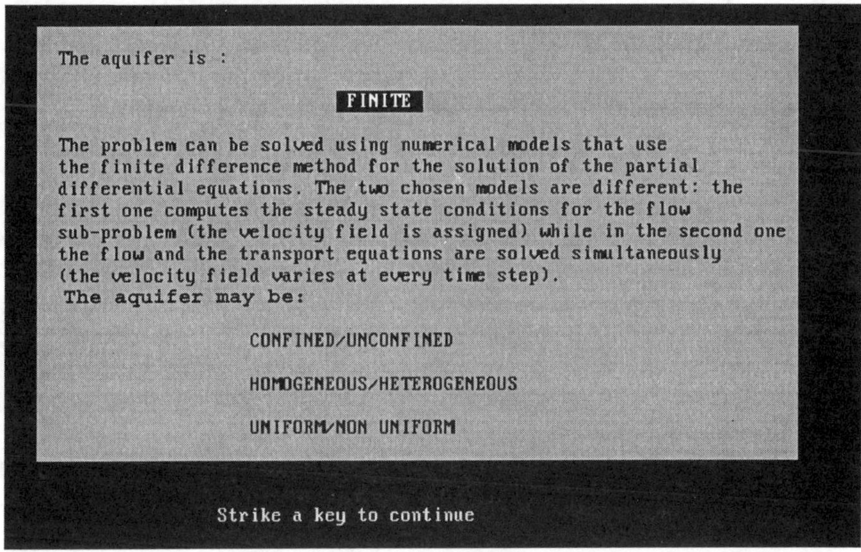

Figure 3: Explanation of model selection

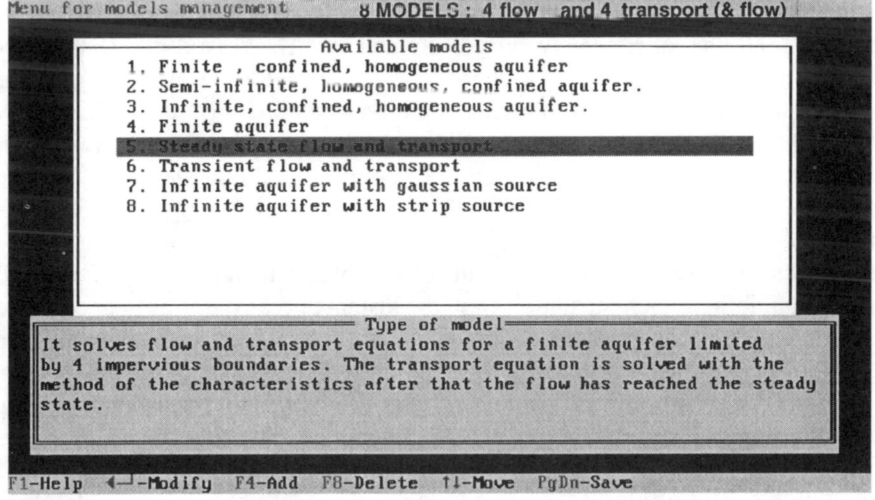

Figure 4: Selection of a model in the model base of GMT

These selected models would be highlighted on the screen together with an explanation of the reasons of these choices (Figure 3). The experienced user has access to the entire set of models and a short description of their applicability (Figure 4), so that the models suggested by GMT can be overridden. For instance, small differences in the porosity of different points in the aquifer results in the conclusion of non-homogeneity and this imposes a strong limitation on the models that can be chosen (only numerical models can be applied to such cases). The user might decide to ignore the differences in porosity and choose an analytical model that is only suitable for homogeneous aquifers.

4.2.3 Frame

A slightly different approach is implemented in FRAME, a package developed by Calori et al. (1993) to help select appropriate air pollution models. There are a wide variety of models presented in the air pollution literature, ranging from simple static models based on the gaussian approximation of the dispersion equation, to complex dynamic particle or puff models. Furthermore, there are many alternative formulations to the simple and widely used guassian approximation for different environmental conditions (Environmental Protection Agency, 1988) and the variety of data and parameters used by these models make it impractical to develop a single database.

The FRAME package comprises three basic modules: a relational database, an expert system shell drawing on various knowledge bases, and a set of files storing detailed descriptions of the models.

The relational database stores the possible attributes of all air quality models (more than hundred) in a large table. Each row of the table correspond to a specific model with the last column of each row containing the name of the file with the complete model description, its application conditions, and related bibliographical references. A consultation with FRAME starts with the description of a specific environmental situation. The user is then presented with a list of all the attributes valid for models which satisfy the first query. The consultation proceeds by defining the characteristics of the desired models more and more specifically. Thus the user cannot define a situation for which no model exists (in contrast to GMT). On the other hand, it is not possible to have complete control over the definition of the problem domain. For instance, if an "urban" condition has been selected, it is impossible to set the terrain to "complex", since there are no models for air pollution problems in "non- flat" cities.

The expert system helps the user consult the database which includes a file name for the set of rules that defines each attribute. Say, for instance, that the user is uncertain whether the value of the "plume" attribute is "buoyant" or "jet". The user can invoke the expert system which accesses a set of rules that define these values. The user is then asked about some meteorological and physical parameters that are necessary to distinguish between the two types of plumes.

The system has been left open to changes and improvements since it is intended for both researchers in the field of atmospheric pollution and model users. New model descriptions and attributes may be entered in the database. Rules relevant to the selection of these models may also need to be added. The system thus contains a "rule base debugger" which helps avoid errors in defining new rules and maintains the integrity and completeness of the knowledge base. A sample session of FRAME is presented in Figure 5.

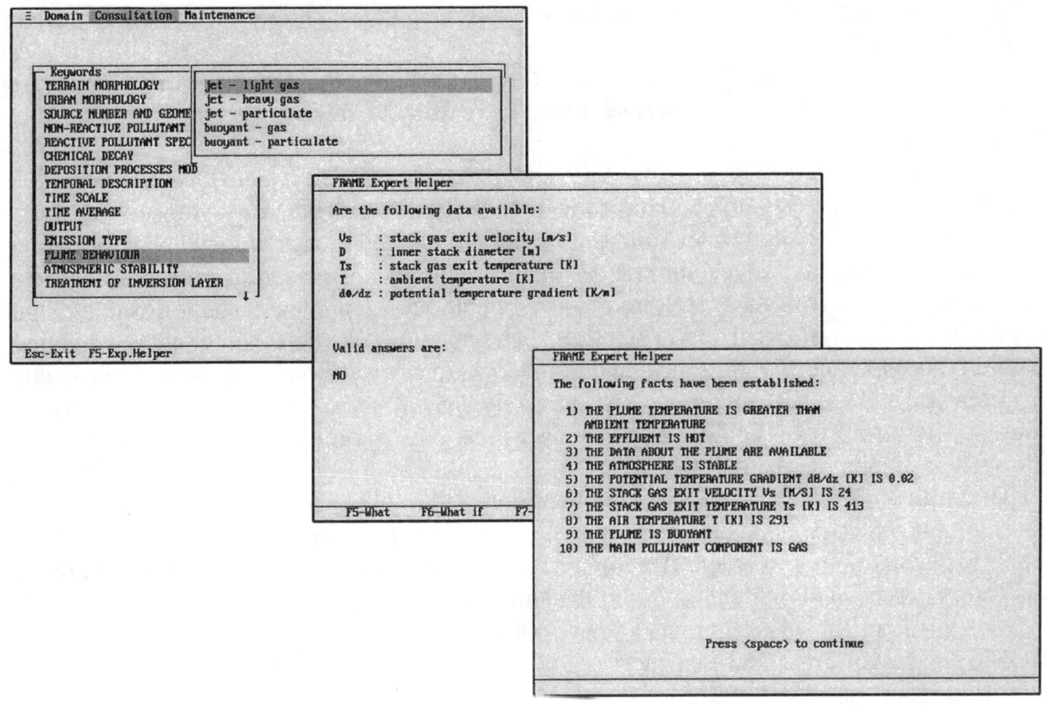

Figure 5: Sample session with FRAME. The user defines the required model characteristics, is supported in the selction of some parameters and can revise the expert system's findings.

4.3 Formulating Input Data

Often considerable experience is needed to formulate the input data files for environmental models. Some expert systems have been produced to shield users from this task. The knowledge to set up these data files usually comes from the developers of the models.

4.3.1 Comax

The COMAX expert system (Lemmon, 1986) was a major effort to develop a system that integrated the Gossym simulation model with an expert system to help in the management of cotton crops. The Gossym model had been developed over 12 years to provide a highly physical simulation of the growth of cotton plants. After extensive development and testing in both laboratory and field conditions the Gossym model was ready for practical on-farm applications. The COMAX expert system was developed to replace the multidisciplinary group that provided Gossym with its input data and interpreted its output in the laboratory.

COMAX is written in the specialised artificial intelligence language, LISP. It uses a forward chaining inference engine and a small rule base to infer new facts from currently known facts.

There are about 70 rules in COMAX's knowledge base. They contain the information needed to set up the input data files for Gossym, execute Gossym and develop recommendations for the farmer from the model's output. The actual weather information recorded from an on-site weather station together with data on stage of crop development and on-site factors (soil data, etc) are recorded in the knowledge base. If, for example, the condition of a rule is true then the rule calls Gossym to determine the date when irrigation would be needed under both typical and hot-dry weather conditions. Gossym is also called to determine nitrogen requirements and the likely date of harvesting the cotton. COMAX presents its output in graphical form by drawing plots of predicted water requirements, nitrogen availability and crop yield.

The rules in COMAX as described in Lemmon (1986) are relatively shallow, being used to assemble input data and call Gossym to calculate the various physiological factors. Their main function is to shield the farmer from the input data formats of Gossym and to interpret the typical numerical output of the simulation program.

McKinion et al (1989) report that cotton farmers experienced in the use of COMAX/Gossym estimated that the system was worth an extra $350 per hecatre but the less experienced users estimated its worth to be only $100 per hectare. The biggest savings were in the reduced use of nitrogen fertilisers and, for one producer, reduced insecticide applications because of the ability to predict the onset of vulnerable fruit and the occurrence of fruit that will not mature before harvest. All users wanted to continue with the system.

4.4 Integration/Substitution

Although we have described the use of expert systems for selecting and setting up environmental models, most environmental expert systems reported in the literature are designed as environmental models, per se, to be used in the design phase of decision making. That is, they employ knowledge about the environmental problem area itself. Usually the rationale for developing these systems is that the knowledge is qualitative and so unable to be represented in equations.

Hayes-Roth et al (1983) describe 10 classes of problems to which expert systems can be applied (Table 2). Not all types are independent; e.g. control problems are seen as being mixtures of interpretation, prediction, repair and monitoring problem types, and some problem types span a number of Simon's phases of decision making. For example, planning problems involve intelligence, design and choice. Diagnostic problems are the most well defined problem type, describing problems where one of more solutions are selected from a small number of possibilities depending upon the match between characteristics of each solution and the characteristics of a problem. In medical problems, the solutions are possible diagnoses (each with identifiable characteristics) and the problem is a patient (with characteristic symptoms). The match may be imperfect. The environmental literature described above contains many examples of expert systems applied to diagnostic problems, especially for diagnosing plant diseases.

Table 2: List of problem types

Interpretation	Inferring system descriptions from sensors
Prediction	Inferring likely consequences of situations
Diagnosis	Inferring system malfunctions
Design	Configuring objects under constraints
Planning	Designing actions
Monitoring	Comparing observations to plans
Debugging	Prescribing remedies for malfunctions
Repair	Executing a plan to administer remedies
Instruction	Diagnosing, debuggung and repairing student behaviour
Control	Interpreting, predicting, repairing and monitoring system behaviour

4.4.1 Tomato Disorder System

For example, Guay et al (1992) describe an expert system built using the PCM-Object shell to diagnose disorders of tomato plants. The program is presently at a prototype stage and contains enough qualitative knowledge to diagnose 10 tomato plant diseases from 175 symptoms. Some symptoms were stored as images in the knowledge base.

The diagnosis problem required the expert system to map multiple symptoms to multiple diagnoses and for the inference engine to undertake uncertain reasoning. The developers chose to use the probabilistic causal model of reasoning (Peng and Reggia, 1990) for this application. This model is noteworthy in that it employs a little used form of reasoning called abductive reasoning. Abductive reasoning reverses the normal order of inference

IF A THEN B

to conclude that if B is observed then A may have plausibly caused it. Each rule has the prior probability of each cause and the causal strength of each cause-effect relationships recorded along with it. Abductive inferences are not guaranteed to be correct, because effects other than A may have caused B. However, the PCM mechanism efficiently searches for the disease that best explains all the symptoms.

The knowledge base contains descriptions of the symptoms and the diseases and the links between each disease and each symptom. New symptoms and diseases can be easily added. An expert at diagnosing tomato diseases provided descriptions of common symptoms and 10 common diseases together with the probabilities of occurrence of these symptoms and diseases for the prototype's knowledge base.

Preliminary experience with the prototype indicates that the knowledge representation, search mechanism and probability calculations are acceptable when applied to closed problems; i.e. those where all necessary symptoms are provided before possible diseases are identified. It is deficient when applied to open problems, i.e. those where possible diseases are tested as symptoms are supplied in response to questions.

4.4.2 Oasis

The expert system developed by the South Florida Water Management District (1986) has several interesting features (Goforth, 1989). First, it was initiated and largely developed by the users themselves to solve the problem of saving the enormous experience developed during 25 years by the Director of the District's Operations Division and his assistant. Second, the decision making process is extremely complex. The District is responsible for more than 200 water control structures over an area of 46,000 square kilometers served by a network of 3,200 kilometers of primary canals. It includes cities such as Miami. Management objectives are very complex. They include flood protection during the wet season (June through October), water supply during the dry season, preventing saline intrusions into aquifers, protecting the environmental quality of wetlands and lakes (the largest is Lake Okeechobee) and supplying sufficient flow to the Everglades. This complex environmental system is monitored by a dense data collection network. Real-time data are collected by field personnel, telephone units and a microwave telemetering network consisting of more than 650 electronic sensors which are automatically interrogated. All this information arrives at the District's control room and forms the basis of the manager's decisions.

The expert system, called OASIS (Operations Assistant and Simulated Intelligence System), has been developed on a highly specialised AI computer machine using the ART[2] shell and has been integrated with other software modules to monitor actual hydrological conditions, to check background alarm situations, and to access the large database. The data network has not been directly connected to the expert system to avoid overloading the computer. It is connected to a central unit which preprocess the data before providing just the required data to OASIS.

The knowledge base contains both declarative and procedural information. The declarative knowledge includes both static descriptions of elements of the hydrological system and dynamically varying data. The hydraulic characteristics, locations and interconnections of canals and water bodies are examples of the former whilst hydrological and meteorological data collected by the District and weather forecasts are examples of the latter. Guidelines for operating single control structures, empirical relations between variables, and legal and technical constraints to be satisfied constitute the procedural elements of the knowledge base.

Developing the knowledge base is widely regarded as being the most difficult task in

---

[2]ART (Automated Reasoning Tool) is a product of Inference Corporation.

developing an expert system (Hart, 1986). It is instructive to describe the four step approach used in forming the OASIS knowledge base (Goforth, 1989). In order to facilitate the first step (familiarization with the domain), a hydraulic engineer and not a computer expert was given the main responsibility for designing and managing the system. He examined past behaviour of the managers and collected all relevant information (manuals, guidelines, exisiting analyses). In the second step (observation of experts in action), the team of developers watched the daily operation of the system and the way decision were taken over a few months. Next, they undertook structured interviews with the managers, taking them through possible environmental conditions to ask how and why they would decide in such circumstances. The knowledge base was first developed on a subarea of the District to provide a representative prototype of the final system. Finally, the emerging knowledge base was verified to check its consistency and completeness. In practice, this last step was undertaken in parallel with the structured interviews.

Several interesting conclusions can be drawn from the OASIS experience. First, an expert system can really be competitive with a human expert in a complex, but well defined, decision making domain. Secondly, the critical part of the process is not the availability of hardware and software, but a human activity - the acquisition of knowledge. Thirdly, the knowledge acquisition process helps the expert understand which data are really taken into account in decision making and why certain decisions are taken. An environmental manager would commonly say that all data are relevant, or that more information would lead to a better decision. But, in practice, all decisions are based on only a limited set of significant data.

Another conclusion from the development of OASIS is that the knowledge, at least in the first three phases, tends to be formalized in IF-THEN rules. Thus a specific action (or a set of actions) tends to be identified with the outcome of a unique set of hydrometeorological conditions. The knowledge base becomes extremely large in order to cover all possible conditions and automatic inference is almost nullified. Each consultation is reduced to matching the description of the situation with the premise clause of single rule. The progressive verification of the knowledge base in the fourth phase helps find a higher level formulation of the knowledge which reduces the number of rules. Situations that have similarities are condensed into single parameterized rules. The higher the level of abstraction that can be encoded into the knowledge base, the more compact and powerful it becomes.

4.5 Postprocessors

Expert systems have also been used to help less experienced managers interpret output from complex environmental models. In this case the domain of expertise recorded in the knowledge base is not the environmental problem area but the interpretation of model predictions.

4.5.1 PigE

Menzies et al. (1988) provide a good example of this sort of expert system. The AUSPIG model (Black and Barron, 1988) predicts the economic productivity of piggeries by using a physiological model of pig growth. The model has proven valuable in predicting ways to improve piggeries but the model's output occupies 40 reports each typically an array of 20x40. These reports are interpreted by the model developers. An expert system was needed to allow pig producers to use AUSPIG directly.

The expert system developers noted that the experts tended to scan the reports generated by AUSPIG and focus on just a few of the numbers. They then derived further information from these selected numbers. The developers distinguished a small number of primitive operations such as identifying runs of consecutively high or low numbers and searching for the highest or lowest numbers in a report. These operations could then be called from production rules which represented the actions of the expert in scanning reports for "interesting" numbers. Menzies et al (1988) report that they encoded 72 rules over 108 person-days of effort. They also report that the experts were forced to examine their ideas carefully in order to form the rules for interpreting AUSPIG's output, and thereby they came to a better understanding of their own beliefs about the factors relevant to pig production. This beneficial side-effect is common to many expert system developments.

These rules were embedded in a specially written PROLOG shell program which used forward chaining. The output reports from AUSPIG are converted into PROLOG facts which are scanned by the rules. "Interesting" numbers are identified by comparing these reports with special facts entered by experts in a separate file. For example, the expert may have noted in this file that a pig is abnormally hot if a particular parameter (the Evaporate Critical Temperature) in a certain report is high for 10 or more consecutive days. If a rule detects such a heat-stressed pig then the rule may cause some actions to be carried out by AUSPIG or messages printed to the pig producer.

Like OASIS, the PigE system outperformed a human expert when applied to two piggeries. The developers suggest that the superior performance of PigE was partly because it was better than humans in making decisions when confronted with multiple factors. However, the test of PigE was conducted some months after the human produced his interpretation and so PigE's knowledge base contained an improved understanding of AUSPIG's output. Thus the test was not entirely fair.

The developers expect that, on average, the AUSPIG and PigE combined system should increase net profits of a piggery by 10-30%, even after taking into account real world factors not included in the knowledge base. The system is now out of the laboratory and in production. The combined system is used by over 35 piggeries in Australia and is being evaluated by piggeries in a number of other countries.

5 Conclusions

Expert systems offer access to a class of knowledge - qualitative information - not normally used explicitly in environmental modelling and so greatly expand the predictive power of models. In addition, there are three valuable features of these systems that are not directly related to qualitative modelling. First, they keep a record of their reasoning and so can produce simple explanations of their answers. Secondly, they normally provide estimates of the uncertainties of their predictions. Thirdly, they are written as interpreters so that the qualitative knowledge can be modified by those unskilled in programming.

As the literature shows, many environmental expert systems have been developed but few have reached full production use. Two that have, OASIS and PigE, are described here. Perhaps one of the reasons is that the most useful role for expert systems in decision making is not yet clear.

This chapter describes five roles for expert systems in environmental decision-making. The first, help in selecting appropriate models, typically draws upon the knowledge of scientists who are versed in the strengths and weaknesses of a variety of models. The second, help in calibrating the chosen model, makes the knowledge of the developer of a particular model or another person familiar with it available to a naive user. The third, prediction of environmental effects, uses knowledge of a problem domain such as plant disease or animal behaviour and so uses the knowledge of a domain scientist (rather than a modeller). The majority of reported systems fall into this category. One trend that is apparent amongst these systems is the development of integrated systems that employ both qualitative and quantitative knowledge to assist environmental decision makers. The fourth, expert systems that help users understand model output, typically employ the knowledge of the developers of a specific model. In this respect they are like the second type.

Finally, there are efforts to generalise model structures although these expert systems are still in the developmental stage. Whereas there is a large body of experience in the use of expert systems for the first four roles, there has been little work done on the use of expert systems for generalizing models. The advantages of this approach are large, for knowledge can be employed more efficiently and models can respond to diverse problems. But the difficulties are also large, because the underlying structure of models will need to be understood in a general way and powerful computers will be needed to implement these declarative representations of models. This remains the most challenging role for expert systems in environmental decision making.

References

Bachant J., and McDermott J. (1984) R1 revisited: Four years in the trenches. *AI Magazine*, Fall, 21-32.

Baffaut C., and Delleur J.W. (1990) Calibration of SWMM runoff quality model with expert system. *ASCE Journal of Water Resources Planning and Management*, 116(2) 247-261.

Barnwell T.O., Brown L.C., and Marek W. (1986) Development of a prototype expert advisor for the enhanced stream water quality model QUAL2E. Internal Report, US EPA, Athens, GA.

Black J.L. and Barron A. (1988) Application of the AUSPIG computer package for the management of pigs. Proceedings, Australian Farm Management Society, 155.1--5.10.

Buchanan B.G. and Feigenbaum E.A. (1978) DENDRAL and MetaDENDRAL: Their applications dimension. *Artificial Intelligence*, 11, 5-24.

Buchanan, B.G. and Shortliffe, E.H. (1984) *Rule-Based Expert Systems: The MYCIN Experiments of the Stanford Heuristic Programming Project*. Addison-Wesley, Reading, Massachusetts.

Calori G., Colombo F., and Finzi G. (1993) FRAME: A knowledge-based tool to support the choice of the right air pollution model, IFIP Conf. on Computer Support for Environmental Impact Assessment, Como, Italy.

Carrascal M.J., and Pau L.F. (1992) A survey of expert systems in agriculture and food processing. *AI Applications: Natural Resources, Agriculture and Environmental Science*, 6(2), 27--49.

Davis J.R., and Clark J.L. (1989) A selective bibliography of expert systems in natural resource management. *AI Applications in Natural Resource Management*, 3(3), 1-18.

Davis R., and King J. (1977) An overview of production systems. In: *Machine Intelligence* 8, E.W. Elcock and D. Michie (eds.), John Wiley and Sons, New York, 300-333.

De Leo G., Del Furia L., Gandolfi C., and Guariso G. (1990) Models and data integration for groundwater decision support. In: *Computational Methods in Subsurface Hydrology*, G. Gambolati et al. (eds.), Computational Mechanics Publications, Southampton, 531-536.

De Simoni G., Fanello R., and Guariso G. (1986) Computer aided planning of a multi-purpose river basin. In: *ENVIROSOFT '86*, P. Zannetti (ed.), Computational Mechanics Publ., Southampton, 421-434.

Duda R.O., Gaschnig J.G., and Hart P.E. (1979) Model design in the PROSPECTOR consultation system for mineral exploration. In: *Expert Systems in the Microelectronic Age,* D. Mitchie (ed.), Edinburgh University Press, Edinburgh, 153-167.

Engman E.T., Rango A., and Martinec J. (1986) An expert system for snowmelt runoff modelling and forecast. Proc. Water Forum, American Soc. Civil Engineers, New York.

Environment Protection Agency (1988) Guideline on Air Quality Models, EPA-450/2-78-027R, US EPA Research Triangle Park, NC.

Fayegh D., and Russell S.O. (1986) An expert system for flood estimation. In: *Expert Systems for Civil Engineering,* C.N. Kosten and M.L. Haher (eds.), American Soc. of Civil Eng., New York, 174-181.

Fishwick P.A. (1989) Qualitative methodology in simulation model engineering. *Simulation*, 5, 3, 95-101.

French M.N., and Bhaskar N.R. (1987) An expert system for runoff hydrograph estimation. Presented at AGU Spring Meeting, Baltimora, MA.

Gashing J., Reboh R., and Reiter J. (1981) Development of a Knowledge-Based Expert System for Water

Resources Problems, SRI Project 1619, SRI International, Palo Alto, CA.

Giarrantano J., and Riley G. (1989) *Expert Systems: Principles and Practice*. PWS-Kent, Boston, Massachusetts.

Goforth G.F. (1989) Advisory system for South Florida water management, Internal Report, South Florida Water Management District.

Goicoechea A., Hansen D.R., and Duckstein L. (1982) *Multiobjective Decision Analysis with Engineering and Business Applications.* John Wiley, New York

Guay R., Gauthier L., and Lacroix M. (1992) An abductive reasoning expert system shell for plant disorder diagnosis. *AI Applications: Natural Resources, Agriculture and Environmental Science*, 6(4), 15-28.

Hart A. (1986) *Knowledge Acquisition.* Kogan Page, London.

Hayes-Roth F., Waterman D.A., and Lenat, D.B. (1983) *Building Expert Systems.* Addison-Wesley, London.

Hwang C.L., and Yoon K. (1982) *Multi Attribute Decision Making: Methods and Applications.* Springer-Verlag, Berlin.

International Ground Water Modeling Center (1992), Software Catalog, Boulder, CO.

Karnopp D.C., Margolis D.L., and Rosenberg R.C. (1990) *System Dynamic: A Unified Approach.* John Wiley, New York.

Keeney R.L., and Raiffa H. (1976) *Decision with Multiple Objectives: Preferences and Value Tradeoffs.* John Wiley, New York.

Klir G.J. (1991) Aspects of uncertainty in qualitative systems modeling. In: *Qualitative Simulation Modeling and Analysis*, P.A. Fishwick and P.A. Luker (eds.), Springer-Verlag, Berlin, 24-50.

Kuipers B.J. (1986) Qualitative simulation. *Artificial Intelligence*, 29, 289-338.

Lemmon H. (1986) Comax: An expert system for cotton crop management. *Science*, 233, 29-33.

Loucks D.P., and Da Costa J.R. (eds.), (1991) *Decision Support Systems - Water Resources Planning.* NATO ASI Series G, vol 26, Springer-Verlag, Berlin.

McCorduck P. (1979) *Machines Who Think.* Freeman San Francisco.

McKinion J.M., Baker D.N., Whisler F.D., and Lambert J.R. (1989) Application of the GOSSYM/COMAX system to cotton crop management. *Agricultural Systems*, 31, 55-65.

Mattson S.E. (1988) On model structuring concepts. Proc. 4th IFAC Symposium on Computer Aided Design in Control Systems, Bejing, 209-214.

Menzies T.J., Dean M., Black J.L., and Fleming J.F. (1988) Combining heuristics and simulation models: An expert system for the optimal management of pigs. Proc. Australian Joint Artificial Intelligence

Conference, Adelaide, November 1988, 176-189.

Milanese M., Vicino A., and Borodani P. (1991) Integration of modeling and AI techniques in KBDSS generators: The EDIPUSS system. *Information and Decision Technology,* 2, 125-132.

Moninger W.R., and Dyer R.M. (1988) Survey of past and current AI work in the environmental sciences. *AI Applications in Natural Resource Management*, 2(1), 48-52.

Muetzelfeldt R., Robertson D., Bundy A., and Ushold M. (1989) The use of Prolog for improving the rigour and accessibility of ecological models. *Ecological Modelling*, 46(1), 9-34.

Peng Y., and Reggia J.A. (1990) *Abductive Inference Models for Diagnostic Problem-Solving*. Springer-Verlag, New York.

Rajagopalan R. (1986) Qualitative modeling and simulation: A survey. In: *AI Applied to Simulation*, E. Kerchoffs, G.Vansteenkiste and B.Zeigler (eds.), Simulation Series 18, 9-27.

Robertson D., Muetzelfeldt R., Plummer D., Ushold M., and Bundy A. (1985) The ECO browser. In: *Expert Systems 85*, British Comp. Soc. Specialist Group on Expert Systems, Coventry, UK, 143-156.

Roesner L.A., Gigure P.A., and Evenson P.E. (1981) Computer program documentation for the stream water quality model QUAL-II. EPA/600/9-81-014, US EPA, Athens, GA.

Shortliffe E.H., and Buchanan, B.G. (1984) A model of inexact reasoning in medicine. In: *Rule Based Expert Systems*, B.G. Buchanan and E.H. Shortliffe (eds.), Addison-Wesley, Reading, Massachusetts, 233-262.

Simon H.A. (1960) *The New Science of Management Decision.* Harper and Row, New York.

South Florida Water Management District (1986) OASIS: The South Florida Water Mangement District's operations artificial intelligence program. District report.

Starfield A.M., and Bleloch A.L. (1983) Expert systems: an approach to problems in ecological management that are difficult to quantify. *Journal of Environmental Mangement,* 16, 261-268.

Stock M. (1988) Planning expert system projects. *AI Applications in Natural Resource Management*, 2(4), 9-16.

Wick M.R., and Thompson, W.B. (1992) Reconstructive expert system explanation. *Artificial Intelligence*, 54, 33-70.

Winograd T. (1975) Frame representations and the declarative/procedural controversy. In: *Representation and Understanding: Studies in Cognitive Science*, D.G. Bobrow and A. Collins (eds.), Academic Press, New York, 185-210.

Zannetti P. (1990) *Air Pollution Modeling*. Computational Mechanics Publications, Southampton.

AUTHORS' INDEX VOLUMES I AND II

SUBJECT INDEX VOLUMES I AND II